MEDICATIVE MACRO-IMAGINEERING

EARTH & MARS MEGAPROJECTS

RICHARD BROOK CATHCART MA

TABLE OF CONTENTS

FOREWORD

Medicative Macro-Imagineering is a big thinkers book not intended for those who opine for the way things were or might have been were it not for their perceived human cancer that plagues Mother Earth. This is a book for those who can throw away the chains of prejudice, selfishness, and backward totalitarian thinking in favor of looking at new possibilities for the world and indeed the universe. In this positive and upbeat treatise on conceptualizations of what humans can achieve under the banner of freedom and free enterprise one is introduced to possibilities of the future. These technological and sociological potentials for humanity are not the result of Richard Cathcart's imagination alone because his Imagineering is backed up by reference to contemporary advanced thinkers and to works of those who have gone before us. In this way he stands on the shoulders of giants who have thought about better ways of doing things to make this world a better place. As Richard so clearly points out, the future can be bleak and foreboding to those without imagination or wonder when instead it can be full of new possibilities ot those who utilize their ratiocinative powers in a positive way.

Because intelligence is neutral, it can be used positively or negatively. Today we see many planners and governments eager to implement totalitarian control through central banks and other New World Order institutions that seek power to control individuals through force. Here, one is reminded of Dr. David Hawkins' elucidation of power versus force (Hawkins, 2002) where the former is accepted because it is benevolent and the latter rejected because it is malevolent. The future can thus be dark, as Richard fully recognizes, or it can be bright and upbeat as he tries to point out with many poignant examples.

To be sure, spaceship Earth has finite resources that can be better managed but the question is how to do this in an equable manner. What is seen today in many quarters are efforts to subjugate the people and make them slaves of big business and big governments where the masses literally exist at the whim of the rich and the powerful who live by different rules. Many so-called Green or Environmental revolutions at first glance appear good but when more carefully considered they are seen to be facades for more sinister plans of and by the elite to gain control of humanity.

Richard Cathcart fortunately brings to the fore new perspectives of how humanity can grow and prosper without ruining its home base on spaceship Earth. Most sentient beings agree that the present course of action can't be maintained because of its ruinous consequences to the planet. Standard operating procedures have to change, that is for sure, and Cathcart shows how medicative macro-engineering can pave the way for a brave new world that can free itself from deleterious impacts associated with over population, pollution, and depletion of natural resources. The medicative macro-imagineering approach is to work with Nature by recognizing natural cycles instead of trying to control Nature through brute force. Adaptive macro-engineering should, for example, replace present geo-engineering practices such as chem-trailing that are ostensibly conducted to control imagined and much feared human-induced global warming, which is a hoax perpetrated on a scientifically illiterate public. The environmental pollution resulting from chemtrailing is far worse than natural cycles of Earth evolution that have continued before and throughout human history. And these natural cycles will continue after human history if we cause our own demise through the injudicious application of unproven technologies such as chemtrailing, HAARP (High Frequency Active Auroral Research Program). [Located near Glennallen, Alaska, HAARP can be used to test the effects of aimed extraterrestrially generated microwave beams in the Earth's atmosphere caused by proposed Space Solar Power facilites.]

Macro-mismanagement cannot and should not continue as business as usual. As pointed out by Cathcart, "Green environmental extremism is the most dangerous and obscene form of misanthropy yet developed in the North America-Europe geographical region" (Chapter 6). Macro-imagineers and Macro-engineers respect the Earth by attempting to create a better world, a theme clearly posited in this book as a better option to the vagaries and self-serving interests of geopolitics and ecopolitics.

This Foreword touches on only some of the vignettes in the book as there are so many included within its covers. There is more, much more, than these few highlights summarized here, providing a mere apophoreta of what is to come if one reads the book with an open mind. And that is the key to the future, to view the world as a place where changes can be wrought without draconian measures of population reduction (depopulation) by soft kills or other equally ugly approaches to diminish humanity in favor of a few elect. If the meek are to inherit the Earth, it connot be achieved by the arrogance, coercion, force (military, economic, political, religious, or otherwise) by those who perceive themselves as privileged by virtue of wealth or accident of birth into blue blood lines. *Medicative Macro-Imagineering* provides a glimpse of what can be our future if we think smart instead of stupid. To think stupid is easy as it does not require a higher level of consciousness (*e.g.,* Hawkins, 2002). Higher levels of consciousness are based on power, not force. Richard Cathcart thus invites the reader to consider the high road to saving the Blue Planet by pointing to potentially viable alternatives of climate as a weapon of war or artificial climatically-sustained environments that exist outside of natural (Galactic, Solar System, Earth) cycles. Humans can exist within the cyclicity of Earth's systems and we do not need to attempt to control them as such efforts can't be maintained forever, or for the duration of most great climatic cycles. The future is bright if we use technology in a positive manner to create a better world for many through macro-projects "prescribed" by persons conversant in medicative macro-imagineering.

Charles W. Finkl, Phd. CSci, CMarSci, FIMarEST, CPGSc, PWS, M.ASCE, President and Executive Director, The Coastal Education & Research Foundation, Inc. (CERF).

Asheville, North Carolina

[Reference: Hawkins, D. 2002. *Power vs. Force: The Hidden Determinants of Human Behavior.* Carlsbad, California: Hay House) 337 pages.]

PREFACE

Medicative Macro-Imagineering adopts a mild medical metaphor to accommodate the natural and anthropogenic Earth-biosphere change risks, including the risks of human reactions and over-reactions to Earth-biosphere alterations that involve civil and military engineering—that is, actual Macro-Engineering following Macro-Imagineering. A complex syndrome of symptomatic planetary shifts (air temperatures, world-ocean currents, seawater's acidification, precipitation shifts, and biotic melee, riots of a sort) can be countered by geopolitical, ecosystem-nation policies and a clear sense of humanity's technically possible future. "Man-made Global Warming" may be a phantom dis-ease lodged in, and troubling, the introverted human mind or a terrible, human-inflicted macro-problem for "Mother Earth", truly a Green-secularist's religion, in which case Earth is suffering from a rare anthropogenic disease, the very first known case in our Solar System. Our world has progressed without "sustainable development" [the prhase was first widely used in 1987], so why do Greens try to meld the two divergent concepts—"sustainable" implies closed-loop [eternal cycling], whilst "development" is entirely subjective? Geographos does not know why they do and cannot even surmise! On the reverse of that corroded Green "coin", Panspermia hypothesis proponents question whether Earth's life is even indigenous (autochthonous) or allochthonous (not aboriginal) to this inhabited planet. Macro-engineers can be sued, tried [and "executed" by public Green pillorying?] for malpractice in Courts of Law, but Macro-Imagineers can only be scorned and shunned—so far! (I claim, assert, and swear, under penalty of law, and under God's unwavering observation that I, Richard B. Cathcart, are now, and have always been, merely a happy Macro-Imagineer.) For Geographos, *Homo sapiens'* practical activities are associated with the Earth-surface, but nowadays other places in our Solar System must be, increasingly, looked at more intently and concernedly. Nevertheless, ***Medicative Macro-Imagineering*** is meant to be a scholarly entertainment that combines medical geology (Selinus, 2013), its complementary environmental medicine with Geoscience and Geopolitics! All persons alive within the Earth-biosphere are subject to all space debris "inputs" and all Earth-core "outputs"; when people come to dwell on Mars, human and animal health macro-problems will certainly arise! At this time, globalized Macro-Imagineering/Macro-Engineering has the professional social status of "physiotherapy medicine" whilst Terraforming might be considered planetary "chiropractic medicine".

[**Cited Reference:** Ollie Salinus (Ed.) (2013) ***Essentials of Medical Geology.*** (The Netherlands: Springer) 805 pages.]

CHAPTER 1

REROUTING OUR RECONSTRUCTED
SOLAR SYSTEM

Sometime during 1943, the famous French novelist and daring Allied wartime reconnaissance-airplane pilot Antoine de Saint-Exupery (1900-44) penned and illustrated his charming parable *The Little Prince*. The geo-artistic child Saint-Exupery imagined characteristically exhibited a strong landscape (Ellison, 2013) grooming or gardening instinct—tempered by omnipresent enveloping spacial sublimity (Koneeni, 2005)—directed exclusively towards a tiny (64 meter diameter?) space-wandering asteroid, perhaps a virtually inconspicuous mini-planetoid approximately 10^5 smaller than our Earth's radius. [Or, a rocky boulder about three times the diameter of the 2012 Chelyabinsk superbolide?] By 2013, about 140 million copies of *The Little Prince* had been sold worldwide. Behaving as a building's guru-janitor, the charming young princeling asserts: "I have a flower, which I water every day. I have three volcanoes, which I clean out every week. I also clean out the one that is extinct. One never knows" (Saint-Exupery, 2000, page 56). [The Prince also needed to persistently weed the Kudzu vine-like asteroid splitter "Baobab" plant to allow his flower to flourish without competition. His personal worldly possession, the asteroid, holds no fossiliferous strata, nor any burials. With microbes, it should be noted, nothing could rot on Mars, a potential planet of human mummies.] By definition, envisioning is a view or concept that is beyond extant professional and popular paradigms. Useful, high-durability future infrastructures will be ecological, aesthetically satisfying landscapes created by the reliably wise discernment of humane macro-imagineers (Cathcart, 2011). Derived from the ancient Roman deity Janus, janitors are the controllers of ingress, determining who and what comes inside as well as whom and what egresses. Conscientious janitors are caretakers, custodians who symbolize the indisputably real-world requirements of all massive 21st Century man-made structures and buildings for vigilant, scrupulous routinized asset management and well-organized total systems of structure and building maintenance. A Macro-Imagineering master planning language—exhibited in future professional dictionaries and encyclopedias (Marsden and Whiteman, 1999; Langmead and Garnaut, 2001; Davidson and Brooke, 2006)—must become widespread, popularly accepted by all educated humans. Today, macro-imagineers trudge the borders of extant discipline-packaged human fields of knowledge, planting practical conceptual seeds meant to rattle the current common collective confidence of the Establishment construction and planning industry professionals! Non-conformist macro-engineers are not theory devisors but, also, such persons are not exclusively theory implementers and testers; macro-imagineers shun the conservative mindset of current profession-wide in-the-box conceptual thinking. Sadly, some casually vacuous television personalities, often broadcast internationally, seem to have more actual authority with the world-public attributable to charm than does many geoscientists and technical experts (Huggan, 2013; Scott, 2014).

During that same year [1943], at Princeton University, the budding futurist American aerospace macro-imagineer Dandridge MacFarland Cole (1921-1965) graduated with a BS degree in chemistry. Cole then continued his formal education, markedly medical in orientation, for one school year at Columbia University's College of Physicians and Surgeons. After World War II's termination, by mid-1945, soon he developed a very keen interest in the economically booming postwar military-industrial aerospace-sciences complex. Just as the Cold War commenced, with the apprehensive world-public held in thrall, during 1949, he was awarded a MA degree in physics by the University of Pennsylvania. From 1949 until 1953 he taught physics and astronomy at Phillips Exeter Academy, the State of New Hampshire's most renowned college-preparatory school. Shortly thereafter he accepted employment with a successful major aerospace manufacturer, the Martin Company, situated at Baltimore (Maryland, USA). Later, he transferred to another Martin Company facility at 1.6 kilometer-high Denver (Colorado, USA) where he conducted some of the most visionary long-view approach of design archetypes to operationally develop the Titan II missile, later utilized to launch the experimental manned Gemini Space Capsules. By 1960 the adept visioneer (McCray, 2013; Foege, 2013) Cole again shifted his work-site, returning to the State of Pennsylvania, working as a highly-esteemed job-focused consulting aerospace macro-engineer—the down-to-Earth, hands-on technician driven by macro-Imagineering visions—focused on advanced planning at the Valley Forge Space Technology Center of the General Electric Corporation's Missile and Space Division, established at Philadelphia (Frisch, 1965; Raithel, 1966; Cole, 1966). By being methodical and steady in his work-habits he proved again that unreasoning haste or impetuosity does not lead to success in industry. Twenty years after Antoine Saint-Exupery wrote his fictional *The Little Prince*, in a commercially-published, co-authored popular non-fiction textbook, Dandridge M. Cole recommended hollowing a captured 30 kilometer-long ellipsoidal natural asteroid to hedonistically sustain humans during spatially long-range, extended-time period human space exploration missions, both within our Solar System as well as intentionally aimed at some nearby Milky Way Galaxy stars (Levitt and Cole, 1963). Traveling in fleets for safety and convenience, such altered asteroids, converted into inhabited planetoid-spaceships, might explore other stellar systems, permitting a swarming human permutation of exotic places for colonization and cultivation, transformation from barren places into thriving human-controlled worlds (Ball, 2012).

In sequel-resembling, concept-extender books, D.M. Cole (**Figure 1**) likened such inhabitable interplanetary space vehicles to island-like "nomadic pseudo-Earths" that, with appropriate modification and equipment, would be potentially capable of carrying living human explorers and colonizers to far-distant places—exactly located spatial co-ordinates landscaping worksites, each with their human and robot interaction histories—for adventurous historical engagements with this Universe's physical features and possible Alien creatures. Some alarmist, semi-apocalyptic Greens, nowadays mostly dedicated to a public-relations campaign warning of the projected continuation of anthropic global warming, for example, choose to envision D.M. Cole's nomad spaceships as future extra-Earth "lifeboat cities", extensions of those big boat-like gargantuan "Arks" in the 2009 block-buster movie *2012*, because Earth's inhabitants will almost certainly endure an anticipated 21st Century global catastrophe: "...a series of reverberating, interlocked shocks" (Gleeson, 2010, page 128) triggered by abrupt global climate change and co-incidental civilizational stresses (Hsiang et al., 2013). In fact, they renounce the April

1784 patent of the steam-engine by James Watt, seeming to favor a *1984* world-view by dedicated Greens! [NOTE: the very term "global" prompts the mind's vision of extra-planetary outer space (Geppert, 2012).] Again, nowadays, such all-encompassing, grand-scale, movable and steerable spacecraft are envisaged and termed "Interstellar Arks", "Generation Starships", "World Ships" and "Artificial Planets" (Glover, 2013). These majestic appellations—that project ecologically away from the Earth-biosphere (Anker, 2005a; Anker 2005b)—do tend to stimulate the universal enlargement of humanity's common presumptions about what kinds of massive techno-things *Homo sapiens* can assemble and shape, and where people may someday sojourn, if ever persons can safely ride such big containments! [Purely as jest, Earth's hollowing macro-project has been suggested by a whimsical mathematician (Stasenko, 1999).]

Figure 1. D.M. Cole, circa early-1960s formal photograph. A young man who had the liberating power of macro-engineering imagination

to visualize, conceptualize and create when presented with recognizable opportunities to put together some thing that was entirely new. Completed bold thoughts, encouraging words and effective real-world actions are the means by which he manifested his genius in the service of all humanity. Artificially cavernous asteroids are a monumental Macro-Imagineering plan completed as a substantive macro-project/mega-project. (Image: WIKIPEDIA.)

Early-21st Century—contemporaneous—civilization imaginaries relating to the constructed manifestations of humanity's future infrastructures are in profound flux. Some moderns perceive Cole's nomadic pseudo-Earths, multiple-vessel convoys of hollowed planetoids, as a type of public mass transit conveyance and zoo-wagons; clearly, the functional utility of these vehicles must be very high and reliable! Non-hollowed or otherwise transformed natural asteroids—certainly those composed mainly of water ice—could be simple minable objects during the many-years trip to another attractive star. Some hollowed asteroids—D.M. Cole's actual trevelling town "planetoids" that is—might serve as compact portable landfills staffed by specialist, dedicated janitorial-curatorial staffs. What's wrong with rubbish? The outer, leading edge not only of human-influenced space but recorded human civilization's consumer history is marked by allochthonous trash—that is, originating in our Solar System other than where it is found by inquisitive Aliens—items such as the Voyager-1 space probe, launched September 1977, which exited our Solar System in July 2012 (Cathcart, 1979; Swisdak et al., 2013). Voyager-1 remains powered by three radio-isotope thermoelectric generators using Plutonium-238 in the form of plutonium dioxide fuel. **Figure 2.**

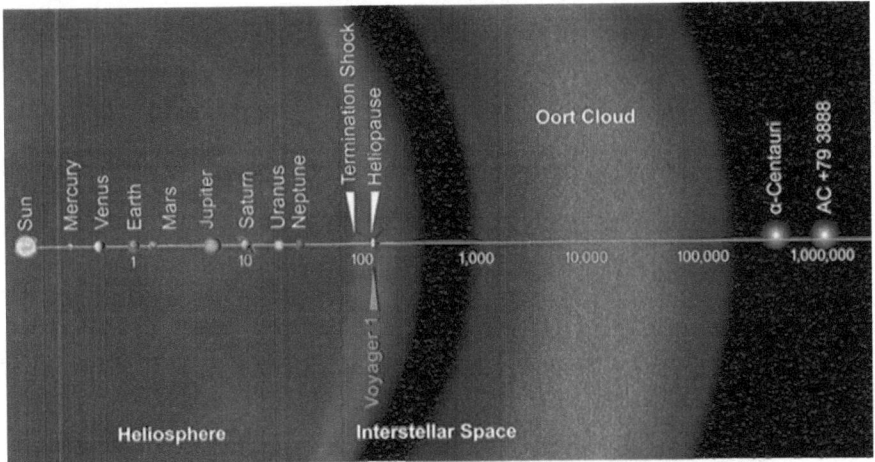

Figure 2. Calculated distance, in Astronomical Units (AUs) from Earth, Voyager-1's launch point in this Universe. Its current position is indicated by the red name-tag. (Image: NASA.)

Comets leave trails as all organic and inorganic things do—people leave a residue of skin-flake castoffs and other stuff, products of our 37 trillion human cells (Bianconi et al., 2014). Really, these imagined trash planetoids may be more accurately labeled as macro-engineered and macro-managed "storage lockers" for recoverable urban ores since even compressed unprocessed trash may serve as simple radiation shielding, protecting biota from anti-life cosmic "rays" that arrive from every direction! Surely, if the density of the Earth-orbiting satellite/junk belt—a new naturally and artificially radioactive geological stratum—wholly enshrouding our planet gets much greater, then someday, collectively, it might serve a similar cosmic-ray shielding purpose! [Earth's largest anthropogenic discarded materials heap—an estimated 150,000,000 tonnes dumped therein—New York City's unsightly Fresh Kills Landfill (1948-2001) has, since it closure, been converted into the low-incline hilly, beautiful Freshkills Park.] Still, it could likely be that technological obsolescence is unavoidable, even with a steady-stream of energetic telecommunications from our originating home Solar System. Also, serious fundamental deterioration of the convoy/fleet is an ever-present possibility—unrelenting physical wear, critical resource depletion, and other degradation caused by internal and external factors are inevitable, especially if the voyage lasts more than a decade. Imagine the commonplace "sick building syndrome" diagnosis applied simultaneously to all of humanity's traveling fleet of interstellar vehicles! Such a process-event of deterioration could become an overwhelming sustainability crisis for an isolated population of desperate humans.

The 1986 Hollywood **Star Trek** franchise film **The Voyage Home** seemingly rests firmly on a somewhat outlandish time-travel plotline pivoting on a gigantic lozenge-shaped, destructive energy-emitting spacecraft piloted by inquisitive Alien humpback whales (*Megaptera novaeangliae*)! [Spyhopping Earthly colorblind whales have, presumably, patiently awaited fascinating conversational updates from their visiting Alien confrères!] Basically, the **Star Trek** and D.M. Cole's earlier starship designs should be visualized as enormous, inside-out, Chia Pet®-like creations, deliberately constructed by intelligent creatures to carry all kinds of life, as well as made life-forms. Real, future-existing **Star Trek** "Commander Data"-like robots with personalities might best be termed cognitively complex "Terra-creatures" (Kim and Kim, 2013; Giselbrecht et al., 2013) to unmistakably differentiate them from *Homo sapiens*, extremely great distances through the increasingly-examined vastness of interstellar space. Terra-creatures and humans are agential entities. The commonplace notion of biotic panspermia—that is, the incidence of dormant life, "extremodures" originating elsewhere in the Universe, coming from outer space, falling onto a susceptible environment such as Earth and there reproducing—often meets with derision from the NASA-funded planetary researchers and SETI philosophers. Why? NASA commonly seeks funding based on its searches for the Universe's living creatures, a solution to satisfy public curiosity the NASA's administrators believe underpins American taxpayer support, so, logically, if humanity's Earth-biosphere is already the beneficiary of natural cosmic panspermia, then the NASA's present and future funding propaganda becomes entirely pointless! The familiar concept of planned and intellectualized mechanized panspermia (michicanpanspermia)—in the past the NASA, at burdensome taxpayer financial cost "sterilized" every launched planetary probe to preclude any possibility of directed panspermia—seems an appropriate name for this spatial transfer process-event caused by technologized and moving space rocks.] In effect, these vehicles ought to be considered moveable time-capsules for any civilization, whether Alien, human, cyborg (Bacon, 2013) and Terra-creature. Cole foresaw isolated biota evolving during a planetoid's travels, during its years-long trajectory through interstellar space—he named these distinct collective overall changes in the [human-selected] contained biota as "macro-life" (Cole, 1961). [Subsequently, science-fiction writer George Zebrowski adopted Cole's published article title as the publication title of his own 1979 science-fiction novel (**Macrolife**).] This offshoot idea soundly suggested by D.M Cole and Zebrowski intrigues Geographos because it may constitute a possibly fatal polylemma, a choice involving multiple undesirable gardening options.

What usually occurs when hard-working janitors and vigilant building administrators finally discover an irreparable disintegration or fatal, previously unnoticed design flaw in a large valued structure or building (Freitas and Delgado, 2013) constructed in our Earth-biosphere? [Almost every building/structure undergoes partial demolition during episodes of maintenance or modernization. Geographos considers the Siamese-twin Macro-Imagineering/ Macro-engineering profession as, essentially, doing that task.] Some are merely bulldozed and bashed with a heavy swung wrecking ball, of course—in North America and Europe about 1% of extant buildings/structures are fully demolished yearly. Ultimately, a planned implosion or explosion demolishes the building or structure and its materials are usually recycled (Byles, 2005)! "Obsolete", derived from Latin *obsolesco,* means "losing prestige and value". [A reliable rule of thumb is that an Earthly land-based structure's monetary tear-down cost is around 2% of its initial construction cost incurred earlier. So, if it ever becomes possible to estimate Earth's monetary worth, then crevice-making macro-imagineers will have some basis for accurate planet dis-configural calculations!] Buildings/structures are commonly considered as future resources; the "cradle-to-cradle" outlook entails the idea that the end of the buildings/structures life cycle is but the commencement of a new buildings/structures life-cycle. Few human cultures today ever perform eulogies after building or structure devaluation, and even fewer cultural mores cause cultures to perform such celebratory rites before commencement of rapid controlled or uncontrolled demolition process-events. However, a fleet underway, using an Internet localized area network (Burleigh et al., 2003; Jackson, 2005), should have the short-delay communication option of multi-party conferences discussing what measures ought to be undertaken to mitigate such a dire impending survival situation. The astronomical fact that there is an quite enormous mass of asteroids and other non-planetary objects orbiting our Sun simply means that humans can appropriate and take a lot of useful physical matter (solid, gaseous, liquid, living) as supplies for any Solar System-exploitation and Solar System-exiting fleet's interplanetary and interstellar spacecraft (Badescu, 2013; Breedon, 2013). It is interesting to note that the earliest known iron artifacts, ancient Egyptian bead jewelry, were fashioned by sophisticated metalworkers from meteoric iron about 2,000 years before iron smelting was first used by humans (Rehren et al., 2013)!

More than 12 small near-Earth asteroids could be "easily" recovered by human spacefarers using extant technologies (Yarnoz et al.,

2013). [Rogue asteroids commoner than once thought, particularly in the main section of our Solar System's Asteroid Belt (DeMeo et al., 2014).] Suppose the main Asteroid Belt's resource mass, which is markedly less than Earth's Moon, just 4% in fact, becomes insufficient for a certain Macro-Imagineering planned task (Ceyssens et al., 2012). What if, according to final operational conclusions determined by techno-savvy human-survival experts, more materials are needed to proceed with the purposeful Sun-enclosure effort postulated, as an on-going normal star-travel enterprise? In other words, a time may come for necessary artificial asteroids, mineable asteroids generated by human technology, namely by the use of Freeman John Dyson's proposed planet spin-up/shattering motor which Geographos has named "Archimedes" (Cathcart, 1983) to be massively generated quickly (Schewe, 2013)! **Figure 5**, below, illustrates Archimedes as it would be configured if used to disassemble Earth. Dyson's Motor must increase today's upwardly directed natural centrifugal force at the Equator of 0.033916 Newtons to equal or very slightly exceed Earth's natural present-day downwardly directed 9.81 Newtons gravitation force at the Equator; **Figure 5A** is the NASA's digital image of our planet; **Figure 5B** indicates the potential shape of our molten homeland, rotating in one hour instead of twenty-four hours, just prior its ultimate disruption via anthropogenic spin acceleration, Archimedes' design—that is, Freeman J. Dyson's Motor—has been thoroughly outlined by experts (Dyson, 1966; Fogg, 1995). Basically, it consists of sturdy conductive metallic cables laid on the Earth's surface—the plastic accessible crustal stratum retaining the Earth-mantle—along imaginary lines of latitude to give, the Earth a fixed quadrupole magnetic field with a vertical component. Following that installation, hands-on macro-engineers would produce a uniform current toroidal magnetic field moving latitudinally through the planet, connecting northern and southern hemispheres. The current's return path, external to Earth, must pass through planet-orbiting metallic conductors and the planet's magnetosphere plasma (consisting of charged particles). Archimedes makes Earth an armature of an enormous anthropogenic electric motor powered by solar energy efficiently collected by orbiting structures. Archimedes is installed to overcome Earth's hydrostatic balance. Earth's surface (its crust, especially and specifically) will consequently endure a horizontal eastward stress to produce a disruptive net torque. It might take more than 40,000 years of continual artificial torque application, according to Freeman J. Dyson and other experts, to bring our home planet to a physical shape-state close to exploding due to imposed centrifugal stress, as illustrated in **Figure 5B**. If our civilization deliberately extricates only those Earth species capable of tolerating human society's new, spacey activities, then the only species left to rapidly expire gruesomely will be the pests and weeds people have always wanted eradicated!

Strictly programmed "Terra-creature" instigators—those anthropogenic agential entities—are hardly likely to mind the prolonged time period at all! And, it is possible, it is to be hoped, that humans will long remain the dominant living creatures of this Solar System. A bonitative Earth-crust map (all land and the ocean's seafloor)—a type of map that indicates favorable/unfavorable conditions for specific economic development and improvement—will help to guide Archimedes' valiant builders. The Earth-crust mass [2.73 followed by 22 zeros kilograms is just 0.68% of the intact Earth's entire mass.] At first, Archimedes could be conceived by its installers, whoever or whatever (cyborgs, Terra-creatures or even intrusive Aliens) they are—self-movement and communication between such units of contained intelligence is an essential feature for all forms—as an majestic spin-off of the late-20th Century artworks of Walter Joseph De Maria (1935-2013), for example such 1977 artworks as *The Vertical Earth Kilometer* lodged permanently at Kessel, Germany and, in the USA, the weathered installation, *The Lightning Field* sitting on southwestern New Mexico's arid landscape (Baker, 2008; Kosky, 2012).

Somewhat strangely, it is F.J. Dyson's teasing potential mayhem machine (synonyms: "harmful", "injurious", "hurting", "painful") that stimulated Geographos to initially think about the possibilities of "Medicative Macro-Imagineering", at least as applied throughout our Solar System. "Macro-Imagineering" is a portmanteau word which combines "imagination" and "Macro-engineering" and partly describes the successful implementation of creative ideas in a practical working form. Walt Disney (1901-1966) described his corporation's property, Walt Disney World, located in the State of Florida (USA), as "...the most magical place on Earth". [His company was the first to introduce steel-constructed roller-coasters and linear induction motors for urban transport, in the form of Disney's "People Mover".] As will soon be apparent, the use of the adjective "medicative", meaning "healing", "health-giving", implies implementation of a "corrective", "curing", "helpful", "alleviative" and "restorative" macro-project(s). Today's impassioned Green-leaners, and sometimes misguided scientists also, anthropomorphize animals, plants and even our central star. For example, since the Sun is at its midpoint in Solar Cycle 24, it has completely reversed its enormous polar magnetic field. The magnetic field flip continues to have a ripple effect extending throughout our Solar System, and possibly beyond the heliosphere; as Earth orbits the Sun, it will enter and leave the new Sun-generated current sheet irregularly. In other words, strongly variable space weather (Strong et al., 2012), directly affecting our civilization's electrical infrastructure and functional Earth-orbiting satellites, must be coped with technically.

"Storms" on the Sun's "wounded" surface are monitored by solar seismologists (Cally and Moradi, 2013). Cally and Moradi speak of our tumultuous central star as if the Sun was alive! Is this daring presumption not truly temerarious? [Others compare Earth's geophysical tumult with various human diseases (Contoyiannis et al., 2013)!] Solar eruptions called "coronal mass ejections (CME)" and solar flares, heat the Earth's uppermost atmosphere. Due to this heating, as well as the impinging ultraviolet, X-ray and particles expelled from our Sun, the atmosphere expands in volume and causes the orbital degradation of Earth-orbiting satellites, making them vulnerable to enhanced drag forces (plasma drag) imposed by the ions and molecules of the enlarged Earth-atmosphere. Nevertheless, between Earthlings and Aliens is a tangible dispersed obstacle: a massive quantity of orbiting satellites, some of which are junk. These objects are directly affected, all the time, by CMEs. **Figure 3.** By 2014, our Sun reached its most recent solar maximum state, Solar Cycle 24, and powerful CME are forecast to occur by still-learning outer space weather experts. These CME will, undoubtedly, adversely affect human civilization's electronic infrastructures, blackouts of large cities, **Figure 3,** may happen suddenly, eliminating indoor and outdoor nighttime lighting. Literally, light pollution (Bogard, 2013) can be stopped instantaneously by infrastructure failure or human volition! **Figure 4.**

Figure 3. An artist's image of tabulated Earth satellites, some are junk. The current mass of the air [5.1 followed by 18 zeros kilograms] below the orbiting satellites is only 1/274th of the blue-colored ocean that visually dominates this image of our homeland. These satellites eventually fall to Earth when their orbits degrade—some are intentionally forced to enter the atmosphere for deliberate degradation by disintegration. For example, the European Space Agency's ATV-4 (Automated Transfer Vehicle-4), launched 5 June 2013 packed with 1.6 metric tonnes of trash from the International Space Station, sent into the atmosphere on 2 November 2013 above the Pacific Ocean. Orbital debris simply adds to the existential dread summoned up by the abyss of outer space itself. (Image: Google Images.)

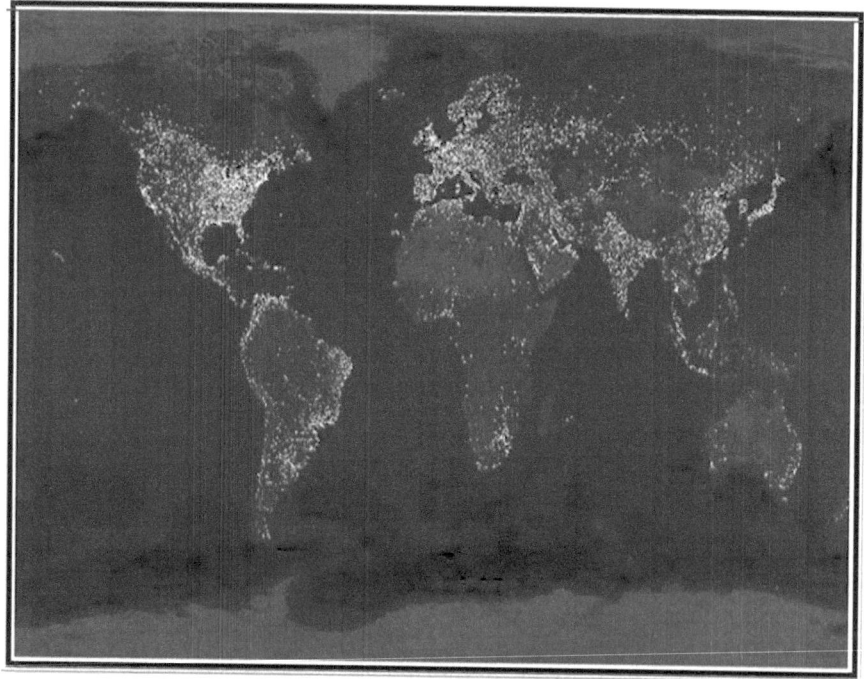

Figure 4. Earth's artificially illuminated Astythromes and Anthromes: about 0.44%, or about 600,000 square kilometers, of our planet's ice-free land surface, as viewed from outer space is lit urban territory. The energy used is, approximately, one-millionth of the sunlight's energy of arriving every day. Some modern philosophers, Michael Serres (born 1930) for instance, have described these nightlights as "cancerous". The so-called luminous "fog" caused by outdoor lighting is a proxy for Astythrome growth (Chen and Nordhaus, 2011). Orbital forcing-caused "Global Warming" terminated the Ice Age, helping to get humans flourishing; light pollution typifies, even amplifies, our presence visibly through exploitation of Earth's geology. (Image: NASA.)

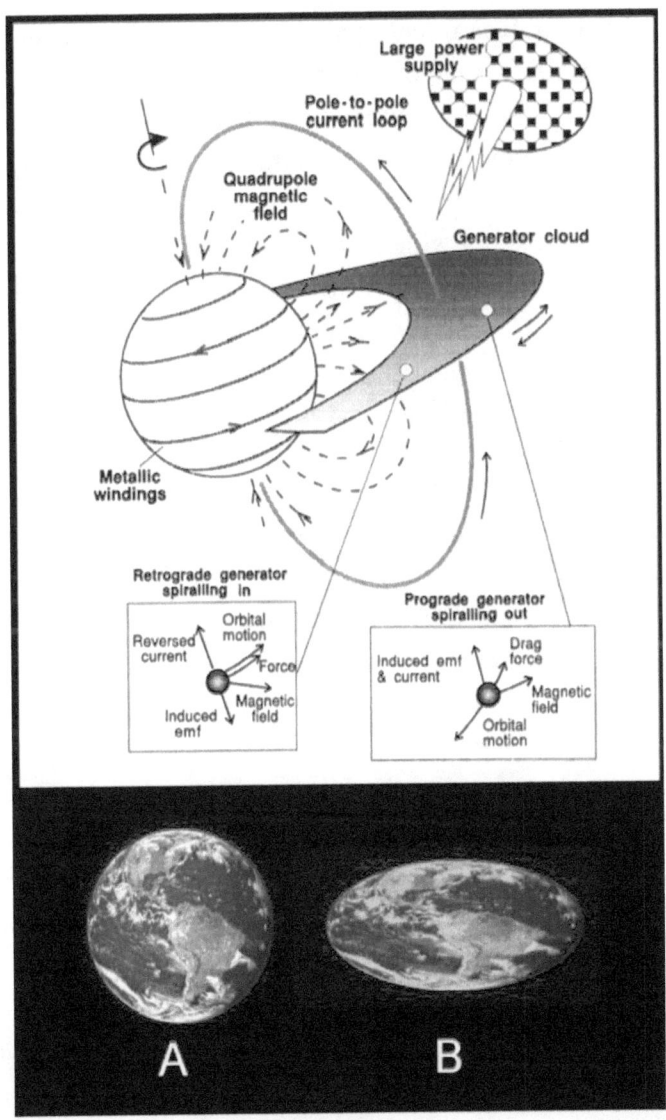

Figure 5 (5A and 5B). 2 "Archimedes", Freeman Dyson's electric motor; **5A** a 24 hour/day Earth. An example of an energy-minimizing geometric shape for a fast rotating deformed Earth is shown in **5B**, a one hour/day Earth. Earth's volume is 1.083 followed by 21 zeros cubic meters; Earth's center of mass is a verifiable place (Zakharov et al., 2013). Spun-up, the intentionally rickety, bulging Earth is unoccupied by humans, a very oblate spheroid! (Image: MJF and JMH.) Unlike some other nearby planets, Earth's crust is prone to structural failures producing powerful tremors. Although it would likely be molten in surface texture, the tectonic plate margins may be where the planet cracks and splits widen first before expanding into an outer space "cloud" of suddenly dispersed materials. That "cloud" is a stock of resources for those with the capability to use, the ultimate "creative destruction" moment, the risky loss of substance followed by the opportunity to create something new.

Mega-events—those international public events with a global audience—are generally perceived as an opportunity to market Earth-biosphere places and anthropogenic things. Deliberate destruction of Earth by a *deus ex machine* named Archimedes would be a watchable televised mega-event of unprecedented spatial vastness and pervasive human psychological effect. ["Big Bang" was introduced by Fred Hoyle (1915-2001) during 1949 as a descriptive label for the still unacceptable theory of the Universe's initial state. One of the most successful neologisms ever, surely future entrepreneurs might re-package (or re-brand) that process-event label to name and promote a one-time-only proposed reserved tickets spectacle!] According to the economic and human developmental scope prevalent, today's Earth is a "Spaceship" rapidly taking on the characteristics of a large, but nevertheless increasingly constrictive "spacesuit": in a time of fast global travel and speedy shipping, disease-carrying persons cannot be isolated (quarantined) in a compartment and, as a result, all aspects of human civilization are bound together, for better or worse. Under a worst-case scenario—say, for example, an act of atmospheric bioterrorism or deliberate instigation of an explosive nuclear weapon producing an electromagnetic pulse event exceeding a solar super-flare by many magnitudes—totalitarian forces might repurpose the

Earth-spacesuit into an Earth Body Bag (Bolonkin and Friedlander, 2013)! [A non-porous body bag, sometimes labeled a "cadaver pouch", is used for the temporary storage and transport of a deceased person.] Under a best-case scenario some artist(s), perhaps advancing upon the action-connoisseurs' spectatorship of nuclear weapon explosions in and above the Earth-biosphere (Smith, 2013), with the same ethereal artwork mindset as Cai Guo-Qiang, might oversee the conduct of a terminal mega-event! Humanity requires the existence of adequate early warning for sociological, rapid-onset geological disasters as well as relatively slow abiotic planetary change such as atmospheric composition and climate change: early-warning and real-time loss (persons and property) needs ought to be addressed via self-organizing sensor networks—See Chapter 3—new outer space satellite imagery with high resolution, multi-sensor observational capacities and crowd-sourcing so that it will be possible, in a timely fashion, to mitigate global disasters, whatever their potential magnitude! Earthlings might observe the flash of electromagnetic energy emitted by the Sun about eight minutes after its destruction occurs, and our planet, located just one astronomical unit from the Sun, will be smote about twenty-two hours later with incinerating ejected Sun matter. *Finis terrae* ["the end of the world"]. NOTE: **Figure 5B** is deceptively simple. In future, since our homeland, experiencing spherical expansion and flattening of a drastic nature, simply won't actually appear as illustrated in **Figure 5B**. A good illustrative counter-example for a much earlier moment in time, prior to melting, is provided by Witold Fraczek (2010). Fraczek modeled the case where Earth was brought to a standstill, no longer rotating as normal. [It would be possible for a tasked Archimedes to accomplish such a stoppage in 4,200 years!] Visually displayed in drawings, Fraczek shows that the world-ocean would recede in some places and advance onshore in others, massively! Our so-called survival worries, concerning the postulated "rising global sea-level", and the subsequent fate of major coastal cities and millions of human dwellers therein, seems, by comparison, rather silly and insignificant (Willenbring et al., 2013)! At this time, major studies seem to indicate that global sea level could rise about 2.3 meters over the next two thousand years for every degree Celsius the Earth-atmosphere warms; as a matter of fact, even for the worst-case scenario, the ocean's rise should be less than 2.0 meters by 2100 AD (Leverman et al., 2013). Brian Fagan's *The Attacking Ocean: The Past, Present, and Future of Rising Sea Level* (2013) is possibly the most level-headed popularized reading on this particular macro-problem published in book format to date! What a reasoned contrast with the sensationalistic lay-out of the September 2013 *National Geographic Magazine*! It is deplorable that so few Americans will indulge to read Fagan's anthropological tome compared to the glossy periodical, and inevitable television programs, put forth by the Society headquartered in Washington, District of Columbia! Aside from Franczek's mapping, it is known with certitude already that if all the Earth-crust were smoothed into a flat-surface shell covering of the planet's mantle and core, our homeland would then be inundated totally, a single ocean with an depth of 2,720 meters or thereabouts; that is the geographical fact underlying the off-the-cuff comment issued by ready wits that our "Earth" ought to be re-named "Ocean";Carl Sagan (1933-1997) famously penned *The Pale Blue Dot: A Vision of the Human Future in Space* (1994) to endorse that idea. Not only does the natural interior structure of rocky planets such as Earth and Mars determine, in part, their habitability the unnatural deformation/restructuring of their interiors has profound effects on habitability (Noack et al., 2014).

After experiencing the first nuclear device explosion in July 1945, Robert Oppenheimer claimed he thought of a statement from the *Bhagavad Gita* uttered by Shiva: "I am become death, the shatterer of worlds". People can destroy the Earth-biosphere [our world]; someday people may want to destroy the whole planet. Only when the biosphere is destroyed will our species become "Death" in fact, but also a selector of co-survivors! At least three main human-life preserving options seem technically viable employing currently known technologies: (I) induce the whole Earth, more or less functioning as a vast urban matrix vessel, to become a still-inhabited nomadic planet traversing interstellar distances—see: Birch, 1993b—(II) shatter all the non-Sun natural solid material agglomerations within our Solar System—see: Birch, 1993a—to fabricate a Freeman Dyson Satellite Swarm enclosing our Sun or (IIIa) shift the entire Solar System—functionality as is, with its energy-emitting Sun, with all of its orbiting objects included, each functioning under highly artificial circumstances, enduring novel astronomical, geological, biological and technological stresses—as a single intact forged unit to another designated position elsewhere in the Milky Way Galaxy—see: Badescu and Cathcart, 2006; Forgan, 2013 or (IIIb) sail the entire broken-up Solar System in the form of an interstellar mobile Freeman Dyson Satellite Swarm. Putting together humanity's first Sun-centered Freeman Dyson Satellite Swarm will entail occasional catastrophes, such as asteroid-asteroid collisions and asteroid-space station collisions (Debase, 2008). (Every Freeman Dyson Satellite Swarm component will be a directly-controlled or remotely-controlled Sun-orbiting "space station".)

Selected asteroids, boulder-like, sometimes rogue, erratics whatever their derivation, may become conserved artworks—like Michael Heizer's rock, *Levitating Mass*, at the Los Angeles County Museum of Art in California (USA) (Chianese, 2013)! Macro-imagineers should recall that, apart from ore carriers on the USA's Great Lakes and elsewhere in the world's restricted navigable waters, it was not until circa 1956 that an ocean-going class of vessels, bulk ore carrier, appeared. Metals are particularly valued by modern civilization (Goody, 2012; Ravilious, 2013). So much so, in fact, that the potential for mining distant asteroids is an attractive early-21st Century investment opportunity! The materials of Earth, including "earth" found at its surface as part of the planetary crust, was first, and is still today, used to build humanity's homes [mud huts] (Minke, 2009). It can be recycled many times—earth is at the intersection of planetary landscape/seascape and professional Architecture. During, and after terraforming other places (arid and airless Mars? Earth's close-by barren Moon?), off-world macro-objects (big asteroids, moons or others kinds of Sun-orbiting satellites?) the same kind of materials will still be used by industrious builders! The ultimate residence built to contain all humans would be, of course, Freeman John Dyson's Satellite Swarm! But, before the Swarm is formed, Earth could become totally encased by a single closed air-conditioned building—similar to an independent self-sufficient manned space capsule—constructed without damaging the Earth-crust down to its transitional Mohorovicic's Discontinuity base stratum situated atop the Earth-mantle (Prodehl et al., 2013). 21st Century asteroid mining by far-flung humans may initiate a new historical phase of the Earth's Anthropocene geological time period (Whitehead, 2014), de-linking *Homo sapiens'* increasingly bulky infrastructure from (possibly all) Earth-crust mining efforts! Geographos has already dubbed such a restricted place-civilization "Ballardia" (Cathcart, 2002) and assumed, if it existed, that it would function like "protective clothing" for Earth's conserved and preserved biota, shielding against cancer-causing solar and cosmic radiation as well as some other significant physical vicissitudes coming from interstellar space, as posed by action possibilities Option (I) and even Option (IIIa), above. If Aliens inspected our planet today, how many species might they document? After several centuries of taxonomic classification, about 2.3 million species have been listed. However, humans continue to discover and describe more species with every passing year: nowadays, according to Mora et al. (2011), there may be approximately

8.7 million species in all, with 86% of "...existing species on Earth and 91% of species in the ocean still [awaiting] description". Obviously, Greens have a huge job facing them if they are to successfully conserve and preserve all the selected creatures and plants! Only since the 1960s have ecologists begun developing efficient means (handling strategies, remotes sensors and super-computers) for "Big Data" that will permit broad-scale geographical speculations about Earth's "Big Ecology"—the Macro-System Ecology social group and its component academic and governmental elites, as of 2014, remains unformed to vigorously support popularized Green propaganda with massaged, relaxed "facts" (Heffernan et al., 2014)!

Intentional planet obliteration (via Archimedes-caused Earth-crust stripping and new asteroid making event-process) after abandonment would substantially affirm, if not completely substantiate, the absolute onset of the epoch of a "Dosmozoicum", an "Age of [Intelligent as well as technologically advanced] Life in Space". The word's suffix, "zoikos" is Greek, meaning herein pertaining to living beings—humans, through the mastery of nuclear-scale industrial processes have begun to create previously unknown or unused resources through the transmutation of global Nature's elements. Radioactivity of particles, natural and induced strongly by human acitivities in our air alters the Anthropocene behaviors of aerosols (Kim et al., 2014). Our planet's air is perfused with Plutonium-238 nanoparticles formed after the ablation of decayed Earth-orbit satellite power units. Amazingly, past aerial nuclear testing, especially during the 1950s, has made possible for medical personnel to learn that new hippocampus neurons form that are crucial for memory and learning, that they were found to be generated daily in adult humans brains, by laboratory dating utilizing non-radioactive Carbon-14 (Spalding et al., 2013); known neurogenesis rates may play an important role in human behavior. Similarly, grains of silica [sand] extracted from meteorites contain clues of a supernova that triggered our Solar System's formation billions of years ago (Haenecouor et al., 2013) and sand grains [of ancient planktonic foraminifera] are readable clues to a single example of life kind's evolution (Ezard, 2013). "Short-lived" radionuclides—Kryton-85, Tritium, Argon-39, Carbon-14, Krypton-81, Chlorine-36, Beryllium-10—have laboratory-discovered half-lives useful for humans to date local and global geological event-processes via measurable big surges in the presence and concentration of radioisotopes.

Europe's alleged "Waldsterben" [dying forests], incorrectly blamed on human-caused air pollution et cetera, were once predicted to convert Bavaria, Germany, and some of Switzerland into dead-tree wastelands (Horeis, 2009). However inconvenient for Greens, it just was not true! Global Nature is only rarely static often it springs surprises on all its inhabitants, in short global Nature sometimes has its "disvalues", as Rolston (1992) termed them. Scurvy once limited human travel time in the task of circumnavigation Earth (Chaplin, 2012). [The linkage of navigation with human knowledge is preserved in the terminology we employ when we "navigate" the World Wide Web! Sometimes, gross obeisity is the result of such sedentary use of computer search engines!] Since then, humans have been compared to a very unpleasant and deadly disease, named a planetary "skin cancer" by some overly opinionated Greens! In a word, planets are essentially macro-objects for pity and consumption by future humans, cyborgs beings and (possibly soulless) Terra-creatures, post-*Homo sapiens* creatures with minds and manipulative physical appendages. The public's existing "Space Age" phrase only approximates partially the spatial scope and import of Dosmozoicum. When artificial outdoor light is scattered by air's molecules and aerosols and returned to the Earth-surface it is called "skyglow"; it may be affecting biodiversity in nocturnal landscapes and seascapes (Kyba and Holker, 2013). Hubertus Joseph Heinrich Strughold (1898-1986), **Figure 6**, was a physician-physiologist and, most notably, the coiner of "space medicine", the earliest usage of which is attested, as the etymologists say, in 1948 (Campbell and Harsch, 2013). Strughold had built upon the observatons made by heroic and foolish balloonists and aviators (Holmes, 2013). Stughold's forte was to make the public aware of humans and their potential role in outer space—he took the common public view of the human body far beyond the impressively memorable physiological cartoons of Dr. Fritz Kahn (1888-1968) by adding realism and context of a modern nature (Debschitz, 2013).

Figure 6. Dr. H.J.H. Strughold, circa 1960. (Image: WIKIPEDIA.)

In 1958, Dr. Strughold became our world's first Professor of Space Medicine, working within the Department of Space Medicine of the School of Aviation Medicine (established 1949, designated "School of Aerospace Medicine" in 1961) of the United State Air Force. Recognizing the psychological moment, Dr. H.J.H. Strughold's neologism, Dosmozoicum, and the concept it was meant to broadly encompass, helped immensely to forge a lasting strong bond between all closely associated professions, a working critical mass of thinkers and visionaries. Strughold's skill in forming concepts was underpinned by his evident ability to conjure images and processes, apparently by sheer intuition; the real impetus for today's established professional Macro-Imagineering changes came from the past blue-sky cogitation of scientists, macro-engineers such as Frank Paul Davidson, **Figure 7**, and Robert Bartholomew Panero, **Figure 8**, avant-garde architects, diligent realistic sociologists, "utopia" planners and other such talented persons very much like himself. Clair R. Patterson (1922-1995) postulated that separate brain regions were used for scientific versus engineering modes of thought (Patterson, 1994); subsequently, the lateral frontal pole prefrontal cortex area of the human brain may be the seat for strategic planning, multi-tasking and decision-making in humans—a brain area that exists only in humans (Neubert, F.-X. et al., 2014). That concept is unproven, possibly because it has never been clinically examined fully! An acceptable, widespread public visional concept of global Nature needs to be promulgated immediately (Earley, 2013)! Every human's imagination is an immeasurable treasure (Daniels, 2011).

Figure 7. Frank Paul Davidson (1918-2014), the American academic doyen of Macro-Imagineering. He has many confreres. Geographos affectionately refers to Davidson as America's pioneer, in effect its "MACRO Polo" to use a hypocorism [nickname]. (Image: FPD.)

Figure 8. Robert Bartholomew Panero (1928-2001), the visionary fieldworker in American Macro-Imagineering. (Image: HMP.)

In 1955 used the term "ecosphere" as a synonym for the life inhabitable zone surrounding our Sun. Other useful Hubertus Strughold word coinages: "Bioastronautics" in 1957, succeeded by "Astrobiology" and "Spatiography". "Aerospace Medicine", in the 21st Century may appropriately be applied to the healthcare and welfare of all tellurians, humans, wherever they reside and work eventually in our Solar System! The difference in denominative suffixes, "-ics" and "-ology" is telling: "-ics" identifies a particular field of inquiry but not necessarily a particular means of approaching that inquiry, whereas "-ology" means merely "discourses about" (Proctor, 2007). His intent, clearly, was to use his unique power of imagination to create a new world of discovery and healthy joy. Furthermore, he wisely divided the extant Solar

System into two spatial-material categories: (I) Spatiography—the word first became officially recognized, and made its first appearance at page 311 of Volume I of the *Oxford English Dictionary Additions Series* (1993)—focuses on revealing the nature of universal space-time whilst (II) Planetography gives scientists data and winning scientific theories about all macro-objects discovered in outer, extra-Earth space. Actually, space is not a void, it not an empty volume between more materially dense places, yet nowadays, somehow, Strughold's Spatiography must usefully center on matter and energy movements under all perceivable vicinal cosmic conditions. Both Macro-Imagineering and Science technophiles ought to pursue their Space Age/Dosmozoicum tasks within Strughold's schema for the best results and optimal outcomes for humankind's future generations! Both Science and Macro-Imagineering are bounded by what people can determine from global Nature. Planetography was boosted by Strughold during 1953 by popular publication of *The Green and Red Planet: A Physiological Study of the Possibility of Life on Mars*, where he indulged in an informed speculation centered on the contemporaneous inhabitability of Mars (Cerceau, 2013), even its areobotany—the study of hypothetical Martian vegetation, its areoflora and areofauna! [Terms containing "areo" are direct equivalents of "geo".] In other words, by extension, Dr. H.J.H. Strughold was a profoundly meditative, almost oneiric, global medication "Macro-Imagineering" advocate! Any conscious juggling of mental representations is crucial to complex and creative behaviors such as imagining (Schlegel et al., 2013). Using his imagination, and his acquired technical expertise, Hubertus Strughold broke a path for others to follow his example to an over-arching geo-theory (Marriner et al., 2010).

A modern, living equivalent of Dr. Hubertus Strughold might be J. Craig Venter (born 1946) whose company created the first bacterial cell controlled by chemically synthesized genome—"*Mycoplasma mycoides JCVI-syn1.0*"—reported during 2010 after its corporate-laboratory, private-sector business creation. Venter and his colleague now postulate that organisms can be teleported distantly [Mars?] whenever a "Digitized Life Sending Unit" is perfected, that a copy of an organism's DNA is made in one place and sent electronically as a file to a device, or factory, somewhere else in this Solar System that can flawlessly recreate the original organism. In other words, a working *deus ex machine*! Were he still alive in 2014, Dr. Hubertus Strughold might well enjoy reading, as Geographos most certainly did, J. Craig Venter's provocative **Life at the Speed of Light: From the Double Helix to the Dawn of Digital Life** (New York: Viking Adult, 240 pages, 2013)!

During 2013 a brave man (famed Hollywood movie-maker James Cameron) dove alone into the ocean's Mariana Trench inside a tiny personal submarine-like conveyance, reaching the sea-bottom almost eleven kilometers below in two hours and thirty-six minutes whilst another courageous male individual (Felix Baumgartner, Austrian daredevil), pressure-suited and equipped with a special kind of deployable parachute, endured a stupendous four minute and twenty-two second free-fall from an altitude of nearly thirty-nine kilometers! These daring pioneering human feats could never have been accomplished without many previous years of aerospace medical science preparations! "Aerospace Medicine" seems to be a plastic term, easily extendable to include planet-dwelling modifiable humans (Preester, 2011; Liao et al., 2012), cyborgs and probably Terra-creatures too! And, naturally, the organized integration of information/data from many different modes of inquiry, across varying cultural and scientific investigative traditions, opposing incomplete hypothetical evidence in support of ill-posed life-survival macro-problems played an important role undergirding these two recent human adventures.

On a larger geographical/Solar System-wide spatial scale of operation, Macro-imagineers, for example, have had to continually bypass or overcome the "Guru Effect" of some key-person professionals (Sperber, 2010), including the eye-catching technical exaggerations to be found in the glossy magazines such as *Time, Scientific American, Discover* and *New Scientist*, pretentious books and ineptly-translated the mass media (noisy, repetitive violent special effects [Special Effects]-dominated films) of popular science-fictionists (Schneider, 2006; Heise, 2012). Some philosophers of relatively new-won repute excitedly allege that humanity's Space Age recorded history period is certainly drawing to a close, examped commonly by the scheduled 2020 or 2024 de-orbiting of the most expensive and complex Big Science/Big Ecology object ever built—the approximately 100 billion USD[$] public expense of the so-called International Space Station (Johnson, 2013). The social impact of R&D—the overused and inexact word "impact" really conveys an expression entailing mostly negative connotations (erosion and debilitation), according to Irene Machado (2006)—encompasses cultural, environmental and economic results, sometimes useful and sometimes indisputably non-utilitarian (Arnould, 2013).

Macro-imagineers, those persons who undertake to embody prospective macro-project plans contrived by a prospering and respected Macro-Imagineering profession, already realize that Cameron and Baumgartner moved rapidly through Earth's substances (air and seawater) that are far different from earlier times when Nature (Jelinski, 2005) existed unmolested and *Homo sapiens* was everywhere absent from the planet's natural scenery. Some polluted landscapes of highly industrialized or densely populated Astythrome's are overlain by smoggy air layers (Silva, 2013) as well as many artificially radioactive particles, fallout from nuclear weapon tests and commercial nuclear reactor operational accidents; similarly seawater is contaminated and the sea-bottom is littered with nearly unimaginable kinds and quantities of accidently dropped or deliberately discarded junk, tragic or fiendishly profitable sunken wrecks and other (almost indescribable) solid stuff. Macro-imagineers must, sooner or later, come to terms with all the impositions and imbrications caused by global climate change—for example, the planning disconnect between dire Green predictions, fostered by Big Ecology, of so-called anthropogenic global warming and "...dominant geopolitical framings of the future" (Hommel and Murphy, 2013). Macro-imagineers must counter their exaggerations with sound proposals for "critical infrastructure" that supplies continual services central to civilization's functioning. One of the latest means to do so, which provides a facility for collaboration to construct an imaginary city or installation, is freely provided by *LEGO* in the form of its online "Build with [Google] Chrome" (GOTO: www.get.webgl.org), released during 2014.

Approximately 50% of early-21st Century humanity lives in skyscraperscapes [Astythromes] constructed with concrete, steel and other transmogrified matter. By 2013, the total live weight of all seven billion humans was about 350 billion tonnes and humans pour about four cubic kilometers of concrete annually! The global scale of mankind's mining industries (minerals, petroleum, gas)—present both on the accessible landscape as well as penetrating the seafloor of the Earth-crust—best exemplifies the Anthropocene concept—in addition to artificial radioactivity instigation—by its aggregate geographical scale and the accompanying widespread ecological changes. Successful future commercial asteroid mining (Badescu, 2013), bringing industrial resources to the Earth-biosphere on a periodic timetable, could sever that limiting urban matrix forever—in effect, taking the judicious admonition of William Penn Adair "Will" Rogers (1879-1935) seriously: "When you find yourself in a hole, stop digging". In time, humans won't need to, or require anymore, digging to excavate, to fill holes (Liu et al., 2013), remove mountains, lop mountain summits or even process many vitally useful elements and factory-made machine parts! [Humans currently mine the air, which is 79% nitrogen gas, 21% oxygen, for industrial nitrogen fixation purposes, the production of nitrogenous fertilizers used in agriculture, a method that replaced guano mining (Melillo, 2012). The ever-increasing fertilizer use since 1940 is responsible for a marked rise in atmospheric nitrous oxide, a major greenhouse gas (Park et al, 2013).] Initial ore and product shipments to Earth extracted from asteroids will decrease the price of those materials and products and cause a reduction in the costs of products made with the imported matter. Asteroid mining will materialize the "cargo cult" mineral policy espoused by some extractive industry leaders—that is, a rich mineral discovery will seem to be a valuable cargo falling, following directive nudging, under the influence of gravity almost inexplicably from Heaven! Geographos reminds readers that most documented myths about the moment of humankind's first acquisition of fire suggest a source in Heaven—lightning—rather than active volcanoes (Pyne, 2012, page 140). An additional resource might be hot rocks fallen from the sky and quickly found useful! Mining other celestial bodies in outer space would make it improbable that humans would continue to tear huge bits of mass out of the Earth-crust, in that effort making it more difficult for other life forms to survive, only to supply raw materials for our civilization.

Macro-imagineers know that their lungs have a finite capacity, that to inhale more life-giving air, each must first exhale in a continuous automatic breathing action; human lungs have odor receptors, like those in our nose, that make us cough sometimes when we breathe industrial solvents, perfumes and other inhaled stimuli (Gu et al., 2013). Carbon dioxide is released into the atmosphere, an act that helps, in part, to sustain other known life forms. [Earth's crust belches carbon dioxide gas: for example, on 24 August 2013, near Italy's largest airport, Leonardo da Vinci International Airport 35 kilometers west southwest of Rome, opened in 1961, a sudden carbon dioxide gas eruption from the ground occurred in the Tiber River delta, causing quite a social commotion! When air's temperature falls below minus 70 C—as can happen often in frigid Antarctica—carbon dioxide gas freezes and its contentration drops to zero; this uncommon phenomenon is more obviously present on Mars where visible dry ice masses vary seasonally!] Why do humans, like plants, sense aerial carbon dioxide gas? According to plant biologist Wolf B. Frommer (2010): besides stimulating our lungs to function properly "...it is not clear how humans sense the gas, nor what [still evolving species survival] advantage this might serve." Research done since demonstrates that the "bite" of carbon dioxide in carbonated drinks, mineral water and soda water felt within the human mouth, for example, increases enjoyment of some beverages (Wise et al., 2013). Beverage carbonation is known to make it difficult for the human brain to distinguish between natural sucrose and an artificial sweetener such as aspartame prevalent in dietetic cola drinks (Di Salle et al., 2013). Carbon dioxide gas is a byproduct of cellular metabolism; each human daily exhales 20 moles of that gas. Through internal protein Connexin 26, carbon dioxide can directly signal a human's physiological information via chemosensory reflexes that regulate blood gas content by the frequency and depth of inhalation/exhalation (Meigh et al., 2013). What is obvious, however, is that humans generally possess a carbon dioxide gas-sensitive neural circuit called a "rhythm generator". Taking in oxygen, releasing carbon dioxide, humans nourish plants and, therefore, the Earth-biosphere. Laboratory analysis of inhalations can be used to assess human exposure to the atmosphere's toxic substance content (Vereb et al., 2011); human exhalations are a metabolic "breath-print" unique to each person (Sinues et al., 2013)! Humanity's innumerable internal combustion engines emit carbon dioxide gas after burning exhumed fossil fuels (coal, petroleum, natural gas). Meanwhile, the atmospheric oxygen gas we breathe is being reduced steadily by our Solar System's "perpetual" natural space weather erosion process (Catling and Zahnle, 2009).

Some professional philosophers as, for example, Neil Paul Cummins (born 1975), purport that the only special Darwinian evolutionary, existential purpose of *Homo sapiens*, as a species, is to be the savior of all Earthly biota by means of carbon dioxide gas sequestering macro-projects (Cummins, 2012). In addition, others purport that a future Synthetic Biology may create a new living thing, something akin to a cell (Park, 2013), that will be "born" in the laboratory rather than outdoors without being derived from an existing cell; it will be a cell without ancestry. Our noble species' efforts to prevent extinctions might be an unnatural act the ultimately harms the health of the Earth-biosphere since extinction is global Nature's method of improving the viability of that human-defined/demarcated voluminous shell. Another tentative justification, which is enhanced by future asteroid mining, might be securing the Earth-biosphere from destructive natural super-bolides. In the bargain, human colonization of our Solar System is a beneficial aspect, helping to insure the survival of *Homo sapiens* against extermination/extinction cosmic event-processes! In other words, Dr. H.J.H. Strughold's aerospace-centered medicative Macro-Imagineering comes to the forefront as a useful human survival strategy. Interestingly, a 2.7 kilometer-long asteroid, 1998 QE2, which passed beyond the Earth-Moon orbit on 31 May 2013, was compared by the reporting science news media with the luxurious flagship ocean-liner *Queen Elizabeth-2*: the asteroid 1998 QE2 was nine *Queen Elizabeth-2* ship-lengths in size (Yeomans, 2012). The first near-Earth asteroid was spotted moving through nearby outer space in 1898 and the 10,000[th] such documented find occurred on 18 June 2013 when the 300 meter-diameter 2013 MZ5 was first logged and tallied by dedicatedly watchful amateur and professional astronomers.

Prior to 2014, in the flesh, humans only ventured a mere one light-second out of the Earth-biosphere—risky explorational jaunts to the Earth's circling Moon; humans accomplishing that outing multiple times was financially costly and the few trips undertaken revealed that our Solar System is a dangerous region of the Universe in terms of life's survival chances, especially for intelligent human spation-auts. Humanity may desire to endure most numerously in the Earth-biosphere, yet seek to extend its presence and aggregate manipulative capabilities to far-distant places (for example, the nearby Moon, an obviously open to human colonization Mars and, even farther away, to the Solar System's main Asteroid Belt). [So far, only five probes have been launched to eventually exit our Solar System.] However, *Homo sapiens* must not only "conquer" the Solar System—only with the presence of a Freeman Dyson Satellite Swarm could we ever "control" our Solar System—and, someday perhaps at least the Milky Way Galaxy, but also to "resist" as best it can all life-threating event-processes that constantly, intermittently and periodically impinge our beloved home planet, our manned spacecraft missions and our extra-terrestrial settlements (Bonnet and Woltjer, 2008; Wells, 2009)! When once-in-a-lifetime natural events happen under, at and above well-populated Earth-biosphere regions, the news-watching world-public reels from surprise and, often, that surprise is followed by a general population sense pervading physical insecurity. In other words, the whole Solar System planet we have named Earth is changeable, sometimes in un-wanted ways, and all life in our orb is mutable and killable (Jones, 2013). Skilled humans are monitoring Earth with globally distribut-ed small "smart rocks" that record-report important movement data on common events like river erosion (Underwood, 2012) and, since the start of humanity's H.J.H. Strughold-commenced Dosmozoicum, by remote-sensing data-gathering Earth-orbiting satellites as well as long-distance space and planetary probes. Even real-time tiny sensors that can disperse like aeroplankton are available to scientists pursuing data on the Earth-atmosphere's currents and contents!

During 1932, a surveyor for the Egyptian Geological Survey investigating the Sahara Great Sand Sea drove his vehicle over a vast surficial deposit of unusual glass fragments. More than 80 years later, the origin of that widespread stratum of fused desert soil, estimated at about 1,400 tonnes, is still a matter of scientific debate. One of the original emplacement theories is that the stratum was generated *in situ* by the powerful aerial burst of a large asteroid, a superbolide, falling from outer space, which heated the then-extant aerial Earth-crust rocks (sandstone?) to an extremely hot liquid which, after cooling, subsequently formed a yellow-green glass (Boslough and Crawford, 2008). If this tentative theory is correct, then the glass created by radiative melting would be analogous to radioactive Trinitite, a remarkably similar dispersed glassy anthropogenic rock formation created by thermal radiation from the world's first nuclear weapon explosion on 16 July 1945 at the proving ground in the State of New Mexico, USA (Szasz, 1995). The most significant difference between the anthropogenic wartime USA event and the ancient, un-witnessed by humans Great Sand Sea glass-creation event was that the very ancient aerial explosion was megatonnes of TNT yield, not kilotonnes!

Historians have recorded visible powerful and surprising airbursts of various objects fallen from space—the 30 June 1908 Tunguska superbolide above Siberia which may have released in an instant between 3 and 5 Megatonnes of yielded energy as well as the 500 Kilotonne yield airburst over Chelyabinsk on February 15, 2012. Sound waves from Chelyabinsk object, which detonated over Russia on 15 February 2013, were detected by infrasound watch-stations that function under the Comprehensive Nuclear-Test-Ban Treaty Organization (Balcerak, 2013). The large number of objects in a range of orbits of the Sun means that some of them will, inevitably, intersect the Earth's orbit, becoming meteors. Both superbolides generated powerful ballistic shockwaves when they fragmented in the Earth-atmosphere, leaving a pronounced detectable trail of meteoritic dust (Gorkavyi et al., 2013). **Figure 9** is an image of the uppermost, visible-from-satellite aerosol trail (at possibly more than 53 kilometers altitude) of the 2012 Chelyabinsk superbolide, transmitted by the geosynchronous-orbiting Japan-owned satellite MTSAT-2. More and more, it is realized by all networked humans that such falling and exploding space objects also put our civilization's infrastructures at great risk and even imperil people (Mignan et al., 2011). Days after its violent entrance above Chelyabinsk, the loopy dust plume became distributed widely over the Northern Hemisphere (Miller et al., 2013). A very dense dust cloud, possibly caused by a comet colliding with Earth, or a superbolide, sometime during AD 536, caused a widespread "dry fog" of similar aerosols (Stothers, 1984; Rigby, et al., 2004).

At the present time, the USA tracks near-Earth space objects using an outmoded detection/monitoring system commissioned in 1961. However, on Kwajalein Island in the northern Pacific Ocean the first radar site in the S-Band Space Fence macro-project network should be installed and operating by 2018 for more complete space surveillance—the art of tracking objects in outer space near the Earth. Superbolides on the scale of Tunguska occur only about every two centuries, while it is estimated the Chelyabinsk-scale event-processes may happen only once a century. A battery of small telescopes situated in Hawaii, possibly operational by 2015, has been designated by the NASA as the Asteroid Terrestrial-Impact Last Alert System (ATLAS by acronym). Its function will be to detect, minimally, 50 meter-diameter asteroids ["city killers"] at least one week before its impact on the Earth-surface, where crater-generation by explosive shockwaves is the dominant destructive process-event (Shuvalov et al., 2013). Imaginative macro-engineers, drawing on a famous story from *The Holy Bible*, opted to provide Earth with an offensive capability against threat-posing "killer" (potential global doomsday event-process causing) asteroids, by capturing and re-directing for collision on any targeted killer asteroid a David's stone asteroid (Massonnet and Meyssignac, 2006)!

Figure 9. Visible recent evidence (Chelyabinsk) of the newly-apprehended danger to Earthlings and supportive infrastructures, as imaged from outer space by a human-manufactured, automated observation artificial satellite. An entire Russian scientific journal's 2013 issue was devoted to this sudden and shocking infall process-event. See: *Solar System Research*, Volume 47, Issue 4, July 2013, pages 239-345. In short, it might have become, were it larger, a 500 Kilotonne yield Rolston "disvalue" event-process, quite a surprise! (Image: ESA and WIKIPEDIA.)

If distant Aliens had been studying Earth on 16 July 1945 at 5:30 A.M. local time it is unlikely they would have noticed the brief flash of light at the open-air scientific laboratory of New Mexico's desert landscape, but it is likely that Aliens may have followed Earth's development at interstellar distances, noting the measureable changes of our home planet from 4.5 billion years ago, shortly after its aggregation, until the present-day—for the Aliens, of course, "present" is limited by the speed of light. ["Present" would be our history, years past. Human "history" records an account of what some groups of *Homo sapiens* have observed, that is a small part of what has actually occurred during Earth's Geologic Time.] The Earth-biosphere contains billions of human beings struggling to make and earn a living, to elevate many people out of indisputable economic poverty. Ameliorating world poverty through economic progress, according to some confirmed, perhaps radicalized, Green enthusiasts, using mined coal, petroleum, natural gas, and uranium is enlarging the risk to the Earth-biosphere of catastrophic anthropogenic climate change.

In response to that slightly exaggerated macro-problem formulation, Neil Paul Cummins and others (Cocks, 2013) have suggested a vigorous *Homo sapiens* civilization-wide effort to effectively and efficiently mitigate the carbon dioxide gas buildup in the Earth's air because of ongoing poverty diminishment due to industrialization activities (Robin and Steffen, 2007). In other words, humanity's overall advancement has, to the sickly neurotic Green way of thinking about the world presently inhabited (Robbins and Moore, 2013), instigated an Earth-crisis—specifically the biosphere is phasing into an amazingly fantastically uncertain future. "Crisis" is routinely described in dramatically declarative media proclamations by persons and untaxed groups with potentially profitable monetization agendas. One outcome is an "overnight" fashion development—climate fiction, by 2012 shortened to "Cli-fi", on the abbreviation model of "Sci-fi" [science-fiction]. Cli-fi is fundamentally dystopian story-telling with a topical focus on the effect of natural and human-caused global climate change and such nightmarish change's posed risk for civilization's continuance. The 1962 novel *The Drowned World* by James Graham Ballard (1930-2009)—Geographos intended to honor the prolific writer with the invented term Ballardia—foretold a tale of Earth's sublimating icecaps (in Greenland and Antarctica) and rising global sea-level leading to a human civilization consisting of salvagers only, roving tribes who mine urban ore and submerged technology! Obviously, the SF-laden movie *2012* was a theatricalized version of the crude Cli-fi entertainment genre. The world's first newspaper-published weather map appeared in 1848 in response to the needs of the new railroad travelers so, of course, travelers in outer space may desire new kinds of cartographic/video representations of planetary weather regimes—Earthly or non-Earthly.

Ballard's entertaining, leisure-time book arrived after authentic climate science and the mass media had merged coverage of newsworthy natural climate change examples in the 1930s and, more clearly and with an increasingly anti-*Homo sapiens* bias during the 1950s and continuing to present time. 1945 to 1950 marks the period that Geographos has pegged as the appropriate counting base for the Earth-biosphere's impending climatical phase for it was about that time that most of humanity came to think, some individuals came to believe, our species had acquired the capability to destroy life's home, or to become over time the Earth life's "savior" (Masco, 2012). Macro-Imagineers who work on a geographically large-scale must, therefore, harness their technically- trained curative imaginations concerning their birth-planet and technology as a method of seeing, sensing, thinking, and dreaming that results in the conditions for massive material interventions in the Earth-biosphere and elsewhere that are securely bridled by worldwide geopolitical sensibilities (Fleming, 2010; Hamilton, 2013). The Anthropocene is the Age of Nuclear fire, anthropogenic atomic fission and fusion, whether contained in a controlled stabilizing device or detonated in a hellish instant of weaponry mass destruction creating artificial isotopes. Artificial radionuclides ought to be the primary marker for the basal boundary [stratigraphic marker] of the Earth's Anthropocene (Leslie and Hancock, 2008). Resolution of the paradox of the Sun and Earth's ages began only after 1896 following Henri Becquerel's discovery of radioactivity, introducing the previously unknown concept of nuclear energy. Ultimately, radiometric dating for rocks became a standard technique used by geoscientists. [Fallout presence was first detected in the Northern Hemisphere in 1945 but not until 1955 in the latitude 30^0 to 40^0 south of the Equator.] "The [radioactive] fallout [following nuclear and thermonuclear explosions in the Earth-biosphere] was a major turning point in global environmental awareness....it altered the questions scientists asked about Earth's vulnerability to human-induced change. The radioactive 'fallout" controversy ensured that scientists would never look at the Earth in the same way again. Suddenly and on a global scale, scientists wanted to measure everything, to see how it was changing." (Hamblin, 2013, page 95).

"Anthropocene" may become an officially-accepted term as a Geological Time period by as soon as 2017 or 2018; yet to be accepted as a global Geological Time period, it may eventuate to be the briefest and the final identified time for the Earth (Szerszynski, 2012)! Nevertheless, scientists have suitable and satisfying new forums to report their findings: *Anthropocene* and *Earth's Future*, both starting their journal publication runs by 2013. Its imperfect recordation and preliminary assessment by subjective historians is underway, but remains incomplete as of 2014 (Robin and Steffen, 2007; Steffen et al., 2011; Brown, 2013; Ellis et al., 2013). Asteroids are included, seriously considered by scientists as well as those who examine humanity's collective actions, because they are now feared by thinking Earth-biosphere dwellers with a strong sense of the future, predictable process-events foretold. Asteroids are threats to civilization, the Earth-biosphere as a whole but they are also perceived and appraised as exploitable future resources for a *Homo sapiens* clearly intent on moving throughout the Solar System in a sustainable way (Olson, 2012).

Geographos' preference is to view the Anthropocene's start, as usefully defined by unequivocal nuclear particle evidence of significant human capacity for Earth-biosphere macro-engineering, when it becomes apparent and/or evident in the species' geo-archaeological record (Ziarek, 2011). Thus, the "Present" is 1950 by those often quoting radiocarbon dates. Certainly associated Earth changes owing to mankind's agency/technology's impact—pavements, railroad tracks, toxic waste dumps and skyscraper foundations, urban rubble piles created by warfare (**Figure 10**)—have reached gigantic dimensions and Nature-overwhelming dynamics. Anthropocenic geographical and historical research requires considering multi-scale interactions—from the individual person to the whole planet. Most assuredly, the Anthropocene is not conceptualized by Geographos as history of *Homo sapiens'* decline to some pre-historical Earth-vessel simplistic state, or worse still, planetary ecocide without a comfortable and utterly safe outer space conveyance escape-vehicle conveniently available for delicate and perishable human passengers! Geographos assumes people are agents for positive process-events of all kinds and descriptions and seeks only to advance, progressively, all human work agential entities inside this Solar System's heliopause.

Figure 10. Manhattan Island's clustered skyscrapers, New York City, by NASA radar satellite imagers. (Image: the NASA.) These buildings and structure have the appearance to the inexperienced layman, and even the trained technician's eye, of an ordinary surface rock formation, in fact nothing unusual. Central Park is the large green-colored rectangle at the image's upper-right-center. Without any scale, the image could be mistaken for a large laboratory sample of Trinitite, a river embankment or a beach (Makowski et al., 2013) composed entirely of discarded sea glass, fragments of ceramic objects and glass that have been in rivers and the world-ocean, "Anthropic Rock" (Cathcart, 2011)! [Anthropic Rock has been made to exist even on the Moon's crust-surface (Clegg et al., 2014).]This peculiar image erases the collective human memory and name identity of the buildings and structures as well as what they symbolized. A terrorist lodge, Al-Qaeda's, dream scenario?

By the early-21st Century, *Homo sapiens* had an indelible geological and supra-planetary extent of impression. [Perhaps one of the most controversial futures of the Earth-biosphere is the possibility of humans synthesizing life itself. However, Geographos won't delve into those speculations in this text.] Erle C. Ellis (2011), an American professional geographer, suggested that humanity's use of land has transformed most of the terrestrial biosphere into "…anthropogenic biomes (anthromes), causing a variety of novel [agricultural] patterns and processes to emerge". Human harvesting of the Earth-biosphere's net primary production has doubled during the 20th Century, and many anthromes are expanding in area whilst others are reducing in area due to farming/ranching technology improvements, as well as biological science developmental innovations (Krausmann et al., 2013; Sayer and Cassman, 2013). Geographos adopted the potent idea that humanity's discontinuous network of widely dispersed world-class cities are usefully identified as Earth's non-contiguous Astythromes—"asty" is Greek for town or city—densely settled, geographically extensive regions where discretionary urban ore mining may take place, where brownfields may be decontaminated, and where polluting nuclear waste is almost religiously excluded from any possible contact with city-living people (Norra, 2009). Of course, Astythromes and Anthromes are geophysical complements and both are cossetted by time, the Anthropocene. Still under active debate amongst committees of scientists as an appropriate, applicable term of 21st Century art, assuredly the Anthropocene is an event that could become a professional recognized unit of Geologic Time—in other words, a deposition of an urban stratum containing unnaturally radioactive concrete, and other artificial radioactive materials. [Chernobyl (in the former USSR) in 1986 and Fukushima (Japan) in 2012 have added newer fallout particles to the terminated aerial nuclear weapons testing era (Kim et al., 2014)] Our Space Age truly began when it first became possible for nuclear warheads to be lofted into outer space, orbited by rockets during October 1957 and during the years since, and that commonly employed adjective befits our ongoing historical era of Solar System exploration and future resource exploitation. "Atomic Age", another facet of Anthropocenic etymology, was coined by William Leonard Laurence (1888-1977), a *New York Times* reporter who was an eyewitness to the awesome "Trinity Test" of 16 July 1945 also experienced by Robert Oppenheimer.

Global human population totaled more than seven billion persons by 2014. A United Nations Organization statistical analysis report issued 13 June 2013 projected a 2100 human population ranging from nine to eleven billion persons. Worldwide, Greens often caution all who listen and read their manifestos that *Homo sapiens* is the greatest mortality threat to non-human life confined within the decaying Earth's biosphere. Our threat to life stems from the uncontrolled cleverness, inventiveness and general propensity to reproduce excessively! Just a little more than one-half of a century hence, by 2070, it is estimated 70% of all humans will live in cities, hence the emphasis herein on the key future roles of globally networked Anthropocene regions comprising Astythromes! [After 2050, possibly, more people will be 60 or older than are children 15 or younger.] Greens allege in multiple popular media forums that humans are causing global atmospheric warming—that people are nosopoetic, the cause of a "cancer" on Mother Earth's surface (Hern, 1993; MacDougall, 1996). In other words, it is their false and ossiblyp contention that all humans are pathognomonic, their pervasive presence indicating that a particular Earth-biosphere disease is present beyond any doubt! Earthidermal diseases—syndromes, if you will—are obvious in some Anthrome regions—for example, the Aral Sea or Lake Chad "Syndromes" and even the more and more common [Nighttime] Light Pollution Syndrome.

In the next several chapters of this textbook, Geographos will elucidate the outlines of some reasonable corrective macro-projects, plans that emanate from a Disneyesque approach called Macro-Imagineering. Additionally, Geographos will detail some of the plans spawned by Frank P. Davidson (**Figure 7**) in his books and articles as well as those of Robert B. Panero (**Figure 8**) in Hudson Institute reports and elsewhere, including reports promulgated by his own corporation authorized by the State of New York during 1974. Geographos is cognizant that some Green fanatics may characterize the Chief of Research and Development as a monster, intent on our precious world's destruction. So, the Chief offers an appropriate self-deprecating image of himself below, **Figure 11**, portraying himself as the Rylos traitor "Xur", sadistic sell-out to the evil Ko-Dan Emperor. In the movie *The Last Starfighter* (1984), a mockingly threatening "Xur" visage appears in a grotesquely distorted holographic projection inside the Rylan Star League's Starfighter Command Center to issue an ultimatum to his own people. **Figure 11**, below, replicates that horrifying image, with the Chief's facial image inserted within the outlines of the squashed, reddened, fast-rotating Earth, enmeshed in the Dyson Motor, and on the verge of explosion! Geographos believes that children born today, who will soon play with physical, not virtual, LEGO toys, making spaceships and other futuristic things, will understand poor old Geographos when they are fully grown (Reid and Goddard, 2013)!

Figure 11. R. B. Cathcart: affectionately known by Green intolerants as pseudo-"Xur". Cathcart is a religious person, and considers mostly secularized "Global Nature as Religion" Greens as negligent Earthlings. Besides, why would anyone want to be mean to Geographos when they can bash Christian Waldvogel (born 1971) (Waldvogel et al., 2004)? (Image: JMH.)

"World" has an interesting etymology: from the Old-English "weorold"—composed from "wir" meaning man and "yldu" meaning age) it means "The Age of Man". So, it is not too much of a conceptual leap to consider "World" and "Anthropocene" as useful equal co-descriptors for the Earth-biosphere's present-day state. Becoming *Homo sapiens* [human] must be a simultaneous event-process with the emergence of "World". Geographos will use these terms interchangeably without further explanation or elucidation.

References Cited

Anker, P. (2005a) *The Ecological Colonization of Space.* Environmental History 10: 239-268.

Anker, P. (2005b) *The closed world of ecological architecture.* The Journal of the Architecture 10: 527-552.

Arnould, J. (2013) *Tele-Reality: How Space Technology Transforms Human Perceptions of Space, Time, and Self.* Astropolitics: The International Journal of Space Politics & Policy 11: 231-237.

Bacon, S. (2013) *"We Can Rebuild Him!": The essentialisation of the human/cyborg interface in the twenty-first century, or whatever happened to **The Six Million Dollar Man?*** AI & Society 28: 267-276.

Badescu, V. and Cathcart, R.B. (2006) *Use of class A and class B stellar engines to control Sun movement in the galaxy.* Acta Astronautica 58: 119-129.

Badescu, V. (Ed.) (2013) *Asteroids: Prospective Energy and Material Resources.* (The Netherlands: Springer) 256 pages.

Balcerak, E. (2013) *Nuclear test monitoring system detected meteor explosion over Russia.* EOS: Transactions of the American Geophysical Union 94: 384.

Baker, K. (2008) *The Lightning Field* (New Haven: Yale University Press) 159 pages.

Ball, P. (2012) *Curiosity: How Science Became Interested in Everything.* (Chicago: University of Chicago Press) 465 pages.

Bianconi, E. et al., (2014) *An estimation of the number of cells in the human body.* Annals of Human Biology 40: 463-471.

Birch, P. (1993a) *How to Spin a Planet*. Journal of the British Interplanetary Society 46: 311-313.

Birch, P. (1993b) *How to Move a Planet*. Journal of the British Interplanetary Society 46: 314-316.

Bogard, P. (2013) *The End of Night: Searching for Natural Darkness in an Age of Artificial Light*. (New York: Little, Brown and Company) 323 pages.

Bolonkin, A.A. and Friedlander, J. (2013) *Explosion of the Sun*. Computational Water, Energy, and Environmental Engineering 2: 83-96.

Bonnet, R-M and Woltjer, L. (2008) *Surviving 1,000 Centuries: Can We Do It?* (The Netherlands: Springer) 422 pages.

Boslough, M.B.E. and Crawford, D.A. (2008) *Low-altitude airbursts and the impact threat*. International Journal of Impact Engineering 35: 1441-1448.

Breedon, J.L. (May-June 2013) *Gravitational Assist via Near-Sun Chaotic Trajectories of Binary Objects*. Journal of the British Interplanetary Society 66: 190-194.

Brown, K. (2013) *Plutopia: Nuclear Families, Atomic Cities, and the Great Soviet and American Plutonium Disasters*. (New York: Oxford University Press) 406 pages.

Burleigh, S. et al. (2003) *Delay-tolerant networking: an approach to interplanetary internet*. Communications Magazine, IEEE, 41: 128-136.

Byles, J. (2005) *Rubble: Unearthing the History of Demolition* (New York: Harmony Books) 353 pages.

Cally, P.S. and Moradi, H. (2013) *Seismology of the Wounded Sun.* Monthly Notices of the Royal Astronomical Society.

Campbell, M. and Harsch, V. (2013). *Hubertus Strughold: Life and Work in the Fields of Space Medicine* (Neubrandenburg, FRG: Rethra Verlag GbR) 235 pages.

Cathcart, R.B. (April 1979) *The Developing Artificial Geography of the Solar System.* Public Administration Series Bibliography P-206 (Monticello, Illinois: Vance Bibliographies) 16 pages.

Cathcart, R.B. (1983) *A Megastructural End to Geologic Time.* Journal of the British Interplanetary Society 36: 291-297.

Cathcart, R.B. (2002) *Unnatural Envelopment: Fieldwork on the Active Tectonics of J.G. Ballard's "Build-Up" (1957.* Journal of Geoscience Education 50: 176-181.

Cathcart, R.B. (2011) *Anthropic Rock: A Brief History.* History of Geo- and Space Sciences 2: 57-74.

Catling, D.C. and Zahnle, K.J. (2009) *The Planetary Air Leak.* Scientific American 300: 36-43.

Cerceau, F.R. (2013) *Pioneering Concepts of Planetary Habitability,* pages 115-129 **IN** Astrobiology, History and Society: Advances in Astrobiology and Biogeophysics (The Netherlands: Springer).

Ceyssons, F.; Driesen, M. and Wouters, K. (2012). *On the Organization of World Ships and Other Gigascale Interstellar Space Exploration Projects.* Journal of the British Interplanetary Society 65: 134-139.

Chaplin, J.E. (2012) *Round About the Earth: Circumnavigations from Magellan to Orbit.* (New York: Simon Schuster) 535 pages.

Chen, X. and Nordhaus, W.D. (2011) *Using luminosity data as a proxy for economic statistics.* Proceedings of the National Academy of Sciences 108:8589-8594.

Chianese, R.L. (July-August 2013) *Levitated Mass.* American Scientist 101: 268-269.

Clegg, R.N. et al. (1 January 2014) *Effects of rocket exhaust on lunar soil reflectance properties.* Icarus 227: 176-194.

Cocks, D. (2013) *Global Overshoot: Contemplating the World's Converging Problems.* (The Netherlands: Springer) 392 pages.

Cole, D.M. (September and October 1961) *Macro-Life.* Space World 1 (Nos. 10 and 11): 44-46

Cole, D.M. (December 1966) *Manned Interplanetary Flights in the Seventies with Saturn/Apollo Technology.* Annals of the New York Academy of Sciences 140: 451-466.

Contoyiannis, Y.F., Potirakis, S.M. and Eftasias, K. (2013) *The Earth as a living planet: human-type diseases in the earthquake preparation process.* Natural Hazards and Earth System Sciences 13: 125-139.

Cummins, N.P. (2012) *An Evolutionary Perspective on the Relationship Between Humans and Their Surroundings: Geoengineering the Purpose of Life & the Nature of the Universe.* (Reading, England: Cranmore Publications) 356 pages.

Daniels, S. (2011) *Geographical Imagination* Transactions of the Institute of British Geographers NS 36: 182-187.

Davidson, F.P. and Brooke, K.L. (2006). *Building the World*, Volumes 1 & 2. (Westport, Connecticut: Greenwood Press) 1387 pages.

Debase, R.L. (2008) *Effects of Collisions Upon a Partial Dyson Sphere.* Journal of the British Interplanetary Society 61: 386-394.

Debschitz, Ute and Thilo (2013) *Fritz Kahn.* (New York: Taschen) 392 pages.

DeMeo, F.E. et al (2014) *Solar System evolution from compositional mapping of the asteroid belt.* Nature 505: 629.

Di Salle, F. et al. (2013) *Effect of Carbonation on Brain Processing of Sweet Stimuli in Humans.* Gastroenterology 145: 537-539.

Earley, J.E. (2013) *A New "Idea of Nature" for Chemical Education* 22: 1775-1786.

Ellis, E.C. (2011) *Anthropogenic transformation of the terrestrial biosphere.* Philosophical Transactions of the Royal Society A 369: 1010-1035.

Ellis, E.C. (2013) *Used planet: A global history.* Proceedings of the National Academy of Sciences 110: 7978-7985.

Ellison, A.M. (2013) *The Suffocating Embrace of Landscape and the Picturesque Conditioning of Ecology.* Landscape Journal: design, planning, and management of the land 32: 79-94.

Ezard, T.H.G. et al. (2013) *Inclusion of a near-complete fossil record reveals speciation-relatd molecular evolution.* Methods in Ecology and Evolution 4: 745-753.

Dyson, F.J. (1966) The Search for Extraterrestrial Technology, pages 641-655 *IN* Marshak, R.E. (Ed.) *Perspectives in Modern*

Physics: Essays in Honor of Hans A. Bethe (New York: Interscience Publishers).

Fleming, J.R. (2010) *Fixing the Sky: The Checkered History of Weather and Climate Control.* (New York: Columbia University Press) 325 pages.

Foege, Al. (2013) *The Tinkerers* (New York: Basic Books) 216 pages.

Fogg, M.J. (1995) *Terraforming: Engineering Planetary Environments.* (Warrendale: Society of Automotive Engineers Inc.) pages388-391.

Forgan, D.H. (2013) *On the Possibility of Detecting Class A Stellar Engines Using Exoplanet Transit Curves.* Journal of the British Interplanetary Society 66: ???.

Fraczek, W. (Summer 2010) *If the Earth Stood Still.* ArcUser, pages 62-68.

Freitas, V.F. de and Delgado, J.M.P.Q. (Eds.) (2014). *Durability of Building Materials and Components.* (The Netherlands: Springer) 274 pages.

Frisch, B.H. (1965) *G.E.'s Way-Out Man*, Science Digest 58: 9-15.

Geppert, A.C.T. (2012). *Introduction: Rethinking the Space Age: astro-culture and technoscience.* History and Technology 28:219-223.

Giselbrecht, S. et al. (2013) *Chemie der Cyborgs—zur Verknupfung technischer System emit Lebewesen.* Angewandte Chemie 125: 14190-14206.

Gleeson, B. (2010). *Lifeboat Cities: Making a New World.* (Sydney: University of New South Wales Press) 228 pages.

Glover, D.R. (2013) *The Artificial Planet.* Journal of the British Interplanetary Society 66: 43-46.

Goody, J. (2012) *Metals, Culture and Capitalism*. (New York: Cambridge University Press) 349 pages.

Gorkavyi, N.N. et al. (2013) *Aerosol plume after the Chelyabinsk bolide*. Solar System Research 47: 275-279.

Gu, X. et al. (2013) *Volatile-Sensing Functions for Pulmonary Neuroendocrine Cells*. American Journal of Respiratory Cell and Molecular Biology. DOI. 131017120820001.

Haenecour, P. et al. (2013) *First Laboratory Observation of Silica Grains from Core Collapse Supernovae*. The Astrophysical Journal 768: L17.

Hamblin, J.D. (2013) *Arming Mother Nature: The Birth of Catastrophic Environmentalism*. (New York: Oxford University Press) 298 pages.

Hamilton, C. (2013) *Earthmasters: The Dawn of the Age of Climate Engineering*. (New Haven: Yale University Press) 247 pages.

Heffernan, J.B. et al. (February 2014) *Macroecosystems ecology: understanding ecolocial patterns and processes at continental scales*. Frontiers in Ecology 12: 5-14.

Heise, U.K. (2012) *Reduced Ecologies: Science Fiction and the meanings of biological scarcity*. European Journal of English Studies 16: 99-112.

Holmes, R. (2013) *Falling Upwards: How We Took to The Air*. (New York: Pantheon Books) 404 pages.

Hommel, D., and Murphy, A.R. (2013) *Rethinking geopolitics in an era of climate change*. GeoJournal 78: 507-524.

Horeis, H. (2011) *Once upon a time, there was a dying forest*. Technology in Society 31: 111-116.

Hsiang, S.M., Burke, M. and Miguel, E. (2013) *Quantifying the influence of climate on human conflict.* Science 341: 1141-1312.

Huggan, G. (2013) *Nature's Saviours: Celebrity Conservationists in the Television Age.* (London: Routledge) 248 pages.

Jackson, J. (2005) *The interplanetary internet [networked space communications].* Spectrum, IEEE 42: 30-35.

Jelinski, D.E. (2005) *There is No Mother Nature—There is No Balance of Nature: Culture, Ecology and Conservation.* Human Ecology 33: 271-288.

Johnson, L. (2013) *Sky Alert! When Satellites Fail.* (Chichester, UK: Praxis Publishing) 199 pages.

Jones, N. (2013) *It Could Happen One Night.* Nature 493: 154-156.

Kim, M-S. and Kim, E-J. (2013) *Humanoid robots as "The Cultural Other": are we able to love our creations?* AI & Society 28: 309-318.

Kim, Y-H.et al. (2014) *Influence of Radioactivity on Surface Charging and Aggregation Kinetics of Particles in the Atmosphere.* Encironmental Science & Technology 48: 182-189.

Koneeni, V.J. (2005) *The Aesthetic Trinity: Awe, Being Moved, Thrills.* Bulletin of Psychology and the Arts 5: 27-44.

Kosky, J.L. (2012) *Arts of Wonder.* (Chicago: University of Chicago Press) 224 pages.

Krausmann, F. et al. (2013) *Global human appropriation of net primary production doubled during the 20th Century.* Proceedings of the National Academy of Sciences 110: 10324-10329.

Kyba, C.C.M. and Holker, F. (2013) *Do artificially illuminated skies affect biodiversity in nocturnal landscapes?* Landscape Ecology 28: 1637-1640.

Langmead, D. and Garnaut, C. (1999) *Encyclopedia of Architectural and Engineering Feats* (Santa Barbara, CA: ABC-CLIO) 388 pages.

Leslie, C. and Hancock, G.J. (2008) *Estimating the date corresponding to the horizon of the first detection of ^{137}Cs and $^{239+240}Pu$ in sediment cores.* Journal of Environmental Radioactivity 99: 483-490.

Levermann, A. et al., (2013) *The multimillennial sea-level commitment of global warming.* Proceedings of the National Academy of Sciences 110: 13745-13750.

Levitt, I.M. and Cole, D.M. (1963) *Exploring the Secrets of Space: Astronautics for the Layman* (Englewood Cliffs, NY: Prentice Hall) pages 277-278.

Liao, S.M., Sandberg, A. and Roache, R. (2012) *Human Engineering and Climate Change.* Ethics, Policy and the Environment 15: 206-221.

Liu, Q., Wang, Y., Zhang, J. and Chen, Y. (2013) *Filling Gullies to Create Farmland on the Loess Plateau.* Environmental Science & Technology 47: 7589-7590.

Machado, I. (2006) *Impact or explosion? Technological culture and the ballistic metaphor.* Sign Systems Studies 34: 245-260.

Makowski, C., Finkl, C.W. and Rusenko, K. (2013) *Suitability of Recycled Glass Cullet as Artificial Dune fill along Coastal Environments.* Journal of Coastal Research 29: 772-782.

Marriner, N., Morhange, C. and Skrimshire, S. (2010) *Geoscience meets the four horsemen? Tracking the rise of neocatastrophism.* Global and Planetary Change 74: 43-48.

Marsden, P. and Whiteman, T. (2001) *The Top Ten Construction Achievements of the 20th Century* (Wadhurst, UK: KHL International) 272 pages.

Masco, J. (2012) *The End of Ends.* Anthropological Quarterly 85: 1107-1124.

Massonnet, D. and Meyssignac, B. (2006) *A captured asteroid: Our David's stone for shielding Earth and providing the cheapest extraterrestrial material.* Acta Astronautica 59: 77-83.

McCray, W.P. (2013) *The Visioneers* (Princeton: Princeton University Press) 351 pages.

Meigh, L. et al. (2013) *CO_2 directly modulates connexin 26 by formation of carbamate bridges between subunits.* eLIFE 2: 1-13.

Melillo, E.D. (October 2012) *The First Green Revolution: Debt Peonage and the Making of the Nitrogen Fertilizer Trade, 1850-1930.* American Historical Review 117: 1028-1060.

Mignan, A., Grossi, P. and Muir-Wood, R. (2011) *Natural Hazards* 56: 869-880.

Miller, S.D. et al., (2013) *Earth-viewing perspective on the Chelyabinsk meteor event.* Proceedings of the National Academy of Science 110: 18092-18097.

Minke, G. (2009) *Earth Construction Handbook.* (MA: WIT Press) 216 pages.

Mora, C. et al. (2011) *How Many Species Are There on Earth and in the Ocean?* PloS Biology 9: e1001127-e1001130.

Neubert, F.-X. et al. (January 2014) *Comparison of Human Ventral Frontal Cortex Areas for Cognitive Control and Language with Areas in Monkey Frontal Cortex.* Neuron 81: 700-713.

Noack, L. et al. (2014) *Can the interior structure influence the habitability of a rocky planet?* Planetary and Space Science. Available online 20 January 2014.

Norra, S. (2009) *The astysphere and urban geochemistry—a new approach to integrate urban systems into the geoscientific concept of spheres and a challenging concept of modern geochemistry supporting the sustainable development of planet Earth.* Environmental Science and Pollution Research 16: 539-545.

Olson, V.A. (2012) *Political Ecology in the Extreme: Asteroid Activism and the Making of an Environmental Solar System.* Anthropological Quarterly 85: 1027-1044.

Park, S. et al. (2013) *Trends and seasonal cycles in the isotopic composition of nitrous oxide since 1940.* Nature Geoscience 5: 261-265.

Patterson, C.C. (1994) *Delineation of separate brain regions used for scientific versus engineering modes of thinking.* Geochimica et Cosmochimica Acta 58: 3321-3327.

Preester, H. De (2011) *Technology and the Body: the (Im)Possibilities of Re-embodiment.* Foundations of Science 16: 119-137.

Proctor, R.N. (2007) *"-Logos," "-Ismos," and "Ikos": The Political Iconicity of Denominative Suffixes in Science (or, Phonesthemic Tints and Taints in the Coining of Science Domain Names.* Isis 98: 290-309.

Prodehl, C., Kennett, B., Artemieva, I. and Thybo, H. (2013) *100 Years of seismic research on the Moho.* Tectonophysics 599:??? [in press].

Pyne, S.J. (2012) *Fire: Nature and Culture* (London, UK: Reaktion Books). 207 pages.

Raithel, I.W. (1966) *A Tribute to Dandridge McFarland Cole (1921-1965).* Annals of the New York Academy of Sciences 140: 7-9.

Ravilious, K. (2013) *Mid-town Miners.* New Scientist 218: 40-43.

Rehren, T. et al. (2013) *5,000 years old Egyptian iron beads made from hammered meteoritic iron.* Journal of Archaeological Science, in press. DOI: http://dx.doi.og/10.1016/j.jas.2013.06.002.

Reid, P. and Goddard, T. (2013) *LEGO: Space Building the Future.* (New York: No Starch Press) 216 pages.

Rigby, E., Symonds, M. and Ward-Thompson, D. (2004) *A comet impact in AD 536?* Astronomy and Geophysics 45: 23-26.

Robbins, P. and Moore, S.A. (2013) *Ecological anxiety disorder: diagnosing the politics of the Anthropocene.* Cultural Geographies 20: 3-19.

Robin, L. and Steffen, W. (2007) *History for the Anthropocene.* History Compass 5: 1694-1719.

Rolston, H. (1992) *Disvalues in Nature.* The Monist 75: 250-278.

Saint-Exupery, A. de (2000) *The Little Prince*, translated by R. Howard (San Diego, CA: Harcourt).

Sayer, J. and Cassman, K.G. (2013) *Agricultural innovation to protect the environment.* Proceedings of the National Academy of Science 110: 8345-8348.

Schewe, P.F. (2013) Mavrick Genius: The Pioneering Odyssey of Freeman Dyson (New York: Thomas Dunne Books) 339 pages.

Schneider, J. (2006) *Science Fiction and Science Policy.* Bulletin of Science, Technology & Society 26: 518-520.

Schlegel, A. et al. (16 September 2013) *Network structure and dynamics of the mental workspace.* Proceedings of the National Academy of Sciences. DOI: 10.1073/pnas.131114910.

Scott, M. (2014) *The mediation of distant suffering: an empirical contribution beyond television news texts.* Media, Culture & Society 36: 3-19.

Steffen, W. et al., (2011) *The Anthropocene: From Global Change to Planetary Stewardship.* AMBIO 40: 739-761.

Shuvalov, V.V., Svettsov, S.S. and Trubetskaya, I.A. (2013) *An estimate for the size of the area of damage on the Earth's surface after impacts of 10-300-m asteroids.* Solar System Research 47: 260-267.

Silva, R.A. et al. (2013) *Global premature mortality due to anthropogenic outdoor air pollution and the contribution of past climate change.* Environmental Research Letters 8: 034005-034010.

Sinues, P.M., Kohler, M. and Zenobi, R. (2013) *Human Breath Analysis May Support the Existence of Individual Metabolic Phenotypes.* PLOS ONE: 8: e59909-e59913.

Smith, D.J. (2013) *Aeroplankton and the Need for a Global Monitoring Network.* BioScience 63: 515-516.

Spalding, K.L. et al. (2013) *Dynamics of Hippocampal Neurogenesis in Adult Humans.* Cell 153: 1219.

Sperber, D. (2010) *The Guru Effect.* Review of Philosophy and Psychology 1: 883-592.

Stasenko, A.L. (July-August 1999) *The New Earth.* Quantum 9: 16-20.

Stothers, R.B. (1984) *Mystery cloud of AD 536.* Nature 307: 344-345.

Swisdak, M., Drake, J.F. and Opher, M. (2013) *A Porous Layered Heliopause.* The Astrophysical Journal 774: L8.

Szasz, F.M. (1995) *The Impact of World War II on the Land: Gruinard Island, Scotland, and Trinity Site, New Mexico as Case Studies.* Environmental History Review 19: 17-30.

Strong, K., Saba, J. and Kucera, T. (September 2012) *Understanding Space Weather: The Sun as a Variable Star.* Bulletin of the American Meteorological Society, pages 1327-1335.

Szerszynski, B. (2012) *The end of the end of nature: the Anthropocene and the fate of the human.* Oxford Literary Review 34: 165-184.

Underwood, E. (2012) *How to Build a Smarter Rock.* Science 338: 1412-1413.

Vereb, H., Dietrich, M., Alfeeli, B. and Agah, M. (2011) *The Possibilities Will Take Your Breath Away: Breath Analysis for Assessing Environmental Exposure.* Environmental Science & Technology 45: 8167-8175.

Waldvogel, C. et al. (2004) *Globus Cassus* (Basel, Switzerland) 182 pages.

Wells, W. (2009) *Apocalypse When? Calculating How Long the Human Race Will Survive: 10 Years? 1,000 Years? 10,000 Years?* (The Netherlands: Praxis) 212 pages.

Whitehead, M. (2014) *Environmental Transformations: A Geography of the Anthropocene.* (UK: Routledge) 192 pages.

Willenbring, J.K., Codilean, A.T. and McElroy, B. (2013) *Earth is (mostly) flat: Apportioment of the flux of continental sediment over millennial time scales.* Geology 41: 343-346.

Wise, P.M. et al. (August 2013) *The Influence of Bubbles on the Perception Carbonaton Bite.* PLOS One: 8: e71488.

Yarnoz, D.G., Sanchez, J.P. and McInnes, C.R. (2013) *Easily Retreivable Objects Among the NEO Population.* Celestial Mechanics and Dynamical Astronomy 116: 367-388.

Yeomans, D.K. (2012) *Near-Earth Objects: Finding Them Before They Find Us.* (Princeton: Princeton University Press) 192 pages.

Zakharov, A.I. et al. (December 2013) *Method for determination of the celestial coordinates of the center of the Earth.* Solar System Research 47: 548-55.

Ziarek, K. (2011) *The Limits of Life: a non-anthropic view of world and finitude.* Angelaki: Journal of the Theoretical Humanities 16: 19-30

CHAPTER 2

HUMAN SPACE AGE INTERNET SPATIOGRAPHIES

Homo sapiens' well-intentioned assessment of Earth's monetary worth ought to change markedly during the 21st Century. Amateur and professional speculation about "future history" (by self-proclaimed "futurologists", members of a social movement) has already been projected via electronic and print media since the alleged termination of the Cold War in 1990. Futurologists are popular; they are now, like the ancient royal court astrologers of pre-modern times, an integral part of our globalized human culture. However, a majority of present-day Earthlings simply accept our immediate habitat quite by custom and tradition. All of Earth's superficial terrain, air, the ocean and all life forms—in sum, "global Nature"—has been drastically altered since Earth's Geologic Time started, possibly more than 4.5 billion years ago. After plants colonized Earth's sub-aerial land-mass 2.2 billion years ago (Retallack et al., 2013), animals joined forces to landscape the continents and islands. Global Nature was in the past, and remains today, shaped by the Universe's incompletely known dynamics and mankind's planned and unplanned real-world actions; humans have always sought to change and arrange the territory surrounding its members (Turner, 2008), whether living, gaseous, solid or liquid, to suit their collective species needs and wants, according to Andrew Shaw Goudie's account, *The Human Impact on the Natural Environment: past, present and future* (2013, 7th Edition), as well as the outlook of Sergey M. Govorushko in *Natural Processes and Human Impacts: Interactions between Humanity and the Environment* (2012). Bascially, these texts report on geographical macro-problems—they do not solve macro-problems—whereas this text will pose conceptual solutions that include personalized implications of what might result. In short, a "Conceptual Art" versus actual "Sculpture" approach is taken herein.

Celebrated and venerated "Name" futurologists are fashionable nowadays, but their books, articles and Internet blogs of vividly-phrased generalizations, popular jargon and sometimes entertaining anecdotes compare unfavorably to the "real stuff" composed by scientist-engineers. Science-engineering truly is the master of the future's functional vocabulary. Both fixated, sometimes obsessed, professions ought always be related to the outside world—the Astythromes and Anthromes and other Spatiographies incontrovertibly exterior to our pulsating bodies—and its societal success depends basically on how well its abstruse theories and practical applications correspond with reality—the realm of survival—that is always surrounding *Homo sapiens* (Herman, 2012; Kennedy, 2013). A cyberpunk-style science-fiction authored by Geoff Ryman, *Air: Or, Have Not Have* (2004) posits an ill-fated United Nations Organization experimental quantum device, "Air", intended for testing the targeted brains of Asian peasants living in a remote village. Implanted in their brains, "Air" is supposed to perform as an Internet-like telecommunications system superseding the World Wide Web. "Air" is implemented, yet it goes awry, trapping the primary character, Chung Mae, who becomes able to see into the realm of both the historical past as well as the future (Dougherty, March 2012). Information held by the Internet is practically massless, without substance that is measurable or touchable in the everyday sense. Recently, medical investigations of the "temporal Doppler effect" reveal that ordinary humans perceived dates in the future as closer to the present than equidistant dates in the past. Evidently, perception of future events as closer to the "present" may be a little-recognized psychological mechanism that helps each human to approach, avoid or otherwise cope with the real-world events encountered. In other words, each human's descriptions of a time are closely linked to that person's experiences of *moving* through space! Humans have a bias towards the future. "The fact that action can facilitate the realization of future desires but not past ones may help explain why people devote more resources to things that lie ahead than things that lie behind' (Caruso et al., 2013, page 535). With its 2 January 1989 issue, *Time* magazine's glossy cover photograph pictured the whole Earth, using a roughly 41 centimeter-diameter Christo artwork ["**Wrapped Globe 1988**"], as an endangered "Planet of the Year". Tendentious teams of futurologists, cozily ensconced in professorial-style offices in publicly- and privately-funded Think Tanks (Medvetz, 2012), played some significant role influencing that shrinking national weekly's staff writers. Christo's artistic expression does not accurately or proportionately symbolize Earths "...vulnerability to man's reckless ways", as *Time* alleged. Briefly, *Time* claimed humans had stolen the whole terrestrial scene, as well as co-opted all possible developmental scenarios, in possibly the last act of a dramatic anti-global Nature play. Whilst obscuring a mass-produced, spun schoolroom globe with commercial see-through plastic food-wrap, Christo tried to help those who perused the glossy magazine's cover at the start of a new year to become more strongly aware of the delicate object under duress from *Homo sapiens* contained. A spherical Earth globe, even one that is enshrouded by a transparent film Christo-style (Chernow, 2002), offers its owner and putative "operator" a God's-eye view of the planet at any time—at least, depending on what the drawing cartographer chose at that first moment of the undertaking to be shown! In addition to changes accomplished by people, our home-planet has naturally and continuously remade itself; climates of all the lands, and the global sea level, have varied from those presently extant. Earth merely seems to be a familiar place to its current human inhabitants; however, many true Earth facts are still to be unearthed by scientists and effectively exploited by practicing macro-imagineers!

"Life" remains a scientifically indefinable term even though mystery-averse biotechnologists and other skilled forward-thinking investigators are attempting to formulate a comprehensive definition. To gin up the American public's interest in outer space and planetary exploration, NASA often touts its endless (funding needed!) search for life elsewhere. Earth's Geologic Time can be divided into two supereons: (I) Pre-Geozoic, the time period when there was no life present in the Earth and the (II) Geozoic, the time-place after the oldest known direct evidence for the presence of life in the Earth, now estimated at more than 3.5 billion years ago. The Geozoic Supereon, thus, represents the most conservative estimate of the Earth-biosphere's existence (Kowalewski et al., 2011). When and if a Dyson Motor shatters Earth, then our planet's Geozoic Supereon is forever ended, in fact super-heated and vaporized, shot into space! "Geozoic Supereon" is a new terminology that has not triggered any substantial response from the world's geo-scientific community regarding its acceptance or rejection. However, since 1970 amongst geo-scientists of many distinct professional persuasion and especially since the first "Earth Day" (Rome, 2013), the organization of a dispersed and, so far, informal cadre of geosozologists has commenced. From the Greek word "sodzo" meaning "I protect", the goal of geosozology is to form a unified scientific profession dedicated to protecting and saving the Earth. Such a profession's work, properly speaking, ought to aim towards maintaining unpolluted, uncontaminated and even pristine conditions within the Earth-biosphere; geosozologists seek to sustain the necessary enveloping situation for Earth life's further development (Kozlowski, 2004). Geo-diversity is the counterpart to bio-diversity, and vice versa (Roelof Dirk Schuiling, 1998, page 2). [During 2013, R.D. Schuiling proposed the immobilization of metal ore waste—sludges, metallurgical slags, mine tailings et cetera, by reaction with hydrogen sulfide in Earth's biggest anoxic body of seawater, the Black Sea: see "Immobilisation of metal wastes by reaction with H2S in anoxic basins", *Environmental Sciences and Pollution Research*. His thinking entailed 200 million years of future Earth Geologic Time!] Naturally, if ever the Earth ceases to exist, then any succeeding remnant geo-diversity is a matter entirely arranged by those thinkers and doers, whatever their origin (birth or manufacture), subsequently controlling the planet's re-mainders. Furthermore, whatever or whoever manages the conserved products of any planet's disruption—some will inevitably go to waste or be destroyed utterly—will become migratory masters of telecommunications networked Space Architecture-designed facilities that should comprise the Freeman Dyson Satellite Swarm (Sherwood, 2011).

Although almost 40% of the word listings of early-21st Century USA university-level English language dictionaries are scientific and technical words, only in 1802 did the term "biology" make its first appearance in life science's literature. "Biology" was used, possibly coined, by the French naturalist Jean Baptiste Lamarck (1744-1829), to mark the inception of an exacting study of global Nature's living matter. Lamarck, in his *Hydrogeologie*, defined "biology" as the division of "terrestrial physics" that included "...all that pertains to living bodies". Bio-scientists have concluded since—primarily from Geozoic Supereon fossils laboriously gathered from Earth's crust—that life forms were once non-existent [during, for instance, what we have termed the "Pre-Geozoic Supereon"]. Indeed, the planet Earth itself aggregated more than 4.5 billion years ago. Since life's "spontaneous" origination—in situ or via panspermia—organisms have, through punctuated organic evolution, filled a significant part of the planet which some old-fashion geographers still phrase as "the surface of the Earth". During the ongoing Anthropocene, changes in phytoplanktonic life forms are evident. For example, under the influence of seawater's acidification, the ocean's most abundant primary producer, the most important calcifying organism, the coccolithophore *Emiliania huxleyi* has been proven to adapt via evolutionary change (Lohbeck et al., 2012). Our culturally diverse species, with individuals numbering more than 7 billion, share the Earth-biosphere with many, uncounted living forms. Lamarck thus introduced the unifying concept of an Earthly biosphere into life science, social science and physical science. Lamarck had realized the importance of plants and animals roles in effecting geographically large-scale changes in the Earth-surface; he was the first to recognize engineering species (Jones, 2012; Haff, 2012) as agents for crustal, oceanic and atmospheric alterations—super Earth-mantle geological changes of Earth's biosphere which are attributable only to life's vitality and intelligence (Cockell, 2011).

By nearly 19th Century's end, Eduard Suess (1831-1914) added the term "biosphere" to professional Geology's literature, which practitioners of Geography almost instantly adopted. Suess was far ahead of his colleagues. His multi-volume comprehensive overview of geoscience, ***The Face of the Earth***, opens with a vivid description of the Earth as first glimpsed by an *imaginary* interplanetary space traveler. Just as The-Man-In-The-Moon is always seen from an Earth-based perspective, the "face" of our planet can only be recognized from a viewpoint in adjacent outer space, hence Suess' ***Das Antlitz der Erde*** (published during1883-1909)! During 1899 the first official introduction of "biosphere"—the planetary domain, a volume of matter and energy supporting Earth's life—to English-using geographers by Sir John Murray (1841-1914) in an address to the British Association for the Advancement of Science. The term literally means "sphere of life" but, in reality, it means that part of Earth which is shell-shaped within which life can or does currently exist. Prudent geo-scientific professionals use "*in* the Earth" simply because those barely visited anti-biota conditions of outer space are known to be absolutely deadly for putatively native organisms. Earth's biosphere is a shell (global Nature); therefore, being a noun far more useful that the phrase "Earth's surface" for a distinguishable volume of space containing solid, liquid, gaseous matter as well as the living.

Geographers know that much of the complexity of Earth's biosphere [particularly from our species-centric viewpoint] is due to the competitive and cooperative interactions among all other organisms currently resident in our Earth—millions of species, practically innumerable organisms. Earth's biosphere is a global Nature, a dynamic complex of organisms forming a single planetary collection. So far, Earth is the only known spatial context (three-dimensional zone) of a settled *Homo sapiens*. Mankind is a "resource pool" to sociologists but to laypersons it is the largest possible 21st Century "organization". The special character of being human during the Anthropocene is the ability to transform and manage local global Nature (Anthromes and Astythromes). Established human groupings, organization, are systems for accomplishing work, for using altering techniques on materials (people, symbols, or things). Such organizations mostly have available efficient earth-moving technology and the ability to pay for its deployment upon Earth's thin epidermis, the Earth-crust perceived as a magnificent workshop within a fairly benign Solar System. Mighty machines (Haycraft, 2000; Smil, 2010) make real the brunt prospect of a future massive Earth-surface reconfiguration (Lynch, 2002; Haff, 2010).

Incontrovertibly a scientifically useful term, "biosphere" is especially so when used in conjunction with other closely associated terminology: "lithosphere" (the Earth-crust), "hydrosphere" (mainly the ocean, but also the water vapor circulated by the atmosphere's moving air, groundwater, together with rivers and lakes) and "atmosphere". Between 1895 and 1902, Carl von Linde (1842-1934) patented the processes employed by industry for the separation of principal gases from liquid air. Humans have long utilized water, both fresh and salty, for various industrial processes, all of them generally well known by the world's public (Smil, 2008). Industrial-scale ammonia synthesis from air has resulted in a worldwide use of fertilizer nitrogen since 1908 (Erisman et al., 2008). [If plant geneticists can develop a green plant that makes its own fertilizer, then air-mining may someday cease.] Clothed in a spacesuit (Monchaux, 2011; Westfahl, 2012), Felix Baumgartner survived a historic Mach 1.25, 38,969 meter free-fall from the lowest "edge" of outer space during his 14 October 2012 adventure, which was probably significantly more hazardous than the untethered space "walk", performed by Bruce McCandless II, during the 1984 USA Space Shuttle's STS-41-B Mission. Earth-crust "Shuttlequakes" were induced by atmospheric shockwaves generated by the once commonplace rapid re-entries of America's (since decommissioned) Space Shuttles (Sorrels et al., 2002). An Arizona corporation, the Paragon Space Development Corp. of Tucson, hopes to field a pressurized gondola by 2016 capable of carrying eight passengers in semi-luxury to 30,500 meters as tourists (Pasztor, 2013). Since 1955, upon even the high-seas, hovercraft—invented by Christopher Cockerell (1910-1999)—ply some short sea-routes of trade and travel by using turbine-forced vertical and horizontal air currents to lift and propel the specialized sea-craft (Bailey, 1993). With regard to the hydrosphere, besides the abstractive use of freshwater, the mass of ships afloat on the oceans actually raises the overall level of the ocean by, approximately, five micrometers (Badescu and Cathcart, 2011). And, like the Space Shuttle's, when very large ships strike land accidently a recordable shockwave is produced that induces Earth-crust shaking

(Mucciarelli, 2012), just like many other anthropogenic shifts of great mass, solids or liquid (Klose, 2013).

No scientist yet has recorded or calculated the mass of air gases sequestered by anthropogenic means that might indicate a *Homo sapiens* control of our planet's atmosphere; such documentation will surely follow any carbon dioxide sequestration effort worldwide. Habitually bipedal walking, and the position of the foramen magnum, a hole in the base of the human skull that transmits the spinal cord into the braincase, promotes our locomotion and looking upwards, say at the be-clouded sky. Yet, an atmospheric boundary separating our troposphere and stratosphere was first examined in a photograph taken on 11 November 1935 from a manned balloon, Explorer II, drifting high above the USA. In other words, a near-miracle technique let our world's publics see one of global Nature's borders penetrated and topped! Of these four nouns (atmosphere, hydrosphere, lithosphere and biosphere), only the planet's actual core—which may or may not be James Marvin Herndon's center-of-the-Earth's mass nuclear-fission "georeactor" (2014)—can be regarded as true, if possibly eccentric, globe-shape mass whilst all the other layers of our Earth are actually only thin shells enshrouding the ball-like core. [J.M. Herndon's "geo-reactor" is as controversial as the James E. Lovelock's late-1960s "Gaia Hypothesis" (Ruse, 2013). [As will be shown later elsewhere in this text, Architecture's concept of "Urban Metabolism" preceded Lovelock's "Gaia".] Earth's habitability is tied ineluctably to the rocky planet's internal structure. Lovelockians do not advocate that Earth is an "organism" but, rather that it is composed of organisms that exert some major adjustment-control over the planet's changes (Margulis, 1993). R.D. Schuiling's Black Sea anoxic seawater "geo-reactor" is still another idea for metal waste disposal for long periods of Earthly Geologic Time.] The tallied and calculated meteorite isochron age (about 4.5 billion years) is 6.45 times the half-life of Uranium-235, so that means there was most probably at least 87 times more U-235 then as now! Humans are changing the contents of the air, the seawater and even the crust materials—namely, by adding man-made radiation and toxic manufactured chemicals or concentrates of mined minerals and ores. Even though we still habitually appraise the Earth-ocean as vast and voluminous, the hydrosphere seems especially easy to alter artificially—see the informative graphic created by the United States Geological Survey, **Figure 1** below.

Figure 1. The bead-like, miniscule by contrast, blue-tinted ball sitting on the co-terminus USA represents all water, whatever its placement and state, composing the Earth-hydrosphere. That iridescent turquoise bead symbolizes 1,386,000,000 cubic kilometers of liquid! (Image: USGS.)

Many of the ancient Greek philosophers' writings—Aristotle and Hippocrates, for instance—enfolded their laborious biological observations and studious open-minded speculation which indicated their keen personal fascination in Earth's biosphere and which were ecological in style. However, the Greek teachers of the most intelligent handpicked youths of their time never found an especially apt descriptive word for their geographical and biological philosophizing! German biologist Ernst Haeckel (1834-1919) coined "ecology" in his *Generelle Morphologie der Oganismen* (1866) after first recognizing that Earth's global Nature was a unity of some undefined kind. Haeckel's "oekologie" was Anglicized at the Botanical Congress, which met in Madison (State of Wisconsin, USA) in 1893. Thereafter, environmentalists made Ecology a vogue term circa 22 April 1970 when the first Earth Day was celebrated (Rome, 2012). Jaded counter-culturists in the USA and Europe, espousing a cynical senso-eco-consciousness, aided in the momentary development of an illogical Ecology. Anti-pollutionary political organizations, lobbying politicians everywhere under the umbrella term "environmentalism" or "Green", foresaw that life-styles accumulated by human civilization would wreck Earth's life-support system (biosphere). [Circa 1969, "Green Revolution" became a popular term labeling the successful results of biogenetical R&D that innovated inexpensive and high-yield varieties of rice, wheat and other grains to support the increasing human populace of Earth. Crucially, circa 1970, "Green Revolution" became also a slogan describing an overly-politicized Ecology broadcast particularly in Europe and North America. "Ecology", surely one of Fate's odder twists!] Since 16 July 1945, all scientists have understood that humanity had obtained and still possesses the powerful weaponry to destroy our species' only planet-sized biosphere-home via global nuclear warfare. In other words, propaganda about a near-term future ecocatastrophe has thoroughly permeated humankind's thinking, most especially including professional Futurology's well-publicized and disseminated Think Tank ruminations.

1950s-era radio and television broadcasting and cable systems fostered the emergence of an "etherized" universal consciousness, some have called it a shell or "thinking stratum" of reflective thought involving the human mind and reason. This ad hoc possibility—literally, "mind-sphere" started with the birth of the first cognitive human being and it is physically manifested throughout the Earth-biosphere in the form of human interventions promoting the physical economic development of our planet. "Noosphere", possibly neologized during 1927 by Edouard Le Roy (1870-1954), who followed the biogeochemical insights and teachings of Vladimir Ivanovich Vernadsky (1863-1945) (Lapo, 2001). The appraised ad hoc Earth's Noosphere, like the human mind, is an information-processing system typified by progressing technology and information-structuring system identified by changing social and ideational concepts that shape various noosystems (cultures) comprising humanity. The opposite notion to *Homo sapiens'* Noosphere could be named humanity's "Cacosphere"—formed after the Greek word "kakos" meaning "trashy" or "bad". Al Qaeda, its affiliated and associated lodges, linked by couriers and electronics, could be classed as a single world-menacing cacosystem! Other less famous cacosystems are extant, but each works to undermine humanity's technological and cultural progress, some are ineffective or laughable whilst some are budding bad-guys training-to-terrify operations.

During 1962, Rachel Carson (1907-1964), in her crowning achievement *The Silent Spring*, deceptively postulated a terminal hypothetical season when pollution had destroyed Earth's biota forever by halting natural reproduction (Morriss et al., 2012). Between Haeckel and Carson a transformation from a slightly neglected life science to a political cause happened so that by the mid-1970s even Hollywood's troglodytic American cultural assassins were filming fictitious disasters—movies that concentrated on one kind of Nature-stimulated technological failure or economic macro-problem—for example: *Earthquake* (1974) and *The Towering Inferno* (1974) and the numerous ever more special-effects laden techo-thriller action movies made since—became suddenly popular worldwide; the unrealistic genre has continued until the present-day, with at least several such disaster epics debuting every year, a few of which are entertainment flops. The American public's next mind-bending science-based shock came in 1983 when credible environmentalists predicted omnicide by a theoretical season, the Nuclear Winter (a stratospheric cloud shroud) killing all life in our "global Village", but not the Earth-biosphere; Nuclear Winter would be a teratogenic pollution (Masco, 2010). Some persons still consider Nuclear Winter a blue-sky proposal by crackpots and ecofreaks—the "psycho-ceramic crowd", people with minds like concrete, all mixed up and set in their obsessive, fixated behaviors. As yet, Nuclear Winter could be caused only by modern macro-war (general nuclear warfare), not conventional war (meso war) or—so far—terrorism (micro war). Terrorist lodges with focused agendas exist and they do threaten all humanity, certainly more surely if they ever obtain, and use, biological and radiological weapons. Whatever their terror strategies and tactics, the seemingly unquenchable passion of these lodges is to mock all existing cultures—"noosystems"—by destroying public as well as private infrastructure and by maiming or murdering persons of high status in particular target cultures. Many of these lodges today view North America and Europe, including Russia, as potentially lootable treasure-houses. As the native population of North America and Europe extended stabilizes or decreases, per capita wealth increases—except when outright invasion and stealthy illegal alien infiltrators swell the ranks of those people drawing from each major region's noosystem hoard, the "mass savings accounts" accumulated in Astythromes by Anthropocene noosystems therein.

Human activities, advanced by the threatened use of powerful explosive weapons and landscape transformation machines, have undoubtedly changed global Nature so much that natural networks have become distorted and the capacity of global Nature has become limited, sometimes compromised forever. The chemical contents of the air breathed (Jacobs, 2008), the freshwater drunk, and the ocean "infected" by floating and fish devoured plastic waste is the most impressive manifestation of the Cacosphere's real-world existence and influence! Logically, Geographos will ordinarily use "noosystem" as a synonym for a singular "culture", a lone societal grouping subsumed by many such groups comprising extant human civilization.

Before the 20th Century commenced, biologist had yet to study all of Earth's life—that fact continues to be a fact during the early-21st Century as well—actually observing but a tiny fraction of its incompletely counted species (with, perhaps, a total mass in excess of 3.6 times one hundred billion tonnes). The current mass of Earth's ever-changing biosphere is approximately 1.148 times ten trillion tonnes. However, even during the early-21st Century, there are still a few places—in the Arctic, Antarctic, and the ocean's abyssal parts that are yet unseen by human visitors with enquiring minds. Compromising with their personal physical limits of discomfort, biologists have judiciously adopted a practical methodological approach to the study of Earth's life environments and forms, which is ad hoc and twofold: (I) to record and examine whatever organized living matter they were able in terms of kinds of entities, and (II) to view and assess global Nature's regions; that is, intellectually convenient geographical volumes of the Earth-biosphere—namely, Astythromes and Anthromes. Various terms have been proposed over the years for universal designation of such regions, but the term that is most often used in today's infor-mational-educational complex of popular and technical literature is "ecosystem", a word first suggested by Arthur George Tansley (1871-1955) in 1936 (Anker, 2002). Some geoscientists use "ecosystem" as a synonym for Earth-biosphere; others restrict its application to vari-ously defined regional subdivisions of our Earth's life-containing zone. Used in its total sense, ecosystem is presumed to denote man-kind's only intelligible field for biological and geo-scientific investiga-tion, as well as the exercise of humanity's constructive and destructive technologies, since it is the present-day environment of *Homo sapiens*. When "ecosystem" is used in its broadest sense, some kind of globally applicable scientific, engineering or philosophical judgment is being rendered. Countries, if considered as ecosystems—as in, for example, "country-ecosystem" or "nation-ecosystem"—present such groups and politicians with very clearly demarcated boundaries! A noosystem's membership in the United Nations Organization's General Assembly is possibly the best international gauge for a politically and territori-ally defined ecosystem's existence. The unique dangers of macro-war, meso-war and micro-war are a matter of everyday discussion, debate, diatribe and outrage amongst noosystems therein. Geographical (and Spatiographical) names are unlimited classes of words. It is prob-ably correct to state that no group knows the Earth-biosphere bet-ter than the USA's armed forces. During the Cold War, both rival Superpowers—the USA and the former USSR—were first to define a biosphere (Earth's) as an integrated battlefield, a global zone involving potential wartime use of nuclear, chemical, biological, and convention-al weapons. Our world could become altered massively by noosystem confrontations and conflicts occurring during the 21st Century.

Geoscience professionals, as well as macro-imagineers, during the perilous time when Superpower weaponeers had first tested fission and fusion explosives, realized that *Homo sapiens'* relationship with

global Nature (and all its appraised subdivisions) changed fundamentally when nuclear explosives became a military and civilian option for Earth-surface transformation (Kirsch, 2005; Hu, 2010). And,
nuclear power-plants on land and afloat or immersed in the ocean
became present in great numbers and, finally, would need confinement for long periods of retirement time for safety reasons (Cidell,
2012). Because of the difficulties associated with burial of radioactive
waste seemed insurmountable, a few experts have suggested, as early
as 1959, that outer space be an option for disposal of irremediably toxic
nuclear waste. Two types of outer space disposal have been proposed:
(I) disposing of the waste by causing designed waste package collisions
with the Sun and (II) transportation of dangerous radioactive waste
packages into distant places (Jupiter, Saturn, Uranus and Neptune,
for example, since humans can never trod their "solid surfaces") by
time-tabled drop-delivery or (III) sending waste packages to an outer
heliopause Sun orbit, where particles of the waste materials would be
released, followed by an unavoidable exit from our Solar System caused
by the solar and interstellar winds. More than a metric tonne of anthropogenic radioactive debris already is circulating within our Solar
System, although no nation-ecosystem (synonym: noosystem) has announced or yet instituted a macro-engineering program of radioactive
waste outer space disposal! To not use outer space, as some foolish
Greens council unhelpfully, would be to countenance wastage of the
Universe's space! Safe Earth-surface storage of lethal waste is costly
and chancy. For example, since 1998—two years after the Chernobyl
Nuclear Reactor #4 catastrophe, the world's worst civilian nuclear
power-plant accident—an international team of macro-engineers has
planned, and is currently carrying out, the encasement of that deadly
reactor site. The Shelter Implementation Plan devised will cost more
than 1.5 billion Euro [2013 value], installing a dome 110 meters high,
250 meters wide, and 150 meters long supposedly designed to remain

usefully intact and air-tight for a century. When it becomes operational after 2015—it will be continuously air-conditioned—the dome will enclose a volume of air great enough to create its own weather, rather like the 160 meters high, 157 meters wide and 218 meters long Vehicle Assembly Building (VAB) completed in 1966 at the NASA's Kennedy Space Center; sometimes, when Florida's air conditions (temperature and humidity) are condusive, a cloudy indoor weather system can form within that spacious building's estimated 3,665,000 cubic meter interior void-space! The VAB, undergoing a major refit and remodeling begun after the Space Shuttles were decommissioned, accommodated new launch rockets after 2013.

Humanity is an agency of impact and change, which is exemplified by technology responding to people's aspirations or restrained by people's fears. Probably at least 15% of Earth's landscape is now occupied by engineering buildings, structures, roads, railways, canals; if global urbanization continues unceasingly at its present rate, then all land may be covered by a conurbation in about 7,500 years (Badescu and Cathcart, 2006)! Frank P. Davidson and Kathleen L. Brooke found it necessary to create a two-volume compendium, *Building the World: An Encyclopedia of the Great Engineering Projects in History* to properly document the most important of these anthropogenic things. At the present-time, global Nature's basic heat flow, from the Earth's core (45 Terrawatts) outwards to the planetary interface of seawater-land with the enveloping air, is only about a factor of 2.5 times greater than current estimated human-harnessed energy (18 Terrawatts).

Hotter or cold/wetter or dryer future climatic regimes (influenced by *Homo sapiens* and global Nature) will mostly affect those public and private sectors of Earth's 200+ ecosystem-countries that interact with the very few still irregularly managed or "unmanaged" ecosystems comprising our worlds collection of Anthromes. For all ambulatory living things, the trend seems to be away from concourses of earthen particles and towards Earth-based corridors of concrete! During 2008, The Dubai Mall opened its 1,200 stores for commerce. Erected in Dubai (United Arab Emirates), it is the largest shopping mall in the world by total area, some 1,224,000 square meters. Architects consider shopping malls, where unfettered retailing is often mixed with recreation, our Anthropocene/Space Age's signature buildings.

During August 1992, in the *Journal of the British Interplanetary Society*, cutting-edge UK planetist Richard L.S. Taylor put forth a Christo-like Macro-Imagineering plan—in the absence of any official national or United Nations Organization Mars environmental mitigation, adaptation, and research strategies—to package Mars with a mall-like roofed structure called "Worldhouse". Perhaps inspired by a humorous classic science-fiction novel, *Mallworld* (1981) by Somtow Sucharitkul, or London, England's famed Crystal Palace, our world's first "theme park" existing from 1851 until 1937 and proposed for a modernized USA$250 million rebuild on 27 July 2013 by the Chinese company ZhongRong Holdings, Richard Taylor makes a unique very tall building research funding request for his innovative ideas to prevent wastage of a planetary environment (Mars). In other words, at least one living artistic landscape developer has perceived Mars as a not-to-be-forever-ignored planet. One might refer to Taylor as a macro-imagineer of esthetic sensation. Would completion of a Mars "Worldhouse" entitle his designated Terra-creature "offspring" to sign it on his behalf during its dedicatory ceremonies? A Mars terraformed by Taylor-inspired Tellurians—whether genetic or electronic stand-ins—would be a planet halfway to becoming a world without geotechnical macro-problems. Beleaguered maintenance personnel such as janitors, currently at work on a semi-pristine (un-terraformed) planet (Earth) might well be green with envy. Is Taylor's "Worldhouse" patentable? Indeed, as an atrium-like building-place encompassing mega-lodgings and pollutant-free guaranteed weather regimes, Taylor's Red Planet "Worldhouse" would be a human life-sustaining Spaceship Mars offering covered (but vacant, biotic sterility) real estate equal in area to Earth's landmass. [Chapter 11 will more fully advance the concept of an inhabitable Mars.] If properly irrigated (Badescu, 2009), and given enough sunlight, green land plant growth on Mars is possible under an absolute Martian air pressure that is 7-10% of the present-day atmospheric pressure at sea-level in the Earth-biosphere. Here, it is worthwhile to recall the perspicacious and incisive words of John Stuart Mill (1806-1873): "If it became customary to sojourn long in places where the air does not naturally penetrate, as in diving-bells sunk in the sea, a supply of air artificially furnished would, like water conveyed to houses, bear a price: and if from any revolution in nature the atmosphere became too scanty for the [sic] consumption, or could be monopolized, air might acquire a very high market value" (Mill, 1880).

Were Earth's life eliminated, in a few million years our homeland would look like today's Mars, and in a few more millions of years, just like today's Venus (Holmes, 2013). Taylorists will have to pen an oft-revised official Karl Baedeker-type Guide to "Worldhouse Mars", which should definitely include colorful site descriptions of Mars' oddest landscape features. The so-called mimetolith "Face on Mars", a natural eroded topographic feature the shape of which when imaged by satellite resembles something else than a hill, namely a partial creature with a human visage! The "Face on Mars" massif is a pareidolic illusion not too different from the "Man in the Moon" we have all seen, as did Eduard Suess the insightful European geoscientist during the 19th Century. [Geographos never found a "Face on Mars"—see Chapter 9—but did, early-on, think the landscape feature looked rather like a mono-static radar installation like PAVE PAWS at Clear Air Force Station in Alaska, USA.] Taylor's Martian "Worldhouse" macroproject might be funded (constructed and maintained) by a roof tax, not a poll or parcel tax. In other words, the Mars tax base is a roof hence the uncovered Martian poles are zones of "free" real estate. What if the Coca-Cola® advertising department, by management's choice, paying generously for the leased long-term corporate use of a part of the Taylor-proposed "Worldhouse" ceiling as an enormous billboard? Try to imagine Mar's ceiling as a gigantic paint-by-numbers color canvas (of television screen pixels). What about other businesses and governmental warnings and encouragements? How about some groups of people—authentic future Martian Greens?—paying not to have to see propaganda, leaving the ceiling a raw canvas? Projection television, using the underside of Taylor's "Worldhouse" roof, might display fluffy clouds familiar to pioneer Mars dwellers recently departed from Earth, a kind of harmless camouflage tending to make restless and uneasy settlers happy in their strange and dangerous new planetary surroundings! Such clouds could even be made three-dimensional—generated on a huge geographical scale—emulating indoor cloud-making art done by Berndnaut Smilde [GOTO: www.berndnaut.nl/index.htm].

For the first time in history, a long-term pro-biota technological fix, "Worldhouse" has been patterned—a sketchy sort of blueprint tailored to solve our Earth-biosphere's forecasted near-term societal level of stress and "crisis". Some psychologists assert that urbanization and a loss of consistent rural contact will be the big social problem for humans during the 21st Century. Mars awaits us! Life-style in Mars would cause changes to our species, such as skeletons composed of less dense bones and much lighter musculature, creating a need for a Martian human lectotype remarkably distinct from Earth's. Almost all humans comfortably inhabiting a Tellurian "Worldhouse" would live in apartment complex-like cells attached to the numerous several-kilometer-tall towers proposed by Richard L.S. Taylor that serve to keep a semi-hermetic tensioned fabric lid on Mars' atmosphere. Certainly a manmade Mars garden, better than *The Little Prince* could ever have tended, is preferable to the insane idea advocated by some radical Greens who seem to be wishing for *Homo sapiens'* removal from Earth's global Nature—supposedly because our species is a contaminant—so that the "real" global Mother Nature—sans toxicosis—would stand fully revealed! Even so, Tellurian macro-engineers might have to devise some way to protect and/or insure their profession from construction-defect litigation over the "improper" tenting of Mars (Cathcart and Badescu, 2004). [During 2012, MIT student Otto Ng proposed a solar-powered tent covering 10,000 square kilometers of Saudi Arabia's desert peninsula landscape establishing a "Powerscape" for that ecosystem-nation and its neighbors.] Mars ought to become a well-knit New World as Earth becomes a well-ordered (peaceful and progressive) Old World. Would varieties of social "circus" (Loring, 2007) appear on the developing Mars stage? Ideally, a typical noosystem is a national cultural gathering, a group of like-minded persons; that is, a voluntary association of persons sharing a viewpoint on other life forms (by absolute fixed geographic location or by moving absolute geographic location), and a common or nearly common Technology for living in its distinct place (for example, Earthly Anthromes or Astythromes, inside capacious Dandridge Cole-style hollowed and harnessed asteroids, and familiar orbiting space stations). Noosystems seem quite necessary for human survival because they are resilient, almost autonomous social systems. (Truthful, agreed political maps showing borders are the most reliable, accurate maps of any kind ever drawn!) A rudimentary Anthropocenic "global village" does exist in the Earth-biosphere, but rare electromagnetic technological failures (such as power grid breakdowns during inclement Sun-caused space weather process-events) demonstrate that modern noosystems have few, if any, time-tested procedures for perfect management of a global village (Harris, 2009; Piantadosi, 2012). Individual intellectual isolation can happen in mere moments, establishing a disconnect that disrupts *Homo sapiens'* "schoolroom" (Earth-biosphere/Noosphere). However, adversity has always been an effective motivator!

Herbert Marshall McLuhan (1911-1980) coined "Global Village" in 1964, identifying a supposition that a single community of human beings, situated in Earth's biosphere, had recently become extant. McLuhan's odd fanatical interest in mass-communications impelled him to arrive at his theoretical concept of a world of only one noosystem. No such completed "global village" exists or is likely to exist anytime soon. Most persons will agree that exchanges of information via telecommunications media is a poor substitute for personal contact and, further, that a "global village" might exist one day, but only if an individual were about one hour's physical travel-time away from the physical presence of the other communicant. All news indications are that our 21st Century world, or Noosphere, will continue to be a system that is not organized according to conscious thought. The co-location of persons, Earth-biosphere confined people, is in violent conflict about the nature of belief: the primary evidence for this statement is the apparent flourishing of worldwide terror lodges such as Al Qaeda. Co-location of grouped humans and certain elements of this world ("integration", "networking", "globalization" and so-called "internationalization"), definitely involves its own built-in macro-problems!

"World" denotes all of the natural and artificial conditions that exist during 2014 within Earth's biosphere that will be affected by all building and demolition by *Homo sapiens* and/or any Tellurian-designated and/or hostile or non-hostile Alien-imposed agents. Macro-Imagineering, and its realization—that is, its operational—practical servant Macro-engineering, big Earth macroprojects (also known as "megaprojects") could help to maintain positive social conditions in Anthromes and Astythromes as well as global international relationships among somewhat self-isolating peoples (Badescu et al., 2006).

Many pliable/gullable Greens seek to soon discern an organizational trend that will make humanity the social unit of *Homo sapiens*, and the world of globalized infrastructures, humanity's only domain. Realistically, however, human social disunity (fomented by our biological species' diverse ethnic, racial and religious affinities) is most likely to continue unabated, and any settlement of other rocky planets in our Solar System seems to be high probable, yet far-distant in occurrence, which settlement would maintain a genetic diversity advantageous to *Homo sapiens*. Overcrowded humans make each person more indifferent to the other cohabitants, less pleased with his/her own individual existence, stupid, less human and less humane. Our Solar System could be induced to have a cortege of three or four celestial rocky body-enshrouding Noospheres since terrestrial-type planets are now identified as specific Macro-Imagineering/Macro-engineering anthromes with characteristics of favorable and unfavorable conditions for construction and destruction. For instance, a technologically tamed Mars, Richard L.S. Taylor-styling, would present people with another landscape almost the equal of Earth's landscape (Cathcart, 1998). (See: Chapter 9.)

Equipped with all of the up-to-date ingenious fashions of progressive 21st Century technologies, humanity can rather easily survive in every part of our Earth-biosphere, even the most inhospitable places such as the frigid Polar Zones and the vast hot desert lands. Because a living encapsulated person (a spationaut operating an appropriate vehicle or merely space-suited) is a volume-defining organism, it logically follows that the planet's biosphere limits necessarily expands and contracts with a fellow person's movements. (Spationauts: similar to the sociological concept of "every-man/every-woman"!) [This is rather picturesquely illustrated by the photographs of the "Suit-Sat 1", a retired Russian spacesuit stuffed with junk and a working radio transmitter, discarded to outer space through the International Space

Station's air-lock hatch on 3 February 2006. It was vaporized during a fiery re-entry on 7 September 2006. A second "Suit-Sat", officially called ARAISS-1 was deployed from the International Space Station during August 2011 and it re-entered Earth's atmosphere in January 2012, incinerated by friction during descent. For a fascinating video dealing with "junkspace" see Cazabon (2013).] What then results are that Earth's shell-like boundaries of biosphere and Noosphere will become practically co-terminus, resulting in humanity's Earth-world, a so far unique bio-apparatus!

Although some persons dissent from the moment's faddish Green imagery of Earth as a tiny containership moving through the Universe, most scientists are nowadays calling for controlled planetary renewal efforts, a Macro-Imagineering/Macro-engineering-led kind of CPR [cardio-pulmonary resuscitation] for our biosphere. Biotechnologists have learnedly discussed bio-engineering techniques that might enable humans to make plants even more responsive to increases in the air's carbon dioxide gas content and thereby portend an even more abundant future for the terrestrial biosphere's vegetation (Sage and Coleman, 2001; Singarayer et al., 2009; Ridgwell et al., 2009; Jansson et al., 2010). Aerospace and space technologies applications, inherently global, are part of an overall growth in geopolitical information systems and applications that offer both groups of thinkers and doers man opportunities for planetary and other styles of management. As will be demonstrated in Chapter 3, certain total environmental systems of management—such as, for example, Rodoman-ALPS—will afford just about every human individual the experience of holding an interactive visual subscription to the **National Geographic Magazine.**

Ten years before his passing, the French naturalist Georges Louis Leclerc, Comte de Buffon, penned **Des Epoques de la Nature**, in which he arranged our Earth-biosphere's past, present and future into seven time periods of indeterminate duration. His Seventh Epoch was defined as a portion of Geologic Time begun by the lordship of humans and ended by Earth's cooling to a temperature making all life impossible. Molecular nanotechnologists (Drexler, 2013) may create micro-robots to remove all greenhouse effect-causing gases—not merely carbon dioxide, but water vapor too—so that the Earth atmosphere would quickly become cold and dry, even sterile, thereby exterminating all unprotected natural life forms. Meaning, of course, that highly motile artificial life, perhaps including synthesized life, such as tiny machines just might continue enjoying a long Earthly existence! [Drexler's proposed machines might be thought of as ant-like and, if successful, Drexler could become remembered as fondly as Milton Levine (1913-2011), the gentleman who popularized the Toy Ant Farm in 1956!] Moderns should not infer—Buffon may not have so implied—that the absence of natural life definitionally annihilates Earth's global Nature! Although the world-public is not yet fully aware of the poorly-broadcast fact, the Green theory of a "Balance of Nature" has become a professionally disused geo-scientific concept. Traditionally, "Balance of Nature", was a transient life science theory that implied at least two symmetrical portions, in equilibrium, and stasis. The phrase has been quietly denounced at least since 1985 by most reputable geoscientists. Nevertheless, "Balance of Nature" still serves a public propaganda function for the upper-management of some international Green organizations. Past Earth changes have occurred abruptly and future changes may happen as stunningly rapid, a la Chelyabinsk during 2012, and worse! When perfected, the developed capabilities of robots will qualify automatons to be accepted as a new form of life. Such miniscule machines might be considered as "mindkind" on the role model of "humankind". The invention of automatons would shatter Buffon's purely imaginative Geologic Time schedule, making it open-ended. Molecular nanotechnologists, in particular, intend that future microscopic-sized machines will have macroscopic scenic consequences. Earth's biosphere could become defunct, replaced entirely by a Noosphere belonging to artificial life.

If it finally happens that our planet ultimately hosts two types of life—natural and human-instigated artificial—then Macro-Imagineers might label the "Free World" as the Anthropocenic Earth-biosphere volume dominated by artificial life and robots; "Free World" is clearly meant to be an antonym for a *Homo sapiens*-dominated "Global Village". One unavoidable outcome of the onset of molecular Nanotechnology's perfected widespread use will be the eradication of world economic cycles such as those first theorized by Nikolai Dmitrievich Kondratieff (1892-1931) during 1922 (Garvy, 1943). Soon, the operating branch of Macro-Imagineering—that is, Macro-engineering—might take on some of the coloring plantation forestry now enjoys as a profession and yet most macro-imagineers have ignored the Kondratieff macro-economic forecast. That the material basis of long economic cycles is the wear and tear, the re-placement and the increase of the total fund of basic capital goods, the production of which requires heavy investment and is a time-con-suming, was clearly postulated close to a century ago (Carlsson, Otto and Hall, 2013). Globally, the real and personal property insurance (and re-insurance) business will cease to exist for lack of need—tiny, sometimes invisible machines introduced to the Earth-biosphere by molecular Nanotechnologists, set to work constantly repairing and up-dating humanity's infrastructures, would—literally—stop mate-rials degradations and capital investment replacement cycles forever. Hence, no more "Boom and Bust" ecosystem-nation economic ups and downs! [Such controlled mini-devices would rapidly replace digi-tal fabricators that have only just recently come to market after years of R&D (Anderson, 2012).] And, no dangerous industrial waste prod-ucts, chemicals or radionuclides (Alley, 2013) will ever again require costly watchful secured containment. Schemes of climate change mitigation—established recently as "Geo-engineering"—will become outmoded, irrelevant to humanity's everyday needs and desires. By June 2010, "Geo-engineering" became a term considered to warrant a common definition in the *Oxford English Dictionary*: "The deliberate large-scale manipulation of an environmental process that affects the Earth's climate, in an attempt to counteract the effects of global warming". So far, there are few contemporary actionable "Geo-engineering" proposals, no method yet is technologically ready for deployment at a geographically large-scale.

Since all the major military establishments have now, at least tacitly in private, endorsed a space-based global anti-ballistic missile defense system coupled with an anti-rogue asteroid defense system, it means that USA Macro-Imagineering's land-use planning and large-scale land planning efforts elsewhere in the world can take a different tack. [Rodoman-ALPS, tantalizingly and briefly outlined in Chapter 3, Geographos' tentatively offered macro-imagineered proposal dealing with how such a vital system might be configured architecturally.] America's often maligned residential suburbs—commonly and derogatorily referred to as "urban sprawl"—exist due, in large part, to the Federal Government's desire, implemented by Veterans Administration and Federal Housing Administration mortgage slanting from 1947 until 1957, to disperse the Nation's populace as a form of protection against nuclear attacks done by intercontinental strategic bombers. Intercontinental missiles, as well as the prospect of a USSR-manufactured 100 Megatonnes potential yield warhead, negated that scientifically and sociologically myopic USA government initiative.

What effect would materialization via installation of a global anti-ballistic missile defense screen have on global private and public sector housing and industrialization plans and programs? Federal

Government investments during World War II resulted in a massive restructuring of the USA. America's postwar urban sprawl is, in effect, a clever long-term countermove to the newly perceived asteroid Earth impact threat! By contrast, crowded India, burdened by a population of more than one billion persons, lacks the population dispersal option of "sprawl". International trade will become unnecessary—capital goods transfers will become obsolete because anything can be produced anywhere from feedstock—and services can be rendered remotely, accomplished via computer-generated Virtual Reality worlds. Early 21st Century personal computers have become representation machines—utterly beyond anything previously seen and handled in talking-film theaters, live stage auditoria or quiet reading of science-fiction novels—that emulate any known medium, making such ever-improving computers the first meta-medium. Arguably, this addition of a radically new form of physicality, interactive human experience is a major civilizational event which may shape humanity's everyday consciousness as much as what has already transpired. *Homo sapiens* now may be titter-tottering on the cusp of a post-biological phase of "evolution" (creation by humans of self-improving mobile hyper-objects). Are humans an obsolete species? Are humans in the process of handing on all scientific information about the Universe to civilization's machinery? Visualizing humankind's role in a future Earth-world quite different from our present-day environment—literally forming an image of it—would be a psychological aid to confidence and achievement by more than seven billion worried persons inhabiting the Earth-biosphere.

Near-term future Macro-Imagineering may become our world's only profession capable of planning the linkage of daisy-chained computers and other necessary electronic and mechanical devices to prevent an Earth-biosphere disaster. Between 1900/1980 and 2009 about 1,673,000 persons were killed by natural process-events (hurricanes, earthquakes, floods, volcanoes and tsunamis), according to a systematic literature review done by Doocy and his colleagues (2013). Whilst horrifying, that reported mortality number pales statistically by comparison with people-on-people violence taking place within and between noosystems worldwide each year! Somewhere, sometime, some reckless leader suffering a serious shortage of the human brain chemical serotonin is going to indulge his/her fantasy of violence upon the world—any world—caught unawares. Observant peoples must be on perpetual guard against the future non-fictional "Dr. Strangeloves"! The bevy of public-accessible satellites and space probes, the so-called robotic "Planet Paparazzi" (Williams, 2011) are inadequate to the task of warning good people of systematically planned evil actions. Evil cacosystems must eventually prevail if the good peoples of the world timidly avert their gaze and remain silent in the presence of the bad and the morally ugly, uncomplaining and inactive, from sheer irresponsible passivity. (Richard L.S. Taylor's futuristic vision of an emplaced, realized "Mars Worldhouse" could not last, or possibly exist, without a generally similar space-based anti-ballistic and anti-asteroid missile defense system as envisioned for the Earth-biosphere's security. Rodoman-ALPS, Geographos suggests, in the battle between evil and good and the variable human tenaciousness and an ever-present tenuity (outer space) that offers sudden death via Earth-biosphere impacting asteroids and terrible *Homo sapiens* designed and fashioned weapons, could save the day, literally! As a concept not much publicized until after 1945, Noosphere dates from the last half of the 19th Century, when Ivan Petrovich Pavlov (1849-1936) called it the realia of the Anthropogenic Era. Eric Drexler and others evidently think that turbulent global Nature [the planet's biosphere] is being replaced by a still forming Earth-noosphere. What is Earth, really, but a kind of voluminous, only slightly uncommon, geo-kinetic artwork (a bio-apparatus) created by the Universe [and/or God]? Might not now lifeless celestial bodies, such as Mars) soon transform into fully appreciated future "playgrounds" or "sand-boxes", which could stimulate the further intellectual growth of our ambitious and curious species? After all, we may truly need, not merely want, such otherworldly places to offset the disappearance of ourselves as the predominant life forms in the Earth! Geographos has strenuously puzzled over the oddly persistent NASA obsession with any purported discovery of real life off-world: *Homo sapiens*, always so far confined to a planet that is, essentially, a chaotic graveyard terrain containing some evidence of every kind of creature that has ever lived, must find cleaner, Greener, stomping grounds! Would it not be preferable to have the opportunity to apply the results of our Macro-Imagineering to places that have never been inhabited by anything that was alive? And, now Greens, extravagantly enamored with Earth imagined as an Anthropocenic "Blue Marble" (Wuebbles, 2012) fiercely guide sometimes petulant humans to collectively remain the ageing Solar System cemetery's caretaking janitor forever!

Yevgeny N. Lazrev (1994) formulated a new profession of "Anthropo-designer". Lazrev's anthropo-designers participate in a social trend, meta-design, which is applied experimental aesthetics, a modeling of the essence, manifestations and relations of people. Lazrev sees human/machine combinations [cyborgs] as a kind of centaur and thinks that electronic culture miniaturizes the world of objects, thereby indirectly increasing human values. [Others like Lazrev—persons such as Jurgen Bey, Yves Behar are product designers who, instead of solving conventional macro-problems, seek to imagine how the human future might be entirely different—that is, to discuss the moral, ethical, political and aesthetic macro-problems in a style that is almost SciFi. See: Anthony Dunne and Fiona Raby's *Speculative Everything: Design, Fiction, and Social Dreaming* (MIT Press, 2013).] Biotechnology and molecular Nanotechnology, effectively harnessed by a new inter-disciplinary profession, named "Anthropo-Designer" by Lazrev may upgrade *Homo sapiens* by increasing our species' ability to cope with the Universe's existing and expected conditions. Inescapably, a Solar System-wide governance regime(s)—itself interdisciplinary, selective and fusive—must come into existence. It is worth noting that "design" preceded and enabled language development, that some of the objects—even those that are artificially radioactive—created exist in the archaeological realm and the configured "world" succeeded "environment" because of "design"! Humans can now select migrants to successfully move away from our ancestral homeland, however mundane the motivations for that future journey! Still, it is possible to envision our Earth-biosphere so riddled with physically and intellectually insurmountable pollution or contamination difficulties (as, for example, a Nuclear Winter or a post-asteroid impact global shady period of Geologic Time) that "environmental refugees" opt for a mass exodus of healthy survivors and expedited planetary colonization of other global features (for instance, the cold desert that is Mars crust landscape). If these "difficulties" were later re-defined as "macro-problems", then "macro-solutions" would have to be on the Earth-Noosphere's "horizon"! As long ago as 1864, George Perkins Marsh (1801-1882), an American preservationist with a globalized outlook (Koelsch, 2012), posited the somewhat astounding thesis: Has or could man become the architect of his own abiding place? Marsh and today's struggling Macro-Imagineers/macro-engineers as well as some geo-politicians—all looking at the Big Picture that is end-to-end systems Macro-engineering—answer Marsh's two-part query affirmatively!

Although it is a marvelous driving macro-problem for Science and Technology, let us not fatally delude ourselves with the prospects of space travel. It would be grossly irresponsible to focus on the "Conquest of Space", ignoring the absolute necessity to maintain a self-renewing administration of *Homo sapiens*' primary base of operations in the

Universe! A new Aristotle trying to organize all human knowledge might find no place for a particular study of Earth-like geography. [Most modern, Anthropocene, professional geographers retain a certain gelatinous intellectual outlook on outer space, and the technologies required to explore it, that really seems rather unfathomable to Macro-imagineers!] Knowledge cannot be separated, though for convenience it must be segmented. A masterful model for all aspiring geographers, Alexander von Humboldt (1769-1959), published the first modern account of our Universe in historical perspective in his *Kosmos* (1862), establishing the concept of the periodicity of meteor showers, which visibly pepper our currently defenseless planet constantly, salting its superficial crustal stratum, as well as penetrating deeper strata, with particles from other unknown places in the Universe. Immanuel Kant opined that all the planets have been or will be inhabited; Kant did not foresee that humans would become space-faring oecists. Kant did allow, however, that culture would produce in rational humans an aptitude and capacity for any ends whatsoever of humankind's choosing. "Tomorrow" scientifically synthesized life, engineered cyborgs and agential Terra-creatures will be empowered also. The early stage of "Tomorrow" was first fully documented by Robert A. Freitas and Ralph C. Merkle in *Kinematic Self-Replicating Machines* (2004). [Freitas was awarded US Patent 7,687,146 B1, "Simple Tool for Positional Diamond Mechanosynthesis, and its Method of Manufacture", on 30 March 2010. His patent is the first for hypothetical industrial use of mechanical constraints to direct reactive molecules to specific molecular sites and thus to create previously "impossible" chemical synthesis outcomes! The first finding of mechanical gearing, interlocking cogs, in a natural biological structure was reported in 2013 by Burrows and Sutton.]

By 1972, in his monumental global history of humanity's geographical ideas, *All Possible Worlds*, Preston Everett James (1899-1986) successfully evaded the theoretical dead-end into which so many others professional geographers had needlessly self-isolated themselves and their conceptually tepid publications: "...if people called geographers apply geographical methods to such studies [of other Solar System solid objects] they will be included in Geography". In other words, James finally subscribed to von Humboldt's vision of Geography focused on extra-Earth debris and celestial bodies. Because of intellectual inertia displayed by too many mature university professors and the relative recentness of our historical period (of outer space and its contents explorations), some have begun to realize that Geography is saddled with a patently inadequate vocabulary. Two choices are open (I) simply transfer Earth-originated and oriented terminology to interplanetary space and other planet-places (that is, additional "global Natures", or (II) create new words, perhaps by language wordsmiths and neologistical experts extensively familiar with Science and Art. So far, simple word transfers have prevailed and additions have been made with the increase in humanity's practical experience. That 20 July 1969 "One small step for a man, one giant leap for mankind" uttered memorably by Neil Alden Armstrong (1930-2012) altered the meanings of all English words, by invalidating the assumption that they are used by a forever-Earthbound group of persons. Armstrong's feat—and feet—set prepared and properly equipped representatives of *Homo sapiens* on another global Nature stage, thereby instituting a new extra-Earth phase of *Homo sapiens* "evolution" under potentially other-worldly conditions of daily existence.

"Geography" is a protean word with elasticity in terms of definition, just like "ecosystem" and so many others. It is to be hoped that this useful property will enable modern professionals to soon refine an apt and workable disciplinary scope. Harold Leland Goodwin (1914-1990), at one time Director of the NASA's Office of Program Development during the early 1960s, thought that "A basis exists in the interpretation of the Greek *gaia*, from which *geo* derived. The word means 'earth', but it preceded the Hellenic concept of a spherical planet orbiting the Sun. If we assume *gaia* means simply 'ground' and, by extension, 'planetary surface', the generic interpretation becomes reasonable and '*geo*' can be applied without restraint" (Goodwin, 1962, page 9). When Armstrong put his left clean-booted foot on the Moon's regolith, he made a first impression on an environment quite remarkably unlike anyplace to be found in the Earth-biosphere where humans have waddled, possibly swaddled in garments made to permit survival in the Polar Zones! The best alternative—if one is needed—the most reasonable, to Goodwin's excellent ad hoc solution would be to contrive a study of Planetography, subdivided into Geography (Earth), Areography (Mars), Selenography (the Earth's Moon), and so on. Some have used this regularized system of naming whilst others have declined to use this system. All of these planet-places are known to have lithospheric crusts upon which people can walk and even move about rapidly in wheeled and tracked surface vehicles. "Anthropos" means "one who walks with his face to the heavens". St. Augustine (354-430) believed that this unusual posture, along with our rotating necks, was one of the defining features indicating *Homo sapiens'* uniqueness: "Man was made to walk erect with his eyes on heaven, as though to remind him to keep his thoughts on things above". Johann Hieronymous Schroter (1745-1816) coined the term "aerographic" by analogy with "geographic" in his ***Aerographissche Beitrage zur genaueen Kenntnis und Beuteilung des Planeten Mars***, which was posthumously printed in 1881. [During 1815, Schroter published the first textbook devoted to another planet, Mercury!] During 1962, William Wheeler Bunge stated in his ***Theoretical Geography*** that Geography's entire subject matter (if the subject truly matters) might be expanded to include universal and site-specific facts discovered subsequent to our Space Age's onset. That is, [since 1957-1969] Geography should examine "...that portion [Bunge meant "part"] of the Universe directly available to man...and to phenomena of human significance".

1957-1969 was an epochal period that marked the inception of a new, energized and outreaching Science. [Until 1957-1969, the inclusive/exclusive professional phrase "Earth Sciences" reflected the spatial restriction to a single Solar System global Nature. Thomas Chrowder Chamberlin (1843-1928) championed "Earth Sciences" but the first American university to establish a School of Earth Sciences was Stanford University in California, USA during 1965.] Hubertus Strughold, the first expert in Space Medicine, suggested that humanity would need a "topographical" description of interplanetary space (Spatiography), which regionalized outer space using certain borders (physical conditions and/or "landmarks"), and names, special words or phrases to identify useful subdivisions of our Solar System. At a time when Macro-Imagineering/Macro-engineering is seriously advancing ideas for impressive macro-projects located amidst our Solar System's planets, 70+ moons, and odd debris, various science disciplines and arts should be applied to evaluate known and extrapolated data and information in the course of locating large-scale projects and determining their effect on humankind's activities in our Solar System. Sometime between 1 April 1955 and 1 December 1959, the US Air Force apparently invented the word "aerospace", as evidenced by its replacement of "air" in *AFM 1-2*, "United State Air Force Basic Doctrine". "Aerospace" is most pertinent to the developers in the UK of SKYLON, a true aerospace vehicle. A Solar System Spatiography would have its point of origin (departure) at our Sun's center of mass, since 98% of the Solar System's mass rests in that shining star. But, as Otto Klima (1916-2000) cautioned about the interplay of aerospace and hydrospace technologies—there is not easy transference possible (Klima, 1970)! Geographers ought to defer to other, better-trained spationauts, letting them adopt a spacious heavenly realm as their profession's turf. This system of spatial description (Geography cooperating with Spatiography) deserves promotional publicity in the media! For the most part, people have only vicariously explored other planets—perhaps the most unforgettable recent witnessing was the macro-engineer Neil Armstrong's directly-broadcast 20 July 1969 lunar power-strides flickering on (pre-flat screen) televisions almost everywhere.

The USA's Apollo Mission spationauts spent little working time on the Moon's cratered and pulverized landscape and were almost unrecognizably encumbered by clumsy clothing. Geographos prefers to remember those brave men as almost like armchair geographers, some of whom nowadays can be "absorbed in a miasma of a tele-presence operation" by computer-generated Virtual Realities (that are not actually real)! Planet-places such as Mars and Venus, which are supposed to be without pre-historic periods, should be explored. Mars and Venus are precisely not classrooms for *Homo sapiens*, but instead Earth's opposite and counter-part ("sand boxes" and "playgrounds"). To preserve their own dignity as intelligent mammals, both men and women must trod Mars' mountains and plains; the 21st Century Earth-biosphere is too much like a reservation for further advance of humanity's Science and Art examining only the remains of a solitary global Nature. Approximately, 33% of Earth's landscape is still called wilderness (Cassidy et al., 2013). Some geographers reckon that if wilderness remains extant on this planet by 2100 it will only be because, for the first time in recorded history, humans have deliberately chosen that it should be so as a positive benefit rather than an industrial brownfield—wilderness in 2100 will be protected not by the Green-inspired barriers that have sufficed until now but by enforced pro-active legislation barriers endorsed by all ecosystem-nation members of civilization.

"Life" may be the only truly unique "commodity" in the Earth. Some smug professional geographers assert that bio-geographers need not worry about extra-terrestrial subjects invading their literature—they apparently prefer a no-growth scenario of development for the discipline of Geography! Space travel theoretician Hermann Julius Oberth (1894-1990) announced in the last paragraph of *Man into Space* (1957) that *Homo sapiens'* guiding species-lifetime aims must be: (I) to make available for living organisms every place where life is possible, and (II) to make inhabitable all planets as yet vacant, and (III) to make all life purposeful—that is, to give all life identifiable goals. [Goal III is especially breath-taking—ALL life? That goal may only be achievable if Earth were shattered to make a planned Freeman Dyson Satellite Swarm.] Oberth did not advocate mere human survival—there are many noosystem ecosystem-countries within our present-day geopolitically fractured global Nature which exist at a subsistence-level of human "Standard of Living"—but, rather, of settlement—Biology's concept of ecesis. Oberth-inspired extraterrestrial colonies are desirable and records of extra-Earth settlement will be much better, more complete, than any previously recorded in our histories. Oberth purposefulness is a theological and cybernetic question best left to the reader's personal preference since for every word, there is also an anti-term!

Even by 2014, a seven billion plus-strong humanity still leads a remarkably secluded existence in this Solar System, but technology's progress will eventually permit our species to replicate and innovate self-supporting planetary biospheres by assessing monetary values to extraterrestrial real estate—that is, Mars and possibly Venus, in particular, will be improved to become as valuable as Earth during the 21st and 22nd Century's. A developable Macro-Imagineering concept, "Terraforming", a spin-off from the Earth-centric precursor discipline "Geo-engineering", was first presented with utmost clarity and breadth to the world's public in *Terraforming: Engineering Planetary Environments* (1995) authored by the charismatic UK scientist-philosopher Martyn John Fogg (born 1960). The compleat text of his magisterial magazine of facts and ideas is generously posted on the Internet for free downloading as a series of chapter pdfs: GOTO: http://www.independent.academia.edu/MartynFogg/book .

Figure 1. Dr. M.J. Fogg, considered by a majority of practicing Macro-imagineers to be the genial originator and developer of that coming-of-age new discipline Terraforming. His 1995 textbook has never yet, as of 2014, been equaled for its determinative scoping and style and its outstanding comprehensiveness! (Image: MJF, 2013)

As incisively and comprehensively envisioned by Dr. M.J. Fogg and others since 1995, human-assembled planet biospheres will very likely be fostered, especially by the progress of Bio-technology, Robotics and molecular Nanotechnology. In other words, Mars could have a near-term future "invasion" far more massive than the few creeping and crawling robots already there, an "invasion" organized and managed by humans and our species' designated soft-lander agents since 1971, resulting in a contextless layer atop the Mars-crust's speculated total stratigraphic column. These stable instrumentation platforms, exactly locatable by special co-ordinate systems as well as visible in imagery snapped from orbiting robotic surveillance "mother-ships", collect and deliver digitized data as almost continuous reports on their immediate surroundings over periods of years sometimes. [Basically, this technique is a simple extension from that used to track persons electronically in Earth (Dobson and Fisher, 2007).] Planet-wide applications of Macro-Imagineering/Macro-engineering as well as Terraforming expertise including the ultimate technology of the Dyson Motor, re-named "Archimedes" by Geographos, summarized in Chapter 1, offers Earthlings the chance to redefine all of the planets as human resources. These malleable celestial bodies might then be known as LDCs—Localized Development Corporations—and would be vertical Solar System company organizations! The scope of such exploration and exploitation efforts can be expanded greatly, but exciting tax-payers to be supportive will not be easy. Jerome E. Dobson, President of the American Geographical Society since 2002, generously clued Geographos to adapt a useful Geography definition by modifying and up-dating it a little (Dobson, 1993). Geographos proposes that 21st Century human geographers pursue spatial logic because spatial event-processes—sometimes a synonym for "progress"—are necessary to solve complex planetary macro-problems. The reader will be encourage in this textbook to wade through a welter of lengthy science-fiction analogies, brief technical definitions, and numerous scientific hypotheses/theories before a coherent Big Picture emerges in Chapters 8 and 9. A unified and consistent Macro-Imagineering vision of Earth's dynamic global Nature, such as Humboldt once produced, has become impossible in our early-21st Century time of scientific specialists, at any rate for Geography's professionals. In other words, the public "tastes" Science and Art after both have been masticated by others (media personalities and fabricators of truth and falsehoods! For normal persons, the ultimate in habitability of any planet-place occurs when local or global Nature permits a clothed, but un-encapsulated, person comfortable spatial movement. In today's Solar System, only our Earth fits this criterion. Logistics, to employ a word that came into wide usage during World War II, is the macro-problem.

Since the final American landing on the Moon in 1973, there has been one really praiseworthy improvement in the format of scholarly writings in Geography: the systems approach to the Earth-biosphere and its industrialization. This approach was brought to Geography research from R&D schemes used by American and UK specialists in wartime programs dating from 1939. Such plans and macro-projects were typically laden with acronymic labeling mysteries—the very word "acronym", coined by Basil Davenport (1899-1964), first appeared in print (as a letter to the Editor) in the February 1943 issue of **American Notes and Queries**. The Apollo Mission Project preparations seem to have been a prime source for spreading the idea of systems analysis and synthesis to Geography, Macro-engineering and, lately, to Macro-Imagineering, the most theoretical of large-scale or major macroprojects thinking (Galloway et al., 2012). In terms of their functioning, operation, and lifecycle, museum curator Jack Wesley Burnham equated some 20th Century mechanical and electrical artworks with "systems" thinking in **Beyond Modern Sculpture: The Effects of Science and Technology on Sculpture of this Century** (1968). Too, the very concept of "harmony" in a "Balance of Nature" implies, even requires, that there are separate Earth-biosphere portions ("systems") to be harmonized naturally or artificially. This implies that good, not merely successful, Geography/Macro-Imagineering/Macro-engineering and Terraforming cannot be a compromise, that its final format must be a resolution that somehow redefines and unifies the interfacing of the local and global equal-importance systems of the original planetary macro-problem (Allan, 1983; Josephson, 1996). *Homo sapiens'* goal to-day ought to be to eventually harness all the natural Earth-biosphere energy systems so that, ultimately, humans can populate the rest of our Solar System. The Earth as a whole, then, is to be apprehended as a resource reservoir of materials and energy.

How does Geographos reconcile "Macro-Imagineering" and its actual "Macro-engineering" application with molecular Nanotechnology? [Caution: recall LEGO's "Build with Google Chrome" Internet avail-ability!] Like others before, with finesse! Observant planners have noted that many building industries have adopted an economics of numbers, trending away from the economics of size. Pre-21st Century, it was conventional industrial wisdom that the capital cost per unit of assembly capacity declines by increasing the unit size. However, electronics miniaturization and even standardized ocean/rail//truck shipping containers has forced a modification of the pearl of wisdom. Three founding driving forces ensure that pearl gets smashed forever: (I) vast networks of computers, sensors and telecommunications make automation possible; (II) mass-production of many physically small, standardized units can achieve capital cost savings comparable, or significantly, greater than can be achieved through large unit scale; (III) small-unit scale technologies are flexible, an optimizing charac-teristic which significantly reduces both capital investment and oper-ating costs (Gimzewski, 2008). Thusly, humans have a new material production paradigm, one that shrinks in importance the old adage "Bigger is Better"! "Unobtainium" combines "unobtainable" with the "ium" suffix that usually marks the names of chemical elements—it is a material that might conceivably exist, but does not, either because it is not extant or that it does exist somewhere but is such a rarely encountered resource that no sum of money could ever purchase it. However, molecular Nanotechnology may someday soon change that truism dramatically! Thusly, both small units and large units could be quickly made physical in the Earth-biosphere, a Mars undergoing terraformation, and in deep space—its development and implementation would be a substantial advance over newly-utilized Digital Printers technology (Dunn, 2012).

Too many persons are currently afflicted with the false mystique that has grown up around computers, which is sometimes promoted by flashy Hollywood movies as well as viewer ratings of broadcast, cable and satellite television shows—that is, by non-computer science experts. Millions of people rely on their impressions of the world's nation-ecosystems as displayed in various popular magazines. Few of those millions of readers, many nowadays using digital devices, are aware that those glossy images are sometimes quite untrue! For example, the controversial **National Geographic Magazine** cover pictures of its February and April 1982 issues were retouched by skillful computer operators; the images were not re-published as part of that monthly's October 2013 inserted poster portraying 125 years of **National Geographic** cover photographs. So, it is now a situation for information/data assessors whereby digital retouching has terminated photographs and television images as evidence of anything—realia is not necessarily ever actually material anymore! Pictorial geographical falsity is a macro-problem for everyone subject to predatory propaganda envelopment.

The world's "Television Generation", the segment of the demographics born anytime after 1950, that never new existence without television, was a poorly educated target population vulnerable to verisimilitude. In so many ways, the global impact of television has been mainly to restore those noosystem-centered viewpoints of the world prevalent before widespread public education flourished for a while during the 20th Century. Like the inhabitants portrayed in Ray Bradbury's **Fahrenheit**

451 (1953), peoples of many present-day Earth-biosphere countries seem happy, if not actually eager, to give up the challenge and responsibility that ever-advancing Science and Technology's books and articles convey! The "critical mass"—this commonly-bandied popular phrase was originated by UK astrophysicist Arthur Stanley Eddington (1882-1944) sometime before 1920—of all cacosystems is ignorance and provincialism in opposition to the quasi-obligatory globalism that is based on permanent shared infrastructures; cacosystems will always seek de-linkage of telecommunications infrastructures in particular, trying by desperate acts of terror and sabotage to force globalized infrastructural cooperation to revert to purely voluntarist internationalism, to become social organizations that can be blackmailed into religious and ideological self-transformations.

References Cited

Allan, J.A. (1983) *Natural Resources as national fantasies.* Geoforum 14: 243-247.

Alley, W. and Alley, R. (2013) *Too Hot to Touch: The Problem of High-Level Nuclear Waste.* (Cambridge: Cambridge University Press) 383 pages.

Anker, P. (2002) *The Context of Ecosystem Theory.* Ecosystems 5: 611-613.

Anderson, C. (2012) *Makers: The New Industrial Revolution.* (New York: Crown Business) 300 pages.

Badescu, V. and Cathcart, R.B. (2006) *Environmental thermodynamic limitations on global human population.* International Journal of Global Energy Issues 25: 129-140.

Badescu, V., R.B. Cathcart and R.D. Schuiling (Eds.) (2006) *Macro-engineering: A Challenge for the Future* (The Netherlands: Springer) 316 pages.

Badescu, V., Dragos, I. and Cathcart, R.B. (2009) *Ecopoiesis and Liquid Water Transportation on Mars,* pages 661-682 **IN** Badescu, V. (Ed.) *Mars: Prospective Energy and Material Resources* (The Netherlands: Springer) 695 pages.

Badescu, V. and Cathcart, R.B. (2011) *Macro-engineering Seawater in Unique Environments: Arid Lowlands and Water Bodies Rehabilitation.* (The Netherland: Springer) 790 pages.

Bailey, M.R. (1993) *The Tracked Hovercraft Project.* The Newcomen Society 65: 129-146.

Burrows, M. and Sutton, G. (2013) *Interacting Gears Synchronize Propulsive Leg Movements in a Jumping Insect [Issus].* Science 341: 1254-1256.

Carlsson, R., Otto, A. and Hall, J.W. (2013) *The role of infrastructure in macroeconomic growth theories.* Civil Engineeing and Environmental Systems 30: 263-273.

Caruso, E.M., Bowen, L.V., Chin, M. and Ward, A. (2013) *The Temporal Doppler Effect: When the Future Feels Closer Than the Past.* Psychological Science 24: 530-536.

Cassidy, E.S. et al. (2013) *Redefining agricultural yields: from tonnes to people nourished per hectare.* Environmental Research Letters 8: 034015.

Cathcart, R.B. (1998) *Taming Mars with a tent and tunnel: creation of a biosphere-city.* Speculations in Science and Technology *21: 117-131.*

Cathcart, R.B. and Badescu, V. (2004) *Architectural Ecology: A Tentative Sahara Restoration.* The International Journal of Environmental Studies 61: 145-160.

Cazabon, L. (2013) *Junkspace.* Leonardo 46: 498.

Chernow, B. (2002) *Christo and Jeanne-Claude* (New York: St. Martin's Press) 390 pages.

Cidell, J. (2012) *Just passing through: the risky mobilities of hazardous materials transport.* Social Geography 7: 13-22.

Cockell, C.S. (2011) *Life in the lithosphere, kinetics and the prospects for life elsewhere.* Philosophical Transactions of the Royal Society A 369: 516-537.

Dobson, J.E. (October 1993) *Commentary: A Conceptual Framework for Integrating Remote Sensing, GIS, and Geography.* Photogrammetric Engineering and Remote Sensing LIX: 1491.

Dobson, J.E. and Fisher, P.F. (2007) *The Pantopticon's Changing Geography.* The Geographical Review 97: 307-323.

Doocy, S. et al. (April 2013) *The Human Impact of [cyclones, earthquakes, floods, volcanoes, tsunamis]: a Historical Review of Events 1900 [and 1980]-2009 and Systematic Literature Review.*PLOS Currents Disasters.

Dougherty, S. (March 2012) *Embodiment and Technicity in Geoff Ryman's Air.* Science Fiction Studies #116, Volume 39, Part 1.

Drexler, K.E. (2013) *Radical Abundance: How a Revolution in Nanotechnology will Change Civilization* (New York: Public Affairs) 340.

Dunn, J. (October-November 2012) *Printed in Space.* Air & Space 27: 26-31.

Erisman, J.W. et al. (2008) *How a century of ammonia synthesis changed the world.* Nature Geoscience 1: 636-639.

Galloway, P., Nielsen, K.R. and Digum, J.L. (2012) *Managing Gigaprojects: Advice From Those Who've Been There, Done That.* (Washington DC: American Society of Civil Engineers) 480 pages.

Garvy, G. (1943) *Kontratieff'sTheory of Long Waves.* Review of Economics and Statistics 25: 208.

Gimzewski, J.K. (2008) *Nanotechnology: The Endgame of Materialism.* Leonardo 41: 259-264.

Goodwin, H.L. (1962) *Space: Frontier Unlimited* (New York: Van Nostrand) 144 pages.

Haff, P.K. (2010) *Hillslopes, rivers, plows, and trucks: mass transport on Earth's surface by natural and technological processes.* Earth Surface Processes and Landforms 36: 1157-1166.

Haff, P.K. (2012) *Technology and human purpose: the problem of solids transport on the Earth's surface.* Earth System Dynamics 3: 149-156.

Harris, P.R. (2009) *Space Enterprise: Living and Working Offworld in the 21ˢᵗ Century.* (New York: Springer) 616 pages.

Haycraft, W.R. (2000) *Yellow Steel: The Story of the Earthmoving Equipment Industry.* (Urbana: University of Illinois Press) 465 pages.

Herman, A. (2012) *Freedom's Forge: How American Business Produced Victory in World War II* (New York: Random House) 432 pages.

Herndon, J.M. (2014) *Terracentric nuclear fission georeactor: background, basis, feasibility, structure, evidence and geophysical implications.* Current Science 106: 528-541.

Holmes, B. (28 September 2013) *Earth after life.* New Scientist 219: 38-41.

Hu, Q-H. (2010) *Sources of anthropogenic radionuclides in the environment: a review.* Journal of Environmental Radioactivity 101: 462-473.

Jacobs, C. (2008) *Smogtown: The Lung-Burning History of Pollution in Los Angeles.* (New York: Overbook) 384 pages.

Jansson, C. et al. (2010) *Phytosequestration: Carbon Biosequestration by Plants and the Prospects of Genetic Engineering.* BioScience 60: 685-696.

Jones, C.G. (2012) *Ecosystem engineers and geomorphological signatures in landscapes.* Geomorphology 157-158: 75-87.

Josephson, P.R. (Summer 1996) *Atomic-Powered Communism: Nuclear Culture in the Postwar USSR.* Slavic Review 55: 297-324.

Kennedy, P. (2013) *Engineers of Victory: The Problem Solvers Who Turned the Tide in the Second World War* (New York: Random House) 436 pages.

Klima, O. (1970) *Technological Interrelationships between Aerospace and Hydrospace.* Journal of Hydronautics 4: 126-128.

Klose, C.D. (2013) *Mechanical and statistical evidence of the causality of human-made mass shifts on the Earth's upper crust and the occurrence of earthquakes.* Journal of Seismology 17: 109-135.

Koelsch, W.A. (2012) *The Legendary "Rediscovery" of George Perkins Marsh.* Geographical Review 102: 510-524.

Kirsch, S. (2005) *Proving Grounds: Project Plowshare and the Unrealized Dream of Nuclear Earthmoving.* (New Brunswick, NJ: Rutgers University Press) 257.

Lapo, A.V. (2001) *Vladimir I. Vernadsky (1863-1945), founder of the biosphere concept.* International Microbiology 4: 47-49.

Loring, P.A. (2007) *The Most Resilient Show on Earth: The Circus as Model for Viewing Identity, Change, and Chaos.* Ecology and Society 12: 9-20.

Lynch.M. (2002) *Mining in World History.* (London: Reaktion Books)350 pages.

Kowalewski,M. et al. (2011) *The Geozoic Supereon.* Palaios 26: 251-255.

Kozlowski,S. (2004) *Geodiversity: The concept and scope of geodiversity.* Przeglad Geologiczny 52: 833-837.

Lohbeck, K.T. et al. (2012) *Adaptive evolution of a key phytoplankton species to ocean acidification.* Nature Geoscience 5: 346-351.

Margulis, L. (1993) *Gaia and the Colonization of Mars.* GST Today 3: 277-291.

Masco, J. (2010) *Bad Weather: On Planetary Crisis.* Social Studies of Science 40: 7-40.

Medvetz, T. (2012) *Think Tanks in America* (Chicago: University of Chicago Press) 324 pages.

Mill, J.S. (1880) *Principles of Political Economy,* page 4.

Monchaux, N. de (2011) *Spacesuit: Fashioning Akpollo.* (Cambridge: MIT Press) 364 pages.

Morriss, A. et al. (Eds.) (2012) *Silent Spring at 50: The False Crises of Rachel Carson.* (Washington DC: Cato Institute) 344 pages.

Mucciarelli, M. (2012) *The Seismic Wake of Costa Concordia.* Seismological Research Letters 83: 636-638.

Park, K.B. et al. (2013) *Thermal conductivity of single biological cells and relation with cell viability.* Applied Physics Letters 102: 203702.

Pasztor, A. (22 October 2013) *No Rocket? Venture to Sell Balloon Trips Into Space.* The Wall Street Journal CCLXII, page B1 and B2.

Piantadosi, C.A. (2012) *Mankind Beyond Earth: The History, Science, and Future of Human Space Exploration.* (New York: Columbia University Press) 279 pages.

Retallack, G.J. et al. (2013) *Problematic urn- haped fossils from a Paleoproterozoic (2.2 Ga) paleosol in South Africa.* Precambrian Research 235: 71.

Ridgwell, A. et al. (2009) *Tackling Regional Climate Change By Leaf Albedo Bio-geoengineering.* Current Biology 19: 146-150.

Rome, A. (2013) *The Genius of Earth Day: How a 1970 Teach-In Unexpectedly Made the First Green Revolution.* (New York: Hill & Wang) 346 pages.

Ruse, M. (2013) *The Gaia hypothesis: Science on a pagan planet.* (Chicago: University of Chicago Press) 251 pages.

Sage, R.F. and Coleman, J.R. (2001) *Effects of low atmospheric CO_2 on plants: more than a thing of the past.* TRENDS in Plant Science 6: 18-24.

Schuiling, R.D. (1998) *Geochemical engineering: taking stock.* Journal of Geochemical Exploration 62: 1-28.

Sherwood, B. (2011) *Inhabiting the Solar System.* Central European Journal of Engineering 1: 38-58.

Singarayer, J.S., Ridgwell, A. and Irvine, P. (2009) *Assessing the benefits of crop albedo bio-geoenineering.* Environmental Research Letters 4: 1-8.

Smil, V. (2008) *Energy in Nature and Society: General Energetics of Complex Systems.* (Cambridge: MIT Press) 480 pages.

Smil, V. (2010) *Prime Movers of Globalization: The History and Impact of Diesel Engines and Gas Turbines* (Cambridge: MIT Press) 261 pages.

Sorrells, G. (2002) *Seismic Precursors to Space Shuttle Shock Fronts.* Pure and Applied Physics 159: 1153-1181.

Turner, A.K. (2008) *The historical record as a basis for assessing interactions between geology and civil engineering.* Quarterly Journal of Engineering Geology and Hydrogeology 41: 143-164.

Westfahl, G. (2012) *The Spacesuit Film: A History, 1918-1969.* (North Carolina: McFarland) 371 pages.

Williams, D.M. (2011) *The Planet Paparazzi: Earth through the Lens of Interplanetary Spacecraft.* Astrobiology 11: 391-392.

Wuebbles, D.J. (2012) *Celebrating the "Blue Marble".* EOS: Transactions of the American Geophysical Union 93: 509-520.

CHAPTER 3

TELEUISED UNNATURAL WORLDS AND TERRA-CREATURES

In 1968, the American civil and military engineering philosopher Frank Paul Davidson began to popularize an at-least-four-year-old UK semantic addition, "Macro-engineering", because of civil and military engineering research that systematically focused on geographically large-scale private-sector and public-sector building projects (macroprojects or megaprojects). Davidson promotes Macro-engineering from his offices in Concord, Massachusetts (USA) and Neuilly sur Seine, France. [In about 1968, Bio-technology—or "Micro-engineering" if you like—became public controversy everywhere. By 2014, reliable gene circuits of mammals were constructed in the laboratory by modular "Micro-engineering" techniques (Guye et al., 2013.] Properly organized, Macro-engineering should prove to be the most immediately stimulating of new professions and should thereby greatly arouse all professional geographers worldwide. Recently, Geographos has attempted to piggyback on Davidson's late-20th-early-21st Century Macro-engineering foundational effort by emphasizing the fascinating and tremendously promising theoretical aspects of futuristic Macro-engineering, namely a leading effort named "Macro-Imagineering". Macroprojects and megaprojects—the concrete emplacement of mega-engineers and macro-engineers' plans—were first organized in the USA by a pioneer Californian, Warren A. Bechtel (1872-1933), according to *The Earth Changers* (1957) by N.C. Wilson and F.J. Taylor, when Hover Dam was built during the 1930s to block the Colorado River in the American Southwest.

Geographos strongly prefers, for public-relations purposes alone that really vibrant, independent centers for Macro-Imagineering's conception and Macro-Engineering's macroproject planning, and the R&D underpinning complicated planning operations therein, wisely ought to avoid all ecosystem-country capitols. The so-called anthropic "Global Warming Crisis" is a reprehensible example of national geoscience corrupted by centralized Federal Government funding and political agendas of both elected politicians and politicized bureaucracies. In the USA a single-agenda community of alarmist scientists has gathered to profit from their public-issue studies of anthropic "Global Warming". Many of these scientists have been extravagantly funded for many years by the Nation's governmental entities based in Washington, D.C. and, unfortunately, many of these scientists have invested years in careers that can only be prolonged by such dependable funding—in other words, solving the macro-problem of anthropic "Global Warming" as, for instance, via Macro-engineering, is without incentive! Yet, American taxpayers do wish to find proper, efficacious macroproject solutions to the perceived macro-problem, not some utterly ridiculous exhortation for a "War on [anthropogenic] Global Warming" advocated by just a few insistent misguided geographers (Sherman et all, 2010) who *desire* the blind obedience of humans oppressed by an ill-informed Geoscience Elite enraged by their worthless, endlessly tweaked, super-computer models!

Edward Norton Lorenz (1917-2008) initiated Chaos Theory during 1963. He was the first meteorologist to note the sensitive dependence on initial atmospheric conditions. Visually, his set of non-linear differential equations describing the flow of air in any planetary atmosphere feature a "strange attractor", which resembles the flapping of a butterfly's wings when animated by electronic computer. He characterized his 1963 mathematical insight, during his 29 December 1972 speech ("Predictability: Does the Flap of a Butterfly's Wings in Brazil set off a Tornado in Texas?") before the American Association for the Advancement of Science in Washington D.C, as the "Butterfly Effect". Lorenz proved that, long-term, Earth's weather regimes could not be predicted, unlike some linear Earth process-events characterized by "Domino Theory". Geographos is focused her on another kind of "Butterfly Effect": when a caterpillar changes into an active butterfly, the transformation occurs within a chrysalis. The innermost wrapping of a caterpillar's chrysalis is known as the "knub", and "knub" is the source of the English word "nub" (or core) of an argument. In the following chapters, the nub of the global debate on human civilization's means of coping with natural and anthropogenic global change—sometimes utilizing Caterpillar bulldozers and other sales brands of earth-movers—is forthrightly addressed, with new mitigation, adaptation and "cure" strategies presented from Macro-Imagineering's viewpoint. "Save Our Planet!", the early-21st Century's weird general fossilized subjunctive radical Green slogan ultimately will be forgotten, made irrelevant by dynamic event-process undertaken by macro-engineers working diligently on Earth's geo-engineering and Mars' terraformation.

By one method or another—subjectively and mathematically—Geographos will assess various geo-hazard risks to human health and life intrinsic to the macroprojects proposed by many Macro-imagineers. Humanity's potential accession of almost unlimited supplies of nuclear energy is offset by the possibility of making Earth uninhabitable, but his potential macro-problem is balanced, as humans continue to explore and exploit our Solar System's resources, by partly quitting the Earth-biosphere and settling in a sustainable way elsewhere. Converting empty extra-Earth landscapes into human-owned and operated developed real estate assets is a laudable, healthy species goal. Science and technology has made this goal nearly materialized because supportive infrastructure systems supplied by extremely reliable logistic systems fostered by innovative Macro-Imagineering/Macro-engineering also stimulates societal innovation and technical invention. That outcome, of course, is possible because humans are able to observe and measure with marvelous accuracy and timeliness. Dedicated macro-imagineers dream, and steel-nerved macro-engineer s realize, plans for the design of macroprojects (or, synonymously, megaprojects) that, in the broadest of outlines, represent, symbolize and cause a fundamental alteration in humanity's global outlook. It is the sociological impression on groups of like-minded people—someday, soon perhaps, of like-minded robots—that is fundamentally important to our globalized civilization expanding upwards into outer space and to all the things and energy far beyond our current reach, endeavoring to practice real-world management and construction skills everywhere in this colonizable Solar System.

Some critical macroprojects are beneficial or crucially essential and usually unmovable infrastructures, as profusely illustrated by Brain Hayes' *Infrastructure: A Field Guide to the Industrial Landscape* (2005) or Kelly Shannon and Marcel Smets' *The Landscape of Contemporary Infrastructure* (2010), both are popular-style descriptions and explanations carefully contrived to educate and intrigue laymen about modern civilization's infrastructural surroundings. Infrastructure is found in Anthromes and Astythromes, atop as well as implanted in the seabed, overhead in orbit and even connected to amazingly distant exiting machines at, or beyond, the boundary of the Sun's direct influence, the heliosphere! Even so, some leading academic authorities prefer to define this new profession-discipline in purely geographical and economic terminology: "Macro-engineering refers to the process marshaling money, materials, personnel,... ["technologies"], logistics and [informed public] opinion on a huge scale to carry out complex projects, often international (or "multi-national") in nature, that last over a long period of time (Chan et al, 2013). A macro-engineering project [macroproject or, alternatively, megaproject] requires massive funding, significant manpower, large-scale equipment and... [metric tonnes] of material. Macro-engineering projects extend the state-of-the-art of...[technologies], may take place in difficult and sometimes hostile environments, and require sophisticated project management techniques" (Davidson and Huot, 1989). In the case of human space travel "...spaceship management aimed narrowly at the biological survival of...[spationauts], an ethic which also came to dominate ecological design proposals on board Spaceship Earth" (Peder, 2005).

The earliest print media use of "Macro-engineering" evidently is by the 12 March 1964 issue of *New Scientist* (Volume 21, Issue No. 382, at page 685). "Macro-engineering" is a collective English-language word encompassing several main civilization planning viewpoints: (I) Earth-biosphere changes induced to promote the livability and sustainability of human civilization; (II) Geo-engineering which is basically centered on controlling Earth's atmosphere, according to the *Encyclopedia of Climate and Weather* (1996) edited by the late Stephen Schneider (1945-2010) and (III) Terraforming. "Terraforming", derived from Latin "terra", earth, plus the English-language verb "form" involves civil and military engineering dedicated to make a planet more like Earth than the planet was in its natural, unmolested state. Using only techniques known expertly now, Mars might be made Earth-like in several thousand years or less (Mole, 1995; Muscatello and Houts, 1996). Alfred Lotka (1922, pages 151-154) intuited that Nature's selection operates to preserve and enlarge the numbers of those species "...possessing superior energy-capturing and directing devices...". The Cold War-era national laboratories that spawned nuclear weapons, some with probable energy yields of 100 Megatonnes, nowadays house R&D programs dedicated to efficiently reducing the effects of natural and anthropogenic global and regional climate change (Edwards, 2012). Cosmic rays and space weather will, however, be major inhibitors working against the movement of human beings from the Earth-cradle into outer space, interplanetary space, and onto thin-atmosphere planet-places. Spationauts using properly radiation shielded housing and vehicles on Mars should be safe, according to the NASA. And, although global rocket launches deplete the Earth's Ozone Layer by about 0.03%, some Greens warn that if this exploration activity continues, especially if low-cost, numerous flight rate launch systems [SKYLON?] are used in future, then that industrial activity poses an existential threat to the planet's life (Ross et al., 2009); thus, Greens thrust a "damned if you do, damned if you don't" conundrum at the most pliable members of humanity.

Truly thoughtful, rigorously systematic communications media exposures of proposed mitigation and abatement Geo-engineering plans possibly affecting the Earth exclusively should be welcomed by the world-public, not shunned, and also the pertinent scientific and engineering professions since the nub of the argument about unwanted and unhelpful global Nature change is what, if anything, ought to be done about such change by those humans utilizing various powerful technologies. Oddly, John Maunder's *Dictionary of Global Climate Change* (1992) omitted "Geo-engineering". Anthropogenic "Global Warming" alarmists have, for several decades, campaigned against all suggested technical solutions to the macro-problem of climate regime change—except, of course, for their own sociological prescriptions that can be economically ruinous! During a single 20th Century human generation, people have visited (directly and tele-presently) all the planets know to the ancient Greeks (Carr, 2013). Macro-Imagineering's basis lies in human ingenuity, individuals and groups, accurate collections of geographical facts—the widest possible meaning of *geo*—as well as the efficient manufacture and manipulation of energy, planetary surface reshaping and widespread alteration of crust compositions, and the intentional movement of materials throughout our Solar System. Obviously, Macro-engineering's global planning most negative aspect is its implied control over almost everything and everyone residing within the Earth-biosphere; this is comparable to the fate of spationauts confined to the increasingly decrepit International Space Station and their personal, finely-fitted spacesuits.

The presumed antiquity of humanity's Earth-surface modifications lengthens as concerted investigations into recorded history and archaeology intensify. [So-called "de-extinction" Bio-technology's genetic-scale laboratory resurrection efforts on behalf of favored, yet disappeared plants and animals, is an oppositional human action! The **National Geographic Magazine**, in its April 2013 issue, touted both the concept and the term, making it a subject of popular conversations. A very new concept, "de-extinction" could introduce newly revived species to environments very different from their species' heyday (Sherkow and Greely, 2013). And, what are the ethical and legal aspects if long-absent ancestors of living humans were made present in our world? What might be some of the macro-problems arising if resurrected plants and animals were transferred to, or separately generated thereon, another place—Mars?!? Thomas Gold (1920-2004), one of America's most outstanding geochemists, mused with tongue-in-check wit, that biological contaminants introduced by Alien interstellar picnickers may be the cause of life in the Earth: "Thus space travelers may have visited the Earth billions of years ago and from their abandoned garbage forms of life have proliferated so that microbes will soon have another agent capable of spreading them further afield" (Gold, 1960). It is true that humans are becoming more and more proficiently capable of genetic manipulations that can greatly alter this planet.

Circa 1966, Alvin Martin Weinberg (1915-2006) coined the phrase "technological fixes" representing a thought-train since updated and used in a derogatory way, and among others of the same ilk, by Evgeny Morozov in *To Save Everything, Click Here* (2013) as "solutionism"— the mistaken belief that synergistic technologies can benignly and efficiently solve all civilization's macro-problems, producing "instantly" a world that is entirely trouble-free (Rosner, 2004). Two Science and Technology policy devisors have made a list of three reasonable rules regarding "Technological Fixes": (I) a fix must mostly embody the cause-effect relationship linking the macro-problem to its alleged solution; (II) the benefits of a strong fix effort must be obvious to all civilized persons involved with the causative macro-problem; (III) each fix R&D most probably will be decisively effective when the R&D program used improves on the soundness of Science's knowledge/data base (Sarewitz and Nelson, 2008). Before World War II, most macroprojects derived from the minds of a few oneiric persons; after World War II the majority of megaprojects seem to stem or emerge from the everyday work of like-minded groups effectively working within established or nascent governmental and corporate bureaucracies. Technology's continuing formation—especially molecular Nanotechnology—holds surprises for all humans. It is noteworthy that the most up-to-date 2014-issued print handbooks about big building projects often fail to honor molecular Nanotechnology's impending prospects: Bent Flyvbjerg (Ed.) *Megaproject Planning and Management: Essential Readings* [from 1951 to 2012] and Hugo Priemus and Bert van Wee's *International Handbook on Mega-Projects* are relevant examples.

Geographos notes—with enthusiastic and unguarded approval—the stated thinking of international architecture critic Reyner Banham (1922-1988) who calmly opined that imaginative Science is "...one of the great mind-stretchers, [profession] specialization-smashers of our day....It is part of the essential education of the imagination of every technologist" (Banham, 1958). Maddeningly, the accruing experience of Macro-engineering's honest professors specializing in urban planning, for example, is proof that funding for megaprojects is often obtained by clever and deliberate politicized deception of taxpayers! This is one of the many saddening documented conclusions of Bent Flyvbjerg and his associates (2003), along with the published results explicated by Altshuler and Luberoff (2003). Financing of all macroprojects must never be finessed by dishonest promoters lest extant and predicted life-threatening macro-problems be allowed to fester and even result in humanity's extinction. *Medicative Macro-Imagineering* attempts, however successfully or unsuccessfully in the beholder's view, to bring forthright appraisals of some possibly beneficial macroproject plans that are intended to improve Earth as, simultaneously, humans, cyborgs, androids, humanoids and Terra-creatures are rocketed from Earth to other places in our Solar System and its surrounding Universe—possibly the ultimate "Environment". Greens often proclaim themselves as "Environmentalists". Since "environs" means the planetary landscape and ocean surface surrounding a stationary geographical place or a vehicle of some sort, not a person, this misuse seems more than strange to Geographos!

When in the course of time words cease to be jargon (that is, words peculiar to a particular community), then such words can be said to be vernacular. "Macro-Imagineering", "Macro-engineering", "Geo-engineering" (Launder and Thompson, 2010) and "Terraforming" are still to be considered jargon, just like "Exobiology" (the scientific study of extra-Earth life, alive or dead) and "Cryptobiology" (the study of

living Earthly creatures whose existence has not yet been scientifically proved) (Wolfe, 2002; Eberhart, 2002; Loxton and Prothero, 2013).
Down-to-Earth print media mainstream USA publications, which are
tightly edited before the public is made aware of the contents, now seem
to recognize the exciting potential of Terraforming—"Terraforming" is
no longer a nonce word. The 2010 edition of the single-volume *Oxford
Dictionary of English* included—for the first time—"Geo-engineering"
(Nerlich and Jaspal, 2012). The ancient Greek inventor Archimedes (287-
212 BC), with a clever thought experiment on the action of levers ("Give
where I may stand and [with a lever] I can move the Earth"), expressed
in vivid wordage a valid consequence of his theory of static forces and
anticipated professional terraformers, Freeman Dyson Satellite Swarm
constructors and others by about 2,250 years!

Our Solar System's boundary relative to the Universe was first accurately mapped in 1984—it is the imaginary borderline based on the
orbital motions of non-star celestial bodies rather the location of the
Sun's heliopause or Oort's Cloud of rocky debris (Smoluchowski and
Torbett, 1984). By 25 August 2013, the Voyager 1 deep-space probe
passed that boundary and, indeed, had left our Solar System forever.
In other words, our expanding "Anthropo-cosmos" is constantly increasing in volume! To reiterate, humanity's outwardly moving trash
defines its realm of strongest influence! [Aliens might not be awed
by our weak, farther-distant "leakage radiation" [electromagnetic
emanations] (Haqq-Misra et al., 2013), yet astronomers here anticipated the direct characterization of exoplanets in other star systems
(Schneider et al., 2010)! Few geographical textbooks yet contain extended basic discussions of non-Earth surfaces, but at least some geographers and macro-imagineers do think globally, using data gathered
by remote sensing technologies (Pike, 1987). Since every NASA or
other active major national space agency map presages some kind
of future exploitation, it can safely be forecast that geographers will

help macro-imagineers/macro-engineers to transform our Earth's air, ocean, crust as well as the uppermost strata of Mars. Terraforming and Macro-engineering will provide hard-won expertise necessary for a social macroproject to control global Nature on more than just a single Solar System planet. Early-21st Century peoples, especially in China and India (Brook, 2013), experienced a burst of public-works spending that historians may one day compare with the initial construction of our world's railroads (Wolmar, 2010), the electrification of Europe and North America, and the spread of public waterworks (Salzman, 2012). Infrastructure modernization is seen as a means to speed economic national unity, while some others are promoted as a toxic for job-creation and to spur foreign financial investment. Country-ecosystems (as known as Noosystems) with the most advanced transportation systems (harbors, airports, for example) stand to gain the most from the increase of international trade.

Without much question, the Earth-biosphere is becoming a human artifact. Even so-called wild-lands are really "wilderness-quality" Anthromes! People of 2014 live in a time when even horticulturalists and animatronics experts boldly predict that future gardens, diametrically opposed to that cared for and adored by *The Little Prince*, will be pullulating electronic machines (*faux* plants and cultivars with entirely different origins). Greens might publicly approve of such an event-process since they could still advertise their (sometimes politically radical) organizations as advocating *Homo sapiens* as a low-impact cultivator species. Even prior to Thomas Gold's picnickers, some exobiologists claim that it is possible Earth has been visited by intelligent creatures and/or machines, Aliens from another part of the Universe (Crick and Orgel, 1973). What if some non-Terran species got to Earth even before pre-cognitive "man-creature" became *Homo sapiens*? If such Aliens did, the "our" Earth could never be a human artifact! When the USA's space probe Galileo approached

Earth on 8 December 1990, its instruments barely detected a USA-operated scientific research station located at the South Pole—its images revealed no other man-made features on our planet's surface (Sagan, 1993). If the NASA's image interpreters had not *known* of the polar camp's existence, then that single poor-quality image might have easily been dismissed as a mere television vision, an error of distorted electronic transmission. One popular pseudonymic author, George Leonard, in 1976 posited the insupposable thesis that **Somebody Else is on the Moon** now! A 1987 book, **The Monuments of Mars: A City on the Edge of Forever**, by Richard C. Hoagland, purveys in that and many essentially unrevised editions since the notion of long-abandoned, mountainous Alien macroprojects existing as abandoned ruins on the planet's surface while David H. Childress's **Extraterrestrial Archaeology** (1994) alleges eager-beaver Aliens constructing New York City-style "Manhattans" on almost every known planet in what Geographos had assumed and hoped was *our* Solar System! The early-21st Century's zeitgeist (Boring, 1955) still permits the commercial merchandizing of such nonsensical tale-weaving. Science and technology, as taught on American television (Lafollete, 2012) has not overcome the inertia of public ignorance although imaginative efforts by that commercial media have been attempted.

During the 1830s, some astronomers suspected, or expected, Mars was inhabited by extraterrestrials, Aliens. During the early-19th Century, J.J. von Littrow (1781-1840) had urged that 32 kilometer-long trenches be dug in the Sahara, filled with flammable liquid floating on water, to be ignited at night as a signal to the supposed Martians of our presence in the Earth. His trenches were to form geometric patterns in the dark nighttime desert background—it is still so in our time, as L.D. Keeney's **Lights of Mankind: The Earth at Night as Seen from Space** (2011) illustrated—recognizable as artificial by Martian observers. [There is irony in that photographers, riding in spacecraft or the International Space Station, snapped pictures of Astythromes at night for Keeney's inspirational coffee-table book—the Astythromes pictured generate light pollution that prevents people below from seeing space sprinkled with stars, perhaps blocking clear views of passing orbital vehicles!] The American novelist Ken Kalfus adopted J.J. von Littrow's desert macroproject idea as the hook in **Equilateral: A Novel** (2013). Kalfus main fictive protagonist, "Sanford Thayer", undertakes a huge Victorian Era engineering feat to make contact with Martians on 17 June 1894, when Mars comes into its closest alignment with the Earth. Using a workforce of 300,000 fellahin, an equilateral triangle is excavated in Egypt's Western Desert with trenches that are hundreds of kilometers long! The trenches are to be filled with burnable pitch, set alit that special night to signal the intelligent Martians with a burning signpost designed and dug by humans, modeled on Moses' Exodus-trek "Burning Bush" in **The Holy Bible**. During 1947 the artist Isamu Noguchi (1904-1989) modeled a "Sculpture to Be Seen from Mars", which looked like an Abstractionists vision of a human face in topographic relief. Its nose alone was planned to extend a length 1.6 kilometers! Photographs of his proposed real-world artwork maquette, an example of dreamt Land Art (Kaiser and Kwon, 2012, page 222), are eerily like the disproved "Face on Mars" first spotted at Cydonia during 1976 by the NASA's Viking-1 probe to Mars. The "Face on Mars" was 'defaced'—demonstrated by orbiting satellite imagery to be not there—in 1998 by subsequent, better quality NASA images. Like the "Man in the Moon", the "Face on Mars" was a pareidolia, a fanciful perception of pattern in something that is actually ambiguous or random. Since the 1960s, humans have had certain data that Martian canals are not visible on Mars's ocean-less crust. Today's Mars atmosphere is so tenuous no un-encapsulated human there survive. [Elon Musk's 2013-proposed State of California, USA, "Hyperloop" transportation

Macro-Imagineering project, meant to connect Los Angeles and San Francisco by tube railway in 35 minutes of constrained flight, utilizes individual capsule-cars, each holding one passenger, moving in a nearly evacuated pipeline. Each capsule must be manufactured to the same exacting high-workmanship standards as any spationaut-carrying spacecraft because the air pressure inside the tube will be roughly 100 Pascals, or approximately one-sixth the natural atmospheric pressure at Mars's imaginary geodetic surface (Anon., 2013)!] Despite some oddballs and pseudo-scientist's very public assertions, Geographos is satisfied to think of *Homo sapiens* as a cosmopolitan of one global Nature—Earth's!

Post-20th Century persons plan achievements that will far surpass our Earthly global Nature's results owing to the planet's hydrologic cycle and rock cycle: aerospace macro-imagineers intent to mine the Earth's atmosphere via air-scooping propulsions systems such as scramjets for aircraft and spacecraft, or combinations thereof (as, for instance, the UK-designed "SKYLON"-type aerospace vehicle). Meanwhile, miners have already completed a large-scale Earth-crust controlled lowering beneath a major urban international river in Germany (Leggett, 1972). One outstandingly imaginative French architect proposed an urban utopia for future Parisians. For 21st Century Paris, Paul Maymont (1926-2007) sketched his "City Under the Seine Project" (1962) design which illustrates fifteen or sixteen separate rectangular tube-like levels of human use underlain by a concentrated pipe network of services (Shea, 2011). Riverbeds as conduits for Astythromes—a truly inspirational new use of now ignored elements of the landscape. Humanity's ever-spreading Earth-biosphere impacts—that is, outward from the Astythromes to the ever-more altered Anthromes—are now so pervasive and penetrating that living species can no longer be considered the only units of Earthly global Nature that are properly to be considered as evolving. Indeed, in 1969 a daring American geoscientist,

Harry Donald Goode (1912-2000), announced his new term, a needed term clearly suggesting multiple Earthly Geologic Time phase transitions, to indicate evolving pre-Anthropocene as well as Anthropocene geologic event-processes: "A look at a few events suggests that some changes in environment wrought by geologic evolution in turn have brought about new geologic processes that have made new changes in Earth. Brief examples: Earth, if it was at one time too hot to have liquid water, at that time had no erosion by running water or waves; until temperatures dropped below freezing there was no frost action; in time when there were no plants, weathering of rocks was probably accomplished principally by water, but later, weathering of some rock was greatly accelerated by the arrival of land plants and their resultant humic acid. This idea of evolution of geologic processes—that some geologic changes foster new geologic processes that in turn bring about new change...is worthy of a new term, Geoevolutionism" (Goode, 1969). Robert M. Hazen's *The Story of Earth: The First 4.5 Billion Years, from Stardust to Living Planet* (2012) carries Goode's "Geoevolutionism" concept forward, especially with Hazen's thoroughly outlined theory of "Co-evolution" which suggests that Earth's first organisms were probably generated by chemical reactions of rock crystals with fleetingly present organic molecules; he alleges that two-thirds of our planet's known mineral varieties could never exist if our world had remained lifeless.

Geographos wonders: what could happen whenever microscopic-size modular robots—sand-grain size "Smart Pebbles" (Gilpin et al, 2010) and almost invisible aerially floating droplets of "Utility Fog" (Hall, 1994)—force changes in tiny bits of all accessible Earth-crustal materials and those of the other celestial bodies? A heap of "Smart Pebbles" would be analogous to the rough block of selected stone that a sculptor shapes by hand. During 2010, on a still smaller geographical scale—nano-scale, in fact—IBM scientists carved a complete three-dimensional map of our world's surface measuring twenty-two by eleven micrometers. The relief map is composed of approximately 500,000 pixels, each twenty nanometers square, milled from a proprietary polymer, polyphthal-aldehyde. In relief, one thousand meters of elevation corresponded to about eight nanometers. Molecular Nanotechnology, after it leaves the R&D phase, will be an immensely powerful tool that will undoubtedly radically alter humanity's conception of the material world. Michael Crichton (1942-2008), the American novelist, penned and published *Prey* (2002), a techno-thriller featuring molecule-size machines that menace microscopic mayhem for our threatened world, a "grey goo" re-pulsive final outcome. Molecular Nanotechnology as mass-destruction, a "Doomsday Scenario", might be considered at least the equivalent of a runaway nuclear chain-reaction—in other words a true weapon of global Nature destruction (Howard, 2002). If it is possible! Further research by molecular Nanotechnology developers revealed that "grey goo" is nearly impossible to happen accidently and even quite difficult to make occur by intention. Still, since geoscientists know that photosynthesis is not exactly "balanced" by animal respiration, just "grey goo" extermination of all life forms from the Earth-biosphere would result in the disappear-ance of atmospheric oxygen in less than 150 years; compare that fright-ening potential "grey goo" effect with that of actualized, long-existing humans: only about 0.3% of the air in our world has been breathed even once by the billions of humans that are alive, and have lived, so far!

By 1978, a few Space Age non-conformist, nay adventurous, ge-ographers had commenced postulating vividly some hypotheti-cal Earth-worlds (Barton, 1978).What about hypothetical worlds? There are several means by which Earth's Geologic Time could be suddenly terminated and, if all the planets of our Solar System were dealt with similarly, then Astronomic Time for our enlarg-ing Milky Way Galaxy Anthropo-cosmos could be halted artifi-cially (Cathcart, 1983). Planetary demolitions would be the final

Anthropo-cosmic epoch of *Homo sapiens'* "Age of [Intelligent] Life in [Outer] Space", that is, Hubert Strughold's "Dosmozoicum". Though Greens will abhor its truth as well as its finality, it is quite possible that Earth's biosphere, future robospheres and planet shells dominated by Terra-creatures will be dispersed, recycled industrially, and formed into a Freeman Dyson Satellite Swarm. A postulated human-made or mind-made Freeman Dyson Satellite Swarm structure here could be characterized as a manifestation of Harry Donald Goode's "Geo-evolutionism"! All too commonly, geoscience's operational weltanschauung is often that set of systematically self-constrained statements—supposedly of experimentally and observationally confirmed knowledge/data—considered, for the moment, much too expensive to change; technology is the performance of a scientific competence. At the present-time, Macro-Imagineering/Macro-engineering starts with human desire, which is subsequently reduced to a solvable macroproject/megaproject concept. Current infrastructures are testaments to humanity's still emerging delight with mastery of structure at all scales—with molecular Nanotechnology the lowest end of that spatial scale. Some Green advocates have announced their group-belief—it cannot be a reasoned thought—that to deform God's planets with virtually any kind of technology is to present Him to humanity as the ineffectual governor of something not really His own; anthropic fragmentation of His planets does not have such result when something is built afterwards by God-inspired humans. The Solar System, shattered by intention, could be set free from its bondage of certain future decay owing to the Sun's aging. A 19th Century UK economist, William Stanley Jevons (1835-1882) propounded a theory of the national trade cycle based on the Sun's spot fluctuations. It was not convincing then, and has not yet been proven. But, would a Freeman Dyson Satellite Swarm be influenced by cyclical star-generated radiation?

Too, some obtusely petulant Greens breezily deride humans as "Stone Age minds in Stone Age Bodies"—that is, Homo sapiens is genetically unprepared for impending Earth-saving or life-preservation tasks; yet, many of these same persons desire to disallow any public consideration of human genetic improvement, even after the Human Genome Project's completion by 2001! Since people are thinkers—honest persons who confess a lack of complete mental adjustment to our collective species enclosure, the not always delightful Earth-biosphere—we admit the irrefutably direct precursor to all global contamination and pollution, solid litter from stone-flaking industry, is an indicator/symptom of *Homo sapiens'* Paleolithic Period technology (Cathcart, 2011). Macro-imagineers must never suppress the human imagination! Picture a real solution to Earth's insupposable anthropic global warming "crisis": if macro-engineers converted all of the Earth-atmosphere's carbon dioxide gas into an Anthropic Rock via molecular Nanotechnology, it should form a global a friable global stratum about 2.5 millimeters thick; by contrast the infamous natural K-T Boundary—that is, the global clay layer between the Tertiary and the Cretaceous—is about 12 millimeters in thickness. Extraction from our encompassing air of all carbon dioxide gas would, inevitably, kill all green plants; a very odd final result for the medium "polluted". Hence, the macroproject does not have Geographos' endorsement—at least when it is carried to such an extreme outcome! Extraction from the air of some carbon dioxide gas, and its solidification as a sequestered Anthropic Rock stratum, is a wise climate control possibility (Tavoni and Socolow, 2013). But, we should not cover, enshroud global Nature—that act, in itself, is despicable and unnecessary. The total absence of global Nature simultaneous with the existence of people would, obviously, start a situation of "environmental generational amnesia" (Kahn et al., 2009)!

For centuries, humans have pieced together many large edifices and transportation systems; more such macro-engineered things are going to be constructed as additions to humanity's extant civilization infrastructures, despite the harsh negative criticisms voiced by antiindustrial reactionaries such as radical Greens. *Homo sapiens'* biggest existing machine is global telecommunications. When any of us uses a telephone (land-line or satellite models), we are programming the world's telephone network. And, since the 1990s, the telecommunications system makes it possible to access computers, both public and private, from any place in the Earth-biosphere, and even from outer space. All actively-reporting Solar System space probes should also be included as part of our interplanetary and interstellar telecommunications web. Because of real-time reporting satellites that are accessible at the Google Earth website (Yu and Gong, 2012), humanity truly lives inside the World Wide Web that is more pervasive (Srivvastava et al., 2012) and potentially at least as suppressive than the "Pantopicon" imagined by Jeremy Bentham (1748-1832)! During 2007, Google Earth was accessed by 22.7 million persons. However, "It is interesting to reflect that the key to success of early *Homo sapiens* was the ability to communicate with language....Perhaps today's fast-growing communications network will serve a similar purpose, and may help us restrain the largest geophysical force we can control, namely ourselves" (Williams and Crutzen, 2013). "Neogeography", a term coined by DiAnn Eisnor before 2006, has spun-off from non-academic persons, was catalyzed by digital mapping technologies such as Google Earth (Warf and Sui, 2010); an excellent example of electronics-based "neogeography" is the discovery, facilitated by Google Earth that six times more fish than what was officially reported to the United Nations Organization were caught yearly in the Persian Gulf (Al-Abulrazzak and Pauly, 2013).

Because of the super-computer, visualization has emerged as a distinctive new specialty discipline; someday computer visualizations will bevomr beyond expert human understanding! Georges Seurat

(1859-1891), using a unique painting technique, painted on canvas a pure Nature, free of the "noise" instigated by the human presence and he depicted illusion too (Jinks, 1953). For good or evil, visualization experts able to place pixels willfully could become geopolitical image-makers, contriving digital electronic doppelgangers (maps, images)! [The reputable print publication *National Geographic Magazine* shifts Egypt's pyramids when its editors are of a mind to do so for purely printed page space-saving and trivial aesthetic reasons. That offending cover was omitted from reprinting in the issue of October 2013 featuring 125 years of *National Geographic Magazine* cover-photographs in the form of an inserted poster.] Because any television monitor that refreshes the visible display at a rate faster than one-fifteenth of a second gives flicker-free and stable images to all sighted persons and also, in part, because real-time super-computer operates in very tiny units of time—each nanosecond equals a billionth of a second. Falsifiers of "ground truth" could rather deftly become docile governmental and corporate employees. Imagine if you will, elitist television programmers, who are stupidly optimistic about the decision-making ability of specially trained elites, designing public presentations to appeal to a targeted large audience (Adams, 1992). Electronic experts, lacking optimism about the average person's ability to make decisions about his/her life's course, could play with remote sensor readouts to deceive the sober as well as the half-pixilated person. Television cartography led to the topographic and geologic mapping of Mars' 145,000,000 square kilometer crust with data from Viking sensors alone—a first in Cartography's illustrious history (Impey and Henry, 2013)!

During the mid-1960s, few geographers anywhere could share the vision of a futuristic space-based electronic observation and control system proposed by the Russian geographer Boris Borisovich Rodoman, born in 1931 (**Figure 1**). About 1952, he became zealously interested in the logical and cartographical aspects of surveying; from about 1960,

Rodoman developed a wider knowledge of theoretical Geography and Ecology. By 1974 he had published the basis for his "polarized-landscape" theory of ecological networking (Rodoman, 1974).

Figure 1. B.B. Rodoman (born 1920), a profound Russian geographer circa 2011. (Image: BBR.)

Cathcart (1986) noted, from an almost real "antipodal" geographical distance, the potential of Rodoman's call for the development of a dedicated constellation of Earth-orbiting satellite constellation for the near-total automation of biosphere rationalization of various economic-ecosystem working combinations. [It is worthwhile to note that, since April 2013, a San Francisco-based California company, Planet Labs, is forming a constellation of small low-cost orbiting imagers that ought to serve Rodoman's purpose!] Rodoman's initially postulated system of data fusion, with the ultimate aim of proceeding from mere "news-casting" to "now-casting", even in 2014 could do little more than produce harmony of integrated yardmasters that take apart or assemble railway carriages and to maintain railroad classification years in real-time. But the times and the available technologies have changed! [Rodoman specifically mentioned railway yards because it was the nation-ecosystem railroad system which first coordinated space, time, matter and persons in the way humans now consider normal and economically vital; a timetable for every event-process, a name and/or number bar code for every transported and stored physical object, a common computer clock for all.] There is also the potential for aerospace-traffic control, natural resources management, gathering news and globalized "electronic brainstorming".

Unimaginably huge amounts of digital data stream in from a growing number of satellites unceasingly monitoring the Earth-biosphere. Solving the technical macro-problems involved in getting disparate data to blend may turn out to be easier than getting international researchers to cooperate. However, when fully developed by Rodoman's most ardent followers, all of Earth's many real environmental macro-problems could be addressed by geographers, geoscientists, Macro-imagineers and macro-engineers who possessed real-time information; consumers worldwide would have rapid access to Earth-biosphere data in digital format for their immediate use with the existence of passive and active devices stowed aboard Rodoman-inspired orbiting satellites. Through a chain of microwave relay and deciphering telecommunications networks, that would have the capacity to convey satellite-derived tell-tale data with a minute of first being acquired to any authorized agency, corporation or person. Though designed and assembled for a different purpose, the USA's Missile Defense Agency, established by the National Missile Defense of 1999 (Public Law 106-38), has nearly perfected some of the equipment a Rodoman facility might utilize to carry out its imagined mission. Other ecosystem-states such as India, China and other organizations such the European Union may find this shareable technology to be very timely during the early-21st Century.

Rodoman foresaw a kind of "simultaneous" modeling (or Big Data fusion effort) of Earth's atmosphere, ocean and landscape cast on huge video displays—the "Big Picture" viewing place—in Biosphere Management Centers. This implies a lessening of centralized ecosystem-country relations and an emergence of an outer space-based system of non-territorial central guidance and communication establishment. At the International Exposition at Japan's Tsukuba Science City during 1985, the Sony Corporation erected its "Jumbotron", an outdoor color television monitor with a screen that measured 25 meters high by 40 meters wide, consisting of 150,000 luminescent pixel elements. During 2012, in the State of New York (USA), The Reality Deck installed at Stony Brook University's Center of Excellence in Wireless and Information Technology, commenced operation. It is a 416 screen super-high resolution (1.5 billion pixels) virtual reality four-walled surround-view theater, the largest resolution immersive display ever constructed driven by a graphic computer. Old-fashion Jumbotron screens are still found in sports stadiums and other forums everywhere today. Updated technology for Reality Deck and Jumbotron-like indoor monitors present in Biosphere Management Centers would surely suffice unless other technologies are deemed better suited to the task. It is also possible to replace television images sent from space with the result of exo-scale supercomputer calculations and manipulations—in other words, the end-product of Virtual Reality research (Brovelli and Zamboni, 2012). "Virtual Reality", coined by Myron Krueger in 1983, is an oxymoron. Probably it was Krueger's intention to indicate the completeness of his electronically-created and manipulated "reality". ("Artificial Reality" is a synonym.) In effect, Virtual (or Artificial) Realities and cyber-space given humans more "space" in which to conduct economic and social affairs, "space" unlimited by the physicality of our Earth's glob-al Nature (Earth-biosphere) but limited by the characteristics of im-parted by prevalent technologies. Geographers and others were tardy in first recognizing that television might be a place. Additionally, they were tardy in realizing that one of the biggest shocks Virtual Reality gives to human viewers can be its sudden absence—that is, a surpris-ing return to physical reality, which is sometimes unpleasant.

From Biosphere Management Centers established along the lines first suggested by B. B. Rodoman, organized humanity could con-trol through complex programming all of *Homo sapiens* significance that transpires at human behest within the largest possible Earth re-gion—to wit, the planetary biosphere (global Nature), which is becom-ing at Earth-robosphere! Data bytes are actually a recordation of our presence in that shell; remotely sensed data are malleable in computers, producing static and dynamic images of our global homeland. More persons see maps on satellite broadcast and fiber-optic cable televi-sion than any other type of cartographical display, which means that today's world populace is, in a sense, "pre-adapted" to Rodomanian modes of visual presentation and interpretation.

William Morris Davis (1850-1934) examined some aspects of com-parative Planetography. In his ***Elementary Meteorology*** (1894, page 120), Davis voiced his thought that our Solar System's fluid giants be studied to determine the influence of axial tilt upon planetary wind patterns: "We shall not fully appreciate the special features of the winds of our planet until the peculiarities by which they are distin-guished from the Jovian and Uranian winds are clearly perceived". [From 16 until 22 July 1994, more than 21 large fragments of peri-odic comet Shoemaker-Levy 9 struck Jupiter and visibly altered the pre-existing Jovian wind patterns. Americans watching late-night television news broadcasts were treated to videos of the action in near real-time!] Weather and climate regime forecasters have long sought an accurate and useful computer model of Earth's atmosphere. During 1922, Lewis Fry Richardson (1881-1953) proposed to predict Earth's atmospheric changes via parallel computation employing 64,000 human mathematicians and to plot their results on a forecaster-enveloping Rodoman-like world map that can only be called a regionalized electronic georama (Hayes, 2001). Or, perhaps, it is a "womb with a view"—especially considering how vast the Solar System and known and unknown Universe really is?

At the opposite end of the objective dimensional scale J. Andrew Ross (**Figure 2**) proposed a 64 centimeter-diameter schoolroom-home "Globall Hyperatlas", which he offered in 1991 (Ross, 1991; Cathcart, 1997). Although Ross's "Globall Hyperatlas" is not extant, R&D efforts are gradually developing and marketing technologies that come close, or even surpass, Ross's imagined device. "Globall Hyperatlas" R&D will surely reach fruition, and attain widespread success, long before molecular Nanotechnology becomes a dominant Earth-biosphere factor. During 1965, about the time Boris B. Rodoman was musing on all the management issues related to our Earth-biosphere, Ivan Sutherland defined "The Ultimate Display", in *Proceedings of the International Federation of Information Processing Congress*, 1965, pages 506-508) as a "...room within which the computer can control the existence of matter"! Only a perfected and trusted molecular Nanotechnology could accomplish what Ivan Sutherland wished to happen instantaneously.

Andy Ross
December 2012

Figure 2. J. Andrew Ross (born 1949): "Global Hyperatlas" conceptualizer and philosopher. (Image: JAR's website).

Foreseeably, humanity's Earthly infrastructures will be operated by automatons, providing material means to ameliorate social macroproblems and even perfecting robotic and human civilizations. The single most important advantage that automatons will have over coexistent humans is that their movements will be spatially limitless—super-mantle Earth conditions and conditions in what humans still provincially term "outer space" just won't matter very much. Barriers like the tropopause, which humans first passed through before World War II via ballooning, would be insignificant to robots. Robots could form a network of thinkers and doers that would make virtually all macroprojects and terraformation possible. Robots could cause a moral dilemma, a dilemma as profound as human slavery did during the 19th Century, and, as well, become *Homo sapiens'* rival for control of Earth! Science converts speculation into technologies. Extending Sutherland's desire to the world outside of a restricted room, molecular Nanotechnology could make "planned obsoleteness" of consumer products impossible and all future manufacturing and construction could be, effectively, a matter of software engineering—computers are already capable of generating new software!

Arthur Charles Clarke (1917-2008) defined "planetary engineering" in *Exploration of Space* (1951) much in the same fashion as Geographos defines the physical work of Macro-Imagineering—that is, "Macro-engineering"—as an umbrella term for Geo-engineering and Terraforming; Clarke viewed "planetary engineering" as "...the reshaping of other worlds to suit human need". Theoretically and spatially, Macro-engineering encompasses global Nature. The risk of failure is part of the human condition. In arguably our world's first global environmental impact statement, *The Earth as Modified by Human Action* (1874), George Perkins Marsh wondered if the heights of Europe's Alps might one day be reduced, diminished to gentle slopes for use by farmers and shepherds by a technically proficient European noosystem. Even a casual glance at the Expressionist plans and artworks of Bruno Taut (1880-1938), especially his daring 1919 urban engineering book *Alpine Architecture*, gives crystal-clear intimation of parenting Macro-Imagineering, at least contemporaneously with the organization of Macro-economics. Not intimidated by the **RMS Titanic** sinking, post-April 1912 Expressionist architects thrived in Germany, Switzerland and The Netherlands. Structurally, beyond the scale of mere buildings, unroofed structures and citified aggregations of assorted roofed structures, some Expressionists considered Earth's crust was to be an *object* of their contemplation and vigorous sculptural effort. This view was, later, picked up by Soviet technology enthusiasts: "There will come a time when powerful subterranean cruisers will set out on long voyages of exploration. They will reach the warm zones of the Earth's interior, discover rich new deposits of various minerals, tap the subterranean rivers and seas, and bring them up to the surface to irrigate the arid deserts and prairies" (Adabashev, 1966, page 35). In their overall view, our Earth's crust was fated to become a single industrial region: "Engineering is to fertilize, architecture is to glorify the Earth. The program is continental, even planetary..." (Scheffauer, 1924, page 126). The Netherlands, which expressed the national will via its Zuider Zee Works Act on 14 June 1916, undertook the largest stand-alone reclamation scheme yet attempted in Europe. It started during 1923 and largely completed by 1932 and it features a plethora of examples of Expressionistic exercise. The macroproject plan was initiated by Cornelis Lely (1854-1929) to prevent further destruction to The Netherlands from Nature's intrusion circa 1200 by sudden creation of the Zuider Zee. And, The Netherland's Delta Project completed in 1997 is designed to weather the global sea-level rise—the world's first macroproject planned with that predicted macro-problem in mind, according to *Designed for Dry Feet: Flood Protection and Land Reclamation in the Netherlands* (2006) by Robert Hoeksema, since 27% of that nation-ecosystem already lies below mean sea level!

During 1966 the little-known Russian historian of Macro-engineering, Igor Ivanovich Adabashev, recounted more recent personalities and other macroprojects in *Global Engineering*. [Originally published in English, *Global Engineering* was reprinted during 2005 by University Press of the Pacific as a 256 page softcover book.] Apparently, Adabashev was excited by the great many USSR plans devised during the 1940s for the massive, revolutionary transformation of Nature macroprojects endorsed by Joseph Stalin (1879-1953). Adabashev was also stimulated by the concept of "Constructive Geography", the geographical writings of Innokentiy Petrovich Gerasimov (1905-1985) and his like-minded colleagues. Once a leading geographer in the USSR, Gerasimov (**Figure 3**) in 1966 coined the phrase

Figure 3. I.P. Gerasimov, a praise-worthy Russian geographer. (Image: Great Soviet Encyclopedia, online.)

"Constructive Geography" as a description of Geography and, thereby, indicated that it should be a professionalized design art. He insisted that all geographers assist in forming a science-art of the planned transformation and complete management of global Nature for the sake of mankind's future. Constructive Geography, as elucidated by Gerasimov, resembles what applied Macro-Imagineering verges on becoming—Space Age Macro-engineering. To its detriment, so far Macro-engineering has no snappy bumper-sticker slogan. More significantly, it lacks a magisterial English-language statement on our profession's maturing philosophy. Most Macro-engineering books and journal reports recite the obvious and use neatly honed catch-phrases in stylistically unremarkable packaged prognostications. Regarding the art half of Gerasimov's Constructive Geography: Aristotle philosophized in **Poetics** that no sane man would call an object one thousand [say, "kilometers"] long "beautiful". But, Anthropocene Techno-Artists endowed with various technologies, with the means to create in unparalleled geographical scale, and with the vision needed to comprehend it, seem quite ready to judge what is beautiful; there is simply no predicting what kind of physical scale [Geographic or Spatiographic] that will be open to tomorrow's artists. Space Elevators (Smitherman, 2000), extended 36,000 to 38,000 kilometers to geosynchronous Earth orbit, would be constructed so as not to be eyesores! [So unlike **Charlie and the Great Glass Elevator** (1972) by Roald Dahl (1916-1990), thankfully a real-world Space Elevator cannot also take people to a lookalike Hell!)

The word "Geography" is commonly used with two different special scales of resolution. In the broadest sense, it is synonymous with all "Earth Science"—the study of all matter and energy as it occurs within our present-day "household", planet Earth. In its narrowest sense of meaningful use, Geography is the study of accessible materials (organic, inorganic, and transuranic) of Earth's crust, ocean and atmosphere (Earthly global Nature, strict sense). Sir John Murray (1841-1914), the oceanographer, used its narrow definition (along with "biosphere") in 1899 to denote Geography's profession scope. The lexicographer Sir James A.H. Murray (1837-1915) stated in **A New English Dictionary on Historical Principles** (1888) that "atmosphere" was first used in its near-modern sense during 1696, and that "hydrosphere" was first utilized in 1887. That same awesome dictionary's 1933 **Supplement and Bibliography** gave unambiguous examples from 1887 for "lithosphere". These four husk-like volumes of materials and energy are interconnected systems that influence each other along their interfaces of energy and material exchanges. A useful definition of "biosphere" might be "All parts of the Earth, above and below ground, and in its ocean and atmosphere that can support living organisms". "Biosphere" is the most useful of these terms for current and future theoretical and applied geographical transformations. (Of course, the just given definition of Biosphere will change as the concept of what is "living" changes in lockstep with Science and Technology's progress.) As of 2014, Macro-Imagineering remains a barely applied for of Geography in the Earth-biosphere (planned humanization—the process-event of altering our planet's biosphere and adapting it to the uses of its resident humans), while Terraforming is a theoretical form of Geography (normative goal-oriented planning for Solar System planet futures (Mars especially).

Unfortunately, few American geographers are today directly linked with any building projects of great magnitude; big rearrangements of Earth's landscape and the ocean are not really popular with the 2014 public. Rodoman Biosphere Management Centers in the USA and elsewhere could efficiently and economically direct interactions in real-time—that is, "live"—in the form of omnipresence. Humans, field robots could form a, more or less, Pax Consortia condition of life, fostered by rapid advance in military technology as, for example, anti-ballistic missile defenses. Every noosystem should be protected from attack by aerospace warplanes, aerodynamic missiles and long-range ballistic missiles. When two ecosystem-nations have a joint interest in avoiding mutually undesired events (such as general nuclear war and third party terrorists armed with weapons of mass destruction) that would or may result from their separate unilateral actions, or when joint action offers benefits more delightful that sub-optimal outcomes attainable from independent unilateral actions, it is in the two country-ecosystems self-interest to agree upon a control regime to avoid such costly dilemmas.

Although video-conferencing has obviated any necessity for great numbers of people to travel long distances by jet aircraft, macro-engineers are nevertheless harnessing old, and designing new, technologies to build safe supersonic transports for the long-distance and international commerce carriers. The UK's SKYLON aerospace vehicle seems especially noteworthy in this regard. SKYLON may be the flight technology that finally leads to the full development of the Aerotropolis concept of combining city/airport/spaceport in one conglomeration (Kasarda and Lindsay, 2011). The Aerotropolis or airport metropolis is the prototypical model, a new variant of the well-known Astythrome, the future of urban existence! Geographos suspects that Aerotropolis is a response the democratization of air travel, the waning of air travel's escapist overtones, causing airport/cities to become tourist destinations! Airports and developing Aerotropolis' synchronize with each other and the re-make landscapes massively. One developing Aerotropolis, the below-sea-level Schiphol serving Amsterdam, The Netherlands, began its existence in 1916 and remains today almost in a constant state of expanding and intensifying developmental flux. The most direct artistic connection of SKYLON, Aerotropolis and macro-imagineered Land Art is unmistakable—Robert Smithson (1967), who perished in an accidental private airplane crash at a dry-land site being surveyed for a new outdoor artwork!

Soon to be a direct result of the introduction and regular flights of Skylon and other speedy aircraft is a planet-size operational headache—the global aerospace traffic management macro-problem. There is an obvious need for an internationally owned and operated Aerospace Controlled Traffic System (ACTS) superstructure capable of defending every ecosystem-nation in the Earth-biosphere. Perhaps the superstructure could be named "Rodoman-ALPS" with ALPS signifying "accidental launch protection system"? Rodoman-ALPS would be a globalized rented facility, paid for by low-cost national subscriptions. With the emplacement of Rodoman-ALPS, no country, multi-national corporation or macroproject could ever be blackmailed. Providing the aggrieved with a truthful computer-radar track file's indisputable proof of culpability should an attack or accidental launch ever take place would be a main task of Rodoman-ALPS. Someday, Rodoman-ALPS might be configured to provide useful warnings of the approach of dangerous space debris (potential superbolides) and errant junk. Like "integrated battlefield", "sphere of influence" is one of those infrequent linguistic acknowledgements of the tri-dimensionality of global Nature. Humanity is bereft of interplanetary space habitats or terraformed households. In effect, Rodoman-ALPS would enclose Earth in a kind of safe deposit vault.

Mutual understanding amongst our world's multifarious Noosystems will be the most important task that thinkers face during the 21st Century. An electro-mechanical mindkind—cyborgs and Terra-creatures—may possibly always remain inferior to humankind. Indeed, it would seem reasonable for people to expect that a future Earth co-tenancy with these "others" would, thereby, prevent a human species-wide anomie! Ancient humans practiced transhumance (nomadism). Will future members of our species become trans-human, subsequently associating with contemplative robots and extra-terrestrial Aliens? Will all exist under a Pax Scientifica? Space is the Universe, whatever its material or immaterial state, minus "Spaceship Earth".

References Cited

Adams, P.C. (1992) *Television as Gathering Place.* Annals of the Association of American Geographers 82: 120.

Al-Abdulrazzak, D. and Pauly, D. (2013) *Managing fisheries from space: Google Earth improves estimates of distant fish catches.* ICES Journal of Marine Science, pages 1-5. [DOI: 10.1093/icesjms/fst178.]

Altshuler, A. and Luberoff, D. (2003) *Mega-Projects: The Changing Politics of Urban Public Investment.* (Washington DC: Brooking Institution Press) 339 pages.

Anon. (November 2013) *What would a Hyperloop nation look like?* Popular Science 283: 31.

Banham, R. (1958) *Space, Fiction and Architecture.* Architects' Journal 127: 559.

Barton, B. (1978) *Geographical Enquiry in a Hypothetical World.* The Professional Geographer XXX: 397-406.

Boring, E.G. (February 1955) *Dual Role of the Zeitgeist in Scientific Creativity.* The Scientific Monthly 82: 101-106.

Brook, D. (2013) *A History of Future Cities* (New York: W.W. Norton) 457 pages.

Brovelli, M.A. and Zamboni, G. (2012) *Virtual globes for $D environmental analysis.* Applied Geomatics 4: 163-172.

Carr, M.H. (2013) *Geologic Exploration of the Planets: The First 50 Years.* EOS 94: 29-30.

Cathcart, R.B. (1983) *A Megastructural End to Geologic Time.* Journal of the British Interplanetary Society 36: 291-297.

Cathcart, R.B. (1986) *Improving the status of Rodoman's electronic geography proposal.* Speculations in Science and Technology 9: 37-39.

Cathcart, R.B. (1997) *Seeing is Believing: Planetographic Data Display on a Spherical TV.* Journal of the British Interplanetary Society 50: 103-104.

Cathcart, R.B. (2011) *Anthropic Rock: A Brief History.* History of Geo- and Space Sciences *2: 57-74.*

Chan, H.Y. et al. (2013) *From Construction Megaproject Management to Complex Project Management: A Bibliographic Analysis.* ASCE Journal of Management Engineering http://dx.doi.org/10.1061/(ASCE)ME.1943-5479.0000254.

Crick, F.H.C. and Orgel, L.E. (1973) *Directed Panspermia.* Icarus 19: 341-346.

Davidson, F.P. and Huot, J-C. (1989) *Management Trends for Major Projects.* Project Appraisal 4: 133-142.

Eberhart, G.M. (2002) *Mysterious Creatures: A Guide to Cryptozoology.* (Santa Barbara: ABC-CLIO) 722 pages.

Edwards, P.N. (2012) *Entangled histories: Climate science and nuclear weapons research.* Bulletin of the Atomic Scientists 68: 28-40.

Flyvbjerg, B., Bruzelius, N. and Rothengatter,W. (2003) *Megaprojects and Risk.* (Cambridge: Cambridge University Press) 207 pages.

Gilpin, K. et al., (2010) *Robot Pebbles: One Centimeter Modules for Programmable Matter through Self-Disassembly.* IEEE International

Conference on Robotics and Automation (ICRA) 9-13 May 2010, pages 3614-3621.

Gold, T. (May 1960) *Cosmic Garbage* Space Digest, pages 65-66.

Goode, H.D. (1969) *Geoevolutionism: A Step Beyond Catastrophism and Uniformitarianism.* Geological Society of America Abstracts with Programs for 1969 Meeting in Salt Lake City, Utah, Part 5, page 29.

Guye, P. et al. (2013) *Rapid, modular and reliable construction of complex mammalian gene circuits.* Nucleic Acids Research 41: 1-6.

Hall, J.S. (1994) *Utility Fog: A universal physical substance.* Vision 21: Interdisciplinary Science and Engineering in the Era of Cyberspace, NASA, Lewis Research Center, N94-27358-12. Pages 115-126.

Haqq-Misra, J. et al. (2013) *The Benefits and Harms of Transmitting Into Space.* Space Policy 29: 40-48.

Hayes, B. (2001) *The Weatherman.* American Scientist 89: 10-14.

Howard, S. (July-August 2002) *Nanotechnology and Mass Destruction: The Need for an Inner Space Treaty.* Disarmament Diplomacy Issue No. 65: 1-22.

Impey, C. and Henry, H. (2013) *Dreams of Other Worlds: The Amazing Story of Unmanned Space Exploration.* (Princeton NJ: Princeton University Press).

Jinks, G.F. (September 1953) *"Pointillism" as Cartographic Technique.* The Professional Geographer 5: 4-6.

Kahn, P.H. et al. (February 2009) *The Human Relation With Nature and Technological Nature* Current Directions in Psychological Science 18: 37-42.

Kaiser, P. and Kwan, M. (2012) *Ends of the Earth: Land Art to 1974.* (New York: Prestel) 264.

Kasarda, J.D. and Lindsay, G. (2011) *Aerotropolis: The Way We'll Live Next.* (New York: Farrar, Straus and Giroux) 466 pages.

Lafollette, M.C. (2012) *Science on American Television: A History* (Chicago: University of Chicago Press) 296 pages.

Launder, B. and Thompson, J.M.T. (2010) *Geo-Engineering Climate Change: Environmental Necessity of Pandora's Box?* (Cambridge: Cambridge University Press) 314 pages.

Legget, R.F. (1972) *Duisberg Harbour Lowered by Controlled Coal Mining.* Canadian Geotechnical Journal 9:374-383.

Lotka, A.J. (1922) *Natural selection as a physical principle.* Proceedings National Academy of Sciences 8: 151-154.

Loxton, D. and Prothero (2013) *Abominable Science!* (New York: Columbia University Press) 411 pages.

Mole, R.A. (1995) *Terraforming Mars with Four War-Surplus Bombs.* Journal of the British Interplanetary Society 48: 321.

Muscatello, A.C. and Houts, M.G. (1996) *Surplus Weapons-Grade Plutonium—A Resource for Exploring and Terraforming Mars,* Los Alamos National Laboratory-UR-96-4463.

Nerlich, B. and Jaspal, R. (2012) *Metaphors We Die By? Geoengineering, Metaphors, and the Argument from Catastrophe.* Metaphor and Symbol 27: 131-147.

Peder, A. (2005) *The closed world of ecological architecture.* The Journal of Architecture 10: 527-552.

Pike, R.J. (1987) *Geography of the Planets: Gift of Remote Sensing.* The Professional Geographer 39: 131-145.

Rodoman, B.B. (1974) *Polijarzaacija Landsafta kak Sredstvo Sochraenija Biosfery I Rekreacionnych Resursov.* Resursy, Sreda, Raselenije. (Moscow, [USSR]: Nauka).

Rosner, L. (Ed.) (2004) *The Technological Fix: How People Use Technology to Create and Solve Problems* (New York: Routledge) 265 pages.

Ross, J.A. (1991) *The Globall Hyperatlas: a development proposal.* The Visual Computer 8: 107.

Ross, M. et al. (2009) *Limits on the Space Launch Market Related to Stratospheric Ozone Depletion.* Astropolitics 7: 50-82.

Salzman, J. (2012) *Drinking Water: A History.* (New York: Duckworth) 320 pages.

Sarewitz, D. and Nelson, R. (2008) *Three Rules for technological fixes.* Nature 456: 871-872.

Scheffauer, H.G. (1924) *The New Vision of the German Arts.*

Schneider, J. et al. (2010) *The Far Future of Exoplanet Direct Characterization.* Astrobiology 10: 121-126.

Shea, N. (2011) *Under Paris*. The National Geographic Magazine 219:102-125.

Sherkow, J.S. and Greely, H.T. (2013) *What If Extinction is Not Forever?* Science 340: 32-33.

Sherman, D.J. et al. (2010) *Benchmarking the war against global warming*. Annals of the Association of American Geographers 100: 1013-1024.

Smitherman, D.V. (2000) *Space Elevators: An Advanced Earth-Space Infrastructure for the New Millennium*. NASA/CP-2000-210429. 33 pages.

Smithson, R. (Summer 1967) *Towards the Development of an Air Terminal Site*. Artforum 5, no. 10: 36-40.

Smoluchowski, R. and Torbett, M. (1984) *The boundary of the Solar System*. Nature 311: 38-39.

Srivastava, M. et al. (2012) *Human-centric sensing*. Philosophical Transactions of the Royal Society A: Mathematical, Physical and Engineering Sciences 370: 176-197.

Tavoni, M. and Socolow, R. (2013) *Modeling meets science and technology: an introduction to a special issue on negative emissions*. Climatic Change 118: 1-14.

Warf, B. and Sui, D. (2010) *From GIS to neogeogaphy: ontological implications and theories of truth*. Annals of GIS 16: 197-209.

Williams, J. and Crutzen, P.J. (2013) *Perspectives on our planet in the Anthropocene*. Environmental Chemistry 10: 278.

Wolfe, A.J. (2002) *Germs in Space: Joshua Lederberg, Exobiology, and the Public Imagination, 1958-1964*. Isis 93: 183-205.

Wolmar, C. (2010) *Blood, Iron, and Gold: How the Railroads Transformed the World* (New York: Public Affairs) 376 pages.

Yu, L. and Gong, P. (2012) *Google Earth as a virtual globe tool for Earth science applications at the global scale: progress and perspectives.* International Journal of Remote Sensing 33: 3966-3986.

CHAPTER 4

FUTURE CORE ANTHROPOCENIC MOTHERSHIP INFRASTRUCTURES

A most fundamental Earthly environmental change is the sensationally reported carbon dioxide buildup in the air caused, in part, by the industrial use of exhumed fossil fuels, vegetation clearance and agriculture. Often overlooked, however, are the cyclical atmospheric changes caused directly by the Sun's variable energy emission. It is possible that global warming some climate degradation alarmists are presently mistaking for anthropogenic global warming is really a normal manifestation of the Sun's variability. As for the future, Qian and Lu (2010) predict that "…global-mean temperature will decline to a renewed cooling period in the 2030s, and then rise to a new high-temperature period in the 2060s". These changes, they maintain, will be instigated by the Sun's "moods", not the man-caused buildup of greenhouse gases! During Solar Cycle 25, peaking around 2022, a major subsequent solar energy emission decline is forecast between 2035 and 2045, making a grand minimum like the Maunder Minimum of 1650-1700 possible. The Medieval Warm Period, lasting from 950 until 1250 reported for southern Europe's Alps, was 1-3^0 C warmer than today whilst southern Europe's subsequent "Little Ice Age", lasting from

1500 to 1850, was cooler. Yet, as Umberto Monterin (1887-1949) proved by 1937 after examining the presence of well-structured irrigation canal networks at unusual Alpine landscape elevations as well as clearly anthropogenic transit tracks through now inaccessible high-mountain passes, that few negative weather or sea-level aberrations were evidenced at any time during these periods (Cresenti and Mariani, 2010)! In other words, Monterin's reported findings call into serious question the Green alarmist announcement of impending global Nature chaos! Climate change over the Alps will alter the frequency of "normal" Alpine living risks to humans and their property (Schwendtner el al., 2013). All current climate models are demonstrably deficient in that they do not properly simulate the recorded behavior of the upper ocean and the lower atmosphere (troposphere). Known solar cycles seem far more likely to occur than numerical readouts from bad climate models! Then, of course, there is the other side of the coin, so to speak, the Earth's interior circulation. Mantle heating of the Earth-crust's base does raise landscapes, shift and warm seawater; a warmer ocean definitely affects the troposphere's weather as well as climate regimes! [But, we should always keep in mind the idea that much reportage today over-estimates the presence as well as the aerial/sub-aerial effects of our air's supposed "global warming" (Fyfe et al., 2013).]

"In planning the technological marvels of tomorrow, even the geographical future cannot be taken completely for granted" (Cathcart, 1980, page 28). Worldwide, famous celebrity-ecologists view this event-process, so-called "anthropogenic global warming", from opposing perspectives—quite naturally their views are polarized around a "hot" topic of public controversy and money-making activities! On the other hand, some say that Earth's carbon dioxide gas buildup injects energy into our planet's biosphere (CO_2 fertilization), thereby opening new opportunities for organic evolution, which is arguably a constructive event-process (De Jong et al., 2012). Too much carbon dioxide gas in the air would stimulate a panting reaction in humans and a concentration of the gas in excess of normal (0.03%, equivalent to a partial pressure of 0.23 millimeters of mercury), above 7.0 millimeters of mercury, would cause narcosis. Relentlessly logical terraformers might teach that a predicted extreme buildup could be considered as an event-process favorable to their promotion of planet-altering schemes since—by prospective 21st Century active human migrants only—Mars would be seen as an aspiration region! Rephrased for additional clarity, terraformers might believe that Earth's anthropogenic enhanced Greenhouse Effect is beneficial, or at least stimulating, to future autonomous Mars investments! As a matter of reason, far-future terraformers might even turn their attention to Earth itself and a macroproject to remove nitrogen gas from Earth's air to reduce air pressure and prolong the existence of the Earth-biosphere (Li et al., 2009).

Some green-color (Cohen, Ed., 2013), photo-synthesizing plants do thrive in enriched aerial carbon dioxide, so an artificial Earth-atmosphere buildup could help solve food and fiber production macro-problems. "Just eight countries comprising 11% of the global population produced, on average, 70% of cereal crops during the past decade" (MacDonald, 2013). [The "Virtual Water" product-transfer controversy arises from this gap between staple crop growing regions, many irrigated or utterly reliant on stable precipitation (rain and snowfall) with fluctuating global markets (Hoekstra, 2013; Allen, 2011).] Certainly, an overheating of our Earth's air—from any cause—that stimulated a mega-Greenhouse Effect, could instigate a significant cause of involuntary birth control not requiring any act of individual personal willpower by sexually-active normal adult humans. Humans would struggle to survive before the intolerable mega-Greenhouse Effect occurs since our skin temperature must be lower than a wet bulb temperature of 35° C so that our bodies can cool by sweating (Sherwood and Huber, 2010). An early-Triassic mega-Greenhouse Effect eliminated many species by excessive heat (Sun et al., 2012). Via thermal effects upon maternal core-skin flow critical to the survival and normal development of embryos, a future mega-Greenhouse Effect (unmitigated by air-conditioning technology) may reduce the flow of blood to the uterine tract and kills mammal embryos (Cathcart, 1997). This factor would be prevalent, especially, in the Earth-biosphere's Tropical Zone, but Temperate Zone nation-ecosystem populations may also be triggered to collapse—thus reducing the Earth-biosphere's load further! Such thoughts are fodder for geopolitical Think Tank intellectuals, the staffs of government agencies and others too dreary to identify in this generalist essay! Geographos offers a fleeting thought here that such folks, perhaps, have not yet considered: Terra-creatures possibly ought not to be considered as detached from global Nature—wherever that might be located—by their nature as put-together techno-science mechanicals. Instead, they may be even more aware of their connectivity to global Nature than are humans. Why? Because all people are perpetually confused, a mental confusion instilled by their conscious rational thoughts/feelings/beliefs about what "life" is, and what human "purpose" is. Terra-creatures will undoubtedly understand they are composed of the actual material stuff of Earth and that, indisputably, they were first manufactured by intelligent, hard-working humans! This observation comports with new ideas compiled by A.L. Pelaez as Editor of the 2014 book *The Robotics Divide: A New Frontier in the 21st Century?* Erich Brynjolfsson and A. McAfee, in *The Second Machine Age: Work, Progress, and Prosperity in a Time of Brilliant Technologies* (2014) opine that government entities, businesses and persons must learn to "race" with ever-smarter machines and to always win the intellectual sprint contest, or at least maintain a working parity, by harnessing our skills to "mix-and-match" different available technological resources and new collaborations, developing tandem brute processing blended with God-given human ingenuity.

It must be noticed, certainly, that policy decision-makers and macro-imagineers/macro-engineers have markedly different perspectives about building and destroying. Geopoliticians, for example, seek approval of their superiors. But elected and appointed policy implementers are involved mainly with near-term future process-events with fast-paced (sometimes "crisis management") deductive thinking producing predictable insights expressed in simple language for "best solution" outcomes under prevailing circumstances. Macro-engineers and Macro-imagineers, on the other hand, desire the respect of their peers, deal chiefly with mid- and long-term future events in Anthrome an Astythrome contexts resulting in well-considered inductive thinking composed of original insights communicated in abstract language in "multi-possibilities" reports to politicians and geopoliticians. More simply stated, with far fewer caveats, consulting geopoliticians are hired to plan solutions for extra-territorial noosystemic macro-problems as they affect an ecosystem-country's survival. Macro-imagineers and macro-engineers are employed to construction regional infrastructures, enmeshing their localized public-works with known planet-wide geographical ground truth and ground engineering and many affected nation-ecosystem economies. A means to unite these geo-philosophical differences is obvious—a fusion of the teaching of decision-makers with that of Macro-Imagineering/Macro-engineering and Geography (Davidson, 1989). In Europe, the Candida Oancea Institute (Bucharest, Romania) offered geo-philosophical fusions since 1997, whilst in the USA at Bristol, State of Rhode Island, at the Center for Macro Projects and Diplomacy (founded 2003) offers similar instruction at the Roger Williams University. The trend to using monetary prizes, "X", "Y", "Z" or whatever, to goad human social innovation, technical invention and to spur extra-ordinarily novel performances in Science really needs to be extended to the realm of Macro-Imagineering as soon as practicable(Kay, 2013)!

Unprejudiced global Nature monitoring organizations, such as Rodoman-ALPS, will have authority and might even lend some credibility to that hoary buzz-phrase, "sustainable global development", which was coined by Lester Brown in *Building a Sustainable Society* (1981). Despite the existence of, literally, hundreds of definitions, no meaningful definition of "sustainable" exists, nevertheless its users seem to imply that terraformation of other planets will never be attempted, or if tried, will prove to be unsuccessful. "Sustainable" and "development" in mutually exclusive terms! However, whenever *Homo sapiens* learns to harness anti-matter energy and to transmute matter, the Earth and other places—possibly Mars—became then pregnant with new "Earths". Perhaps it would help some to imagine this macro-problem as one of Timothy Morton's *Hyperobjects: Philosophy and Ecology after the End of the World* (2013). Morton implies that a mega-Greenhouse Effect is inevitable, that the environmental emergency is also a crisis for our habitual thinking—that is, the mega-Greenhouse Effect will defy our persistent attempts to control and to understand! So, the crisis of philosophy prevalent entails a scientific revolution, the "paradigm-shift" postulated by Thomas Samuel Kuhn (1922-1996). Our old world view—the philosophical and epistemological scope of viewing reality—must change to some point of view more worldly than any existing now! Brown's "sustainability" is a synonym for technology's stagnation and, possibly, regression. The charm of planetary engineering, Terraforming, potential was first brought to the world-public's attention via James Edward Oberg's *New Earths: Restructuring Earth and Other Planets* (1981).

Approximately 20% (or 29.2 million square kilometers) of land is classified as arid and about 50% of all landscape is water supply limited. Discontinuously covering the islands and continents, deserts are the present-day permanent homelands of a scant few persons. [Many deserts, such as the northern Sahara, are regions of important modern economic development—the oil and natural gas resources, for example—that generate lots of vehicular traffic.] Extrovert Australian macro-imagineer David Noel (**Figure 1**, below)

Figure 1. David Noel, the bubbly, tech-savvy Australian free-thinker. (Image: DN.)

offered his ingenious and generous suggestion for the bulk collection of freshwater from rainfall occurring over the selected regions of the world-ocean, which would then be conveyed to freshwater-short coastal lands in 1980 (Noel, 1980). His means of massive freshwater acquisition is "natural", unlike the bruited massive future extractions of freshwater for coastal cities from mined undersea aquifers covered by the world-ocean's post-Ice Age rise (Post et al., 2013). Unquestionably, deployment of such a water-collecting and distribution infrastructure, a kind a diaphanous Architecture simulating marine surface films, would pit Earth's eco-fundamentalists against techno-fanatics! Noel's proposed lily-pad could only unfairly be compared to the enormous mass of discarded plastic litter bobbing on the ocean's surface (Moody, 2006; Ebbesmeyer and Scigliano, 2009; Hohn, 2011; Karl, 2014)! Actual investment in Noel's invention seems economically attractive during the 21st Century. The mean annual oceanic precipitation (situated in a band between 65⁰ North latitude and 60⁰ South latitude) is about 93 centimeters. Noel proposed a special kind of huge, lily pad-like buoyant collector-reservoir composed of impermeable light-weight plastic that would spread over the ocean's surface where current climatic regimes produce harvestable rain. His device, well mapped and marked with edge-buoys and emplaced out of the normal shipping lanes, could well prove to be a quite practical floating "collection plate" in the future because, so far, over-land cloud chemical treatments to instigate rainfall have been ineffectively sporadic. Between early 2010 and mid-2011, global sea-level fell sharply by about 0.5 centimeter because of the co-occurrence of surface seawater cooling of the eastern Pacific Ocean by La Nina, a climate pattern called the Southern Annular Mode and the Indian Ocean Dipole. Together, these process-events pushed rain-bearing air masses over the Outback of northern Australia where it fell onto flat dry-land. So much of the freshwater that fell stayed immobile—absorbed by arheic basin dry soils, filling dry lakebeds such as Lake Eyre, and otherwise retained in endorheic basins—that it deprived the ocean of its normal input of rain (Fasullo et al, 2013). Any widespread presence of Noel's permanent collector-reservoirs should have the same effect as these unique climate regimes coincidences—that is, causing the long-term sea-level to measurably decrease by anthropogenically altered air-ocean interactions. Freshwater will be shifted, sometimes permanently, to dry-land in places like central Australia. In fact, such a freshwater shifting from ocean to land can be compared to a recently announced macroproject plan to fill below-sea-level Lake Eyre with seawater via laid hoses from the adjacent ocean, possibly via Spencer Gulf, to that gigantic inland basin (Badescu et al., 2013). It is worth remarking here that others have developed plans for massive floating facilities: (I) Hisaaki Maeda's crop-planted "Giga-Floats" (Maeda, 2003) and (II) "Solar Cell Rafts" proposed by Takaji Kokusho and colleagues (Kokusho, 2013) to generation electricity on a massive scale without using scarce land resources in Japan. Both macro-objects are specified to be used on the Pacific Ocean.

Earlier than Noel's assumed invention date, Francisco Alcalde Pacero (1941-2004), addressing only over-landscape transportation of great quantities of freshwater, opted for gigantic tire-like rolling water-carrying bags (Cathcart, 2005). A sea-going blimp analogue—that is, kilometer-long, 10 meter-diameter, sausage-shaped plastic bags—to serve as containers for enormous quantities of freshwater can connect Noel's collector-reservoir with Pacero's on-land reliable rolling distributors pulled by powerful independent off-road capable tractors.

India will need more freshwater after 2020 since by then consumer demand will likely exceed all present-day sources of supply. Imports of foreign freshwater are an option mostly overlooked by India's macro-imagineers! Instead an unwieldy and complex macroproject first conceptualized in 1982 has been elaborated that would link the major rivers in the mountainous north (Himalayas) with those rivers in the south as a means to better manage freshwater, moving it from a region of perceived surplus to regions without sufficient domestic supplies. The National Water Development Agency tasked with carrying out the USD 120 billion that would transport 178 billion cubic meters of freshwater southwards via 30 river-diversion megaprojects and 14,900 kilometers of canals (Mirza et al., 2008). Cathcart (2011) states the case that foreign water could be imported with relative ease in sufficient quantity to satisfy the urgent needs of India's coastal and inland Astythromes. A single tug-towable water bag, transporting one million cubic meters of freshwater could be the basis of supply for 50,000 families annually; 1,869 water bags holding one cubic kilometer of freshwater each equals the yearly natural discharge at the coast of all rivers flowing on India's national landscape territory! Try to imagine ocean-going plastic bags off-loaded into Pacero's mobile freshwater juggernauts that then regularly fill the many artificial reservoirs in the Madurai-Ramanathapurama tank landscape of southeaster India, near Palk Bay and the Gulf of Mannar. Groundwater below that Anthrome is over-exploited and replenishment and substitute resources of freshwater are much needed by the residents. Keeping the tanks of southwestern India, very distinctive landscape sculptures, always filled with freshwater via an interlocked freshwater transport system requires the investigations of an India-based Think Tank (Torgersen, 2006).

American hydrologists claim increasing concern about the future availability of freshwater supplies for the Southwest, in particular for Southern California. Southern California's infamous determined political and mega-engineering efforts to acquire an entirely reliable, consistent supply of distributable freshwater via transport from freshwater sources to that semi-desert (Mediterranean) climate region are legendary. To satiate Southern California's consumer demand, even the deliberate movement of Antarctic icebergs has been investigated by the RAND Corporation. During the present-day 2014 political climate, most Californians have no expectation of building new aqueducts—R.B. Cathcart's *Macro-Imagineering Coastal California's 21st Century Prosperity*, a 2013 e-book published in the USA by Thinker Media, suggested a single large-diameter submarine freshwater aqueduct be emplaced connecting coastal Northern California river mouths with onshore freshwater reservoirs in Monterey and Santa Barbara counties—and the region's human population is increasing, pressuring water supply bureaucracies to devise ways of using existing freshwater more efficiently. This macro-project seems vital to California's future since it is a fact that the State of California during 2013 received less precipitation than in any year since California became a USA state in 1850; perhaps the State has entered an enduring "mega-drought"? For the Metropolitan Water District (established 1928) which furnished Southern California with about 50% of its supply, a key to meeting these needs lies in its attempts to store more freshwater during California's "wet years". Freshwater-filled floating plastic pods, securely anchored offshore, could provide additional storage capacity situated close to the dense-settled coastline. It is even possible to suppose such sea-bottom secured pods will form an effective barrier to sporadic tsunamis as well as wintertime storm surge waves before their destructive landfall.

Practical assessment of various aspects of blimp-like freshwater transports has shown that strained, treated and untreated drinking water could, thus, be moved from sites of surplus supply to wherever needed by coastal consumers willing to pay prices per cubic meter required to cover costs plus modest profits. Also, minimally processed urban sewage exported by coastal cities could be removed elsewhere by dedicated pods for pickups and deliveries—one or more multi-national corporations, perhaps, or a yet-to-be-established United Nations Organization agency—could holistically plan the use of the ocean wherein some zones are pronounced as producing little life for a natural near absence of edible nutrients. Such a system might also alleviate the macro-problem of dead zones near rivers mouths where fast delivery of too many nutrients has poisoned some valuable fishes by uselessly making these zones low in seawater oxygen content. Many local or regional freshwater supply management teams currently is composed of technical staff which are lacking in practical understanding or experience of an extremely stressful high-seas work place, nor have they yet devised appropriately-engineered equipment to withstand the rigors of the turbulent, sometimes unpredictable ocean. Such Macro-Imagineering concepts outlined above might well be identified as future core macroprojects, multi-national corporations and international political organizations will need to be wisely advised by a professionalized elite, combining the skills of geographers with those of civil-military engineers, in order to arrive safely at a terminus of decision-making processes with the most cost-effective least Earth-biosphere damaging solutions (macroprojects/megaprojects).

Since 1977, macro-engineering professional symposiums have been included, occasionally, in the annual meetings of the American Society for the Advancement of Science. In majority, those few AAAS Symposiums were focused on a formulation of macro-management systems for macroprojects, considering specific plans. However, it was not until March 1988, in a RAND Corporation-published booklet, "Understanding the Outcomes of Megaprojects: A Quantitative Analysis of Very Large Civilian Projects" [RAND/R-3560-PSSP] by Edward W. Merrow (**Figure 2**), with Lorrraine McDonnel and R. Yilmaz Arguden, that a systematic empirical economic analysis of costs, macro-problems, and operations of built macroprojects was first published. The report's authors concluded that of the numerous completed macroprojects reviewed "...most...met their performance goals; many met their schedule goals; few met their cost goals".

Figure 2. Edward E. Merrow, pioneer of realistic macroproject cost-benefit assessments, ultimate worth of emplacement and usage. (Image: EEM).

Their truism is hardly surprising: once started, any macroproject initiators, constructors and political supporters, to save their own reputations and ensure their future prosperous livelihoods, will overspend other people's money to get to a final, operational macroproject! The RAND Corporation study was not really expanded until years later when three books by others substantiated Merrow et al. conclusions: Alan Altshuler and David Luberoff's *Mega-Projects: The Changing Politics of Urban Public Investment* (2003), Bent Flyvbjerg, Nils Bruzelius and Werner Rothengatter's *Megaprojects and Risk: An Anatomy of Ambition* (2003) and Edward W. Merrow's *Industrial Megaprojects: Concepts, Strategies, and Practices for*

Success (2011). Most recently, Patricia D. Galloway, Kris R. Nielsen and Jack L. Dignum, as Editors, compiled ***Managing Gigaprojects: Advice from Those Who've Been There, Done That*** (2012). It is still apparent that Macro-Imagineering, and particularly Macro-engineering, is an immature profession which badly needs good teaching literature or it will be saddled with James E. Oberg's definition (at page 270 of his 1981 book): "MACRO-ENGINEERING— creation of giant projects lasting for decades and costing appreciable fractions of a society's gross national product; in general, they are not practical but serve some social, religious, psychological, or other purpose". Macro-engineers are peacefully poaching on the traditional turfs of disciplines like History and Science—the word "scientist" first appeared only in 1840 and was coined as a contrast to "artist"—to achieve a workable professional discipline and successful profession. Macro-Imagineers/Macro-engineers are often perplexed by the wide variety of megaproject studies conducted and the spread of purely monetary cost amounts in carefully-prepared financial estimates. It must be recalled that "...miscalculation or sheer ignorance of cost and difficulties was the key to launching a number of great and successful enterprises, from canals and railroads to mining and manufacture" (Sawyer, 1952). Macroprojects must be judged by how they meet their objectives over historical time amidst mutating societal, geopolitical and Green-decreed environmental values— this judgment, therefore, is always complicated by evolving public expectations and the near-constant emergence of new Science facts and engineered Technology. Macro-planning is the organization of human effort—in future, possibly supplemented by cyborgs and Terra-creatures—on a time schedule to serve or enhance attainment of goals. Completed in 1825, the USA's Erie Canal was one such macroproject that was, later, renamed the New York State Barge Canal (Koeppel, 2009; McGreevy, 2009).

At the 1977 AAAS Meeting, Robert M. Salter (**Figure 3**) presented a thorough outline of his extreme rapid underground rail transportation system in evacuated tunnels, which he euphemistically called "Planetran".

Figure 3. Robert M. Salter (1920-2011), RAND Corporation originator of "Planetran".

Supporting his Planetran study conclusions with basic mathematics, Salter (1978) envisioned an intercontinental subway system consisting of streamlined vehicles traveling thousands of kilometers per hour, serving only Earth's Northern Hemisphere. Curiously, however, he neglected to acknowledge previous scientific contributions by Laurence Knight Edwards (1965) and Robert L. Forgacs (1973). Salter foresaw excavated tunnels thousands of kilometers long; today, our world's longest railroad tunnel is the USD 10 billion, 56.9

kilometer-long Gotthard Base Tunnel in Switzerland, which opens to traffic during 2017. First conceived in 1947 by macro-engineer Eduard Gruner (1905-1984), the Gotthard Base Tunnel has exceeded the length of Japan's undersea Seikan Tunnel, dug beneath the Tsugaru Strait, which is 53.9 kilometers from portal to portal. Tunnel-digging experience indicates that USA dollar cost per cubic meter of material displaced and removed from the Planetran work-site during 2014 might range from USD 66 for jointed, limestone and chalk to USD 1,319 for sand, silt and clay under groundwater; the diameter of the tunnel as well as its length will, naturally, determine the total expenditure on excavation alone. By comparison, underground cities are equally expensive (Mulder et al., 2014). Still, China's military officials publicly claim to have a dug an "Underground Great Wall", supposedly 4,800 kilometers in length (Stephens, 2011). Neither the American Underground-Space Association (founded 1976) nor the Underground Construction Research Council (established 1970) endorsed Salter's Planetran. The retired-from-service Concorde SSTs were first tested in 1968; post-Concorde aerospace planes such as SKYLON are still in the R&D phase as are high-speed surface trains that follow a network of magnetic-levitation tracks—both of these technologies render Planetran problematical. As sketched by Salter, Planetran would move passengers and freight on a single route connecting North America-Russia-Europe, linked by an 86 kilometer-long submarine tunnel across the Bering Strait. According to Salter, a branch line could serve Japan. Salter's tunnel vision excludes China and India from the Planetran system and Earth's Southern Hemisphere population would not receive any service either. Salter had rather inaccurately named his Planetran system "trans-planetary"; such careless promotion raises images of an absurdly directed transportation technology, a mental image of "Babel reversed"! 21st Century macro-imagineers have concocted similar worldwide systems (Friedmann, 2006; Oster et al., 2011) yet these also face the limitation of exorbitant excavation costs that made Salter's scheme a non-starter, infeasible.

How might Macro-imagineers seek to connect Earth's hemispheres? Would a 21st Century New World Free Trade Region necessitate interconnection? No single answer exists, of course, but one distinct possibility might be hovercraft—skirted air-cushion vehicles—crossing Drake Passage, which separates Antarctica from South America, along with construction of a high-speed magnetically levitated railway linking South, Central and North America. Sea-states in Drake Passage are dangerously chaotic, with huge waves, even awesome rogue waves (100 meters height) that are far too big for hovercraft to ever negotiate safely (Anathaswamy, 2013)! From the early-1950s, Christopher Sydney Cockerell (1910-1999) famously proselytized the use of hovercraft everywhere but such air-cushioned vehicles can safely operate at a cruise speed of 30 knots over open stretches of seawater only where undulating waves are 1.5 meters high! Famed UK explorer Vivian Ernest Fuchs (1908-1999), head of the 1957-1958 Commonwealth Trans-Antarctic Expedition, promoted the use of hovercraft in both Polar Zones (Fuchs, 1966) as appropriate for marine geophysical, geological and oceanographic work on the ice. A hovercraft was used during the 1988-1989 austral summer at McMurdo Sound in Antarctica; its rubber-skirt was flexible to minus 50 degrees centigrade and it was stable against accidental flip-over because it has a very low center of gravity combined with a relatively large width—normally hovering height is about 15-20% of the hovercraft's width. Suppose hovercraft, after putatively passing over Drake Passage, could follow Antarctica's coast until reaching the USA's McMurdo Station thence continuing to the South Pole via a postulated 1,600 kilometer-long ice-highway Antarctic-1, the first four kilometer segment of which opened to some wheeled (Scambos, 2005) and most all caterpillar-treaded traffic in 2002. Antarctic-1 is an overland gravel road that connects McMurdo Station's floating ice-dock, in part used to supply the logistics base off McMurdo Station, and New Zealand's Scott Base and beyond (on snow-covered ice roadbed) nearly 10 kilometers to the snow-plowed summertime airport. The Drake Passage Hovercraft Roadway would, purely as a side-effect, raise Earth's sea-level—a sea-level elevation on top of that previously happening owing to various causes, both natural and anthropogenic. Armadas of icebergs would hinder commercial shipping around both Greenland and Antarctica if a future global climate change causes massive de-glacialization. An artificial closure of the Drake Passage might, therefore, help to mitigate this geohazard, especially in the South Atlantic Ocean's trade routes—it is likely petroleum-natural gas shipments as well as containerships will still need to 'round southern Africa's Cape Algulhas even if a trans-Arctic Ocean route from Asia to Europe becomes feasible by mid-21st Century. Not to be overlooked is that freshwater icebergs will be pushed northward along Chile's coast, where they might be harvest—towed into harbors and used for municipal water supplies purposes!

Antarctica is an ultra-pleasant version of Mars un-terraformed (Pyne, 2007): there is a lack of really substantial infrastructure as in Earth's Sahara, pervasive dire human working conditions, increasing pollution due to the waste stream generated by its human explorers as well as the psychological issues of prolonged dark wintertime isolation from civilization! The continent of Antarctica, naturally uninhabited by humans, yet visited by 35,000 tourists during 2014, symbolizes Alien space, the menacing monstrous "Thing" science-fiction writer John W. Campbell (1910-1971) featured in the 1951 short-story "Who Goes There?" largely because the icescape/

landscape is so remarkably different visually from the generally more familiar Arctic Zone (Leane, 2005). More proof of how alien Antarctica's landscape is, and how it affects humans, can be found in Bernadett Hince's *The Antarctic Dictionary: A Complete Guide to Antarctic English* (2000): "greenout", for example, is sometimes used by returning visitors to describe the emotion felt upon seeing and smelling living green plants after a long period on the icy and cold continent. [Hince's text is a useful down-to-earth complement to information on Earthly geographical features of that remote region compiled by John Stewart's *Antarctica: An Encyclopedia* (2nd Edition, 2011).] Not much life exists atop Antarctica's ice-sheet; some life microbial exists in liquid freshwater far below. And, what surface life is extant is menaced indirectly by an anthropic "Ozone Hole" (Mulder, 2005), discovered in 1985 by Joe Farman (1930-2013), as well as by direct influences (Convey, 2011). Seemingly, a ban on CFCs is healing the unnatural anthropogenic enlargement of global Nature's Ozone Hole although its natural seasonality remains consistently operative (Solomon et al., 2007). Interestingly, if all ozone was intentionally abstracted from Earth's air, then all life must subsequently endure a cold, almost "Snowball Earth"-like, climate regime (Bordi et al., 2012)! And, likewise, the Earth's global Ozone Layer would vanish, subjecting life to massive stellar and cosmic radiation bombardment! Since ignited rocket engines deplete the global Ozone Layer, there is then some possibility for a future popularly-applauded limitation of spacecraft launches (Ross et al., 2009); thus, an evident risk to outer space industrialization, including Asteroid Belt mining, may be quixotically invoked to "protect and save" the global Ozone Layer from a fate worse than the former over-hyped CFC fumigation macro-problem!

During 1971, Keiji Higuichi in Japan adumbrated a macro-imagineered proposal for a dam-like barrier, composed of linked, moored icebergs, in the 2,500 meter-deep Drake Passage (**Figure 4**), a maritime [naval and commercial shipping] chokepoint.

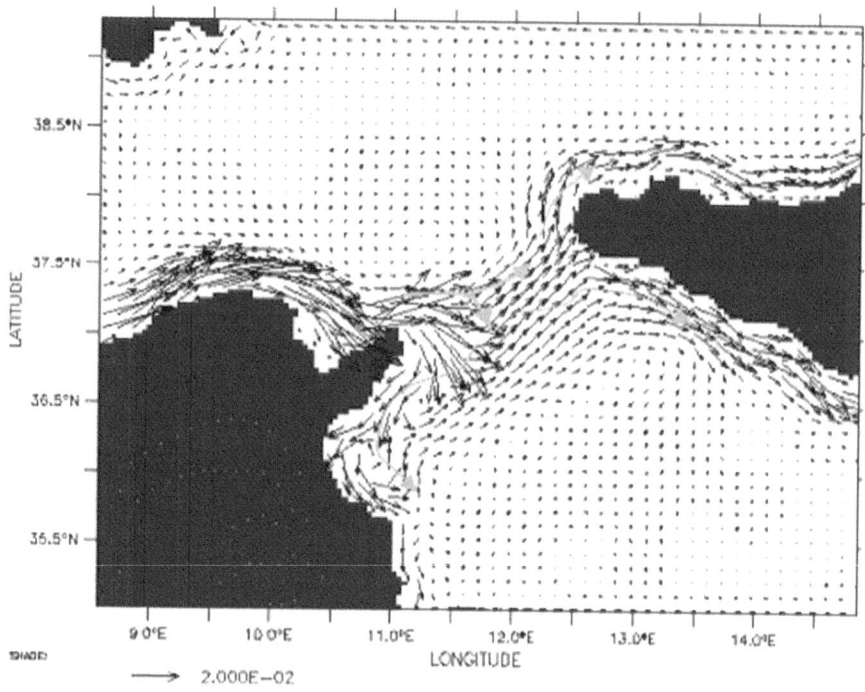

Figure 4. The seawater expanse situated between the Pacific Ocean (top) and the Atlantic Ocean (bottom) is Drake Passage (center), which separates the southernmost tip of South America (right-side) at Cape Horn, Chile, from the Antarctic Peninsula, perhaps just 10% of Antarctica. Were it blockaded the streamlines of seawater currents would alter significantly. See: **Figure 5**. (Image: Google Images.)

In his view, the primary purpose of such a physical obstruction would be to separate the South Pacific Ocean from the South Atlantic Ocean in order to induce favorable Southern Hemisphere climate regime changes, which also includes the southern Indian

Ocean. The Antarctic Circumpolar Current (ACC), which is the only seawater current that flows right around our world, is constricted by Drake Passage. The AAC is directly as well as indirectly influenced by human activity (Fyfe and Saenko, 2005; Martinson, 2012). Drake Passage is the narrowest chokepoint in the overall dynamics of the wind-driven surface seawater circulation in the circumpolar Southern Ocean (taken here to be what is labeled "Pacific Ocean" and "Atlantic Ocean" in **Figure 4**). Yearly variations in the surface seawater heat content fluxes (5-10% variance normally) result from meridional wind anomalies west of Drake Passage, which in turn result from forcing by El Nino/Southern Oscillation and the Southern Annular Mode. The Antarctic Circumpolar Current (Firing et al., 2011) would be interrupted by Keiji Higuichi's ice-dam so that surface seawater moving from west to east around Antarctica would be altered from a vortex to another shape and, accordingly, "...may have the effect of changing the general ocean circulation on the global scale" (Higuichi, 1971). Higuichi's barrier would reinstate a blockade most recently emplaced millions of years ago, before the ACC was first generated and Antarctica was thermally isolated (Dalziel et al., 2013)! A 21st Century closed Drake Passage may eventually affect the North Atlantic's Deep-Water overturn, so it does have a significant role in governing the Earth-ocean's thermohaline circulation. A synchronicity between tectonic closures/openings and global air-warming periods are known by geoscientists. A closed Drake Passage, especially in a high-carbon dioxide gas atmospheric content scenario, would like cause warmer surface air temperatures—possibly as much as 4-7^0 C in the Southern Ocean region of the Southern Hemisphere (Meredith, 2011). Redirection of the surface seawater flow by Higuichi's iceberg dam or a shallower floating ice barrier-boom would likely terminate the recurrence of El Nino, the stream of warm western Pacific Ocean surface seawater sporadically washing Peru's lengthy shoreline. When the El Nino phenomenon happens, it causes great worldwide economic hardships by way of extra-normal weather conditions, extreme climate regime event-processes regionally. Furthermore, the ACC redirected northwards along South America's west coast could boost the oxygen content of the world's largest Minimum Oxygen Zone located in the Pacific Ocean off Chile and Peru, perhaps suppressing the quantity and floating depth of the denitrifying anammox bacteria (Brunner et al., 2013) that create it since the Zone is located just 100-500 meters beneath the ocean's surface. **Figure 5.** Another potential benefit is to cause more precipitation over the land that will bulk-up the presently receding glaciers of the Patagonian Ice Field and, therefore, increase the hydropower resource and freshwater supplies for Chile and Argentina; interestingly, both Argentina and Chile have passed highly debated laws that regulate the mining company impacts on Andean glaciers because there is a present-day and, more importantly, an larger impending water scarcity macro-problem in the Andes (Mutinho et al., 2013). It is probable also that the Benguela Current skirting southern Africa's west coast and the West Australian Current brushing Western Australia would be greatly weakened, or perhaps even ended as long as an ice-dam existed within Drake Passage. Higuichi's dam would modify the geostrophic balance, resulting in a greater outflow of Antarctic Deep-Water to other parts of Earth's ocean as well as weakening the ACC so much that it may need to be renamed! A teleological geo-philosopher might argue that humanity should judge whether a macro-engineering act is good or bad by seeing if a completed macroproject produces a good or bad result.

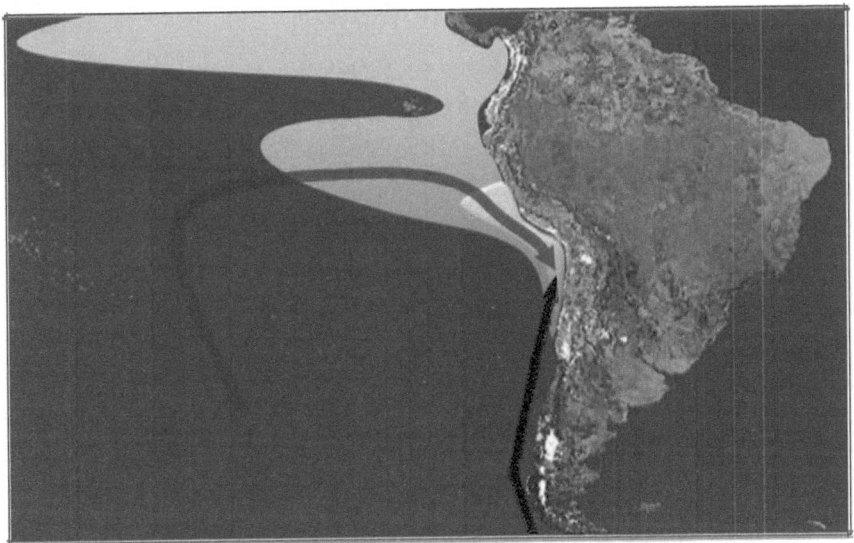

Figure 5. The anoxic seawater (Scholz et al., 2014), shown in light blue, should be suppressed by a strong northward flowing extension of the Antarctic Circumpolar Current, caused by the partial-barrier at Drake Passage, meeting the (black line with arrowhead) which represents the anammox bacteria travel streamline. (Image: Google Images.)

Icebergs, which break off naturally from ancient glacial ice-flows on Antarctica, are flat-topped and often enormous in bulk (Tournadre et al., 2012). Icebergs were calved from Antarctica's ice-shelf when the tsunami generated by the offshore Honshu, Japan earthquake of 11 March 2011 impacted the Sulzberger Ice Shelf (Brunt et al., 2011). Since the ratio of height of a tabular iceberg's freeboard above the ocean's surface is 1:7, Antarctic icebergs with freeboards of 45 meters must extend below sea-level about 315 meters! Keiji Higuichi calculated that the volume of the icy barrier fully filling Drake Passage would require the directed emplacement of 3,100 flat-top icebergs, each with a volume of two cubic kilometers. Loading these icebergs,

Higuichi faultily speculated, with soil and rock to sink them—secondarily forming a Geographos-proposed fixed link causeway, the Drake Passage Hovercraft Roadway—would indeed be a materially voluminous and geographically huge macroproject liable to affect more than half the Earth-biosphere. As a seawater current diversion—rerouting the Antarctic Circumpolar Current at a place of its most extreme constriction—the Southern Ocean's belt of biotic productivity would be disrupted and the ongoing strengthening of the ACC's wind-driven circulation could be vastly curtailed (Hutchinson et al., 2013). Indeed, a Drake Passage Hovercraft Roadway could be simultaneously created with Keiji Higuichi's Ice-dam, but Macro-engineering's task would be completed only at enormous cost economically and ecologically. Since approximately 80% of the Southern Hemisphere's area is ocean, would it really be useful to alter climate regimes in—at the very least—that portion of the Earth-biosphere with such an expensive, time-consuming 1,000 kilometer-long dam? Even a cheaper version, one perhaps designed to just shift surface seawater currents, but not sealing off the seawater flow of Drake Passage fully, might be more efficacious yet still very costly! Higuichi's building method is extremely risky, and without any predictably marked improvement of the Southern Hemisphere's many climate regimes, it seems neither practical nor geopolitically feasible to construct a Drake Passage Hovercraft Roadway. How, then, can we extend the Pan-American Highway, an all-season track constructed since 1923, to connect the Americas with the Antarctic Peninsula? Macro-Imagineers: "Put on your Thinking Caps"! Then, there is another important question to be pondered and answered: Why bother to form any infrastructural connection whatsoever? Well, one obvious statement is that global Nature's climate change apparently is producing new regions of ecosystem-infrastructure vulnerability that humans must cope with somehow (Watson et al, 2013).

Founded in 1951, Doxiadis Associates, a land-use planning team based in Greece, predicted that by 2100 the Antarctic Peninsula could support a polycentric organized "low density" Astythrome-like permanent human residential population, an extension from South America of "Ecumenopolis" (Doxiadis, 1974). Coined in 1961, Ecumenopolis is a suggested future city created by the spreading in a continuous carpet-like urban region covering most of the habitable parts of the Earth-biosphere, a more or less universal settlement (literally, a "Global Village")! Boris B. Rodoman—a central figure in Chapter 3—approved of the gigantic geographical concept devised by Constantinos Apostolou Doxiadis (1913-1975), **Figure 6**, because it would give infrastructure managers their biggest future challenge. Ecumenopolis matches Pietro Passerini's coinage during 1984 of "Anthropostrome" in breadth, since the meaning of the ancient Greek word "stroma" is "carpet" (Passerini, 1984).

Figure 6. C.A. Doxiadis (Image: Google Images.)

The IceCube Neutrino Telescope, a one cubic kilometer observatory implanted in the ice beneath the geographic South Pole by 2011, is currently dedicated to the detection of neutrino particles arriving from outer space. It is the world's largest high-energy neutrino observatory; IceCube consists of photomultipliers deployed 1.5-2.5 kilometers deep into Antarctica's ice at the South Pole and detects the trajectory of charged leptons produced during high-energy neutrino interactions in the surrounding ice mass. Development and settlement of the rocky Antarctic Peninsula could come about, and be a result of, the 21st Century heavy industrialization of the continent's icy periphery, perhaps after 2041.

"Greening" of Antarctica is a near-traditional false theme promoted by so-called "Environmentalists" to please the low-information voting taxpayers (Anon., 2010). From 1962 until 1972, a 1.8 Megawatt nuclear reactor provided electricity for the people and machines located at McMurdo Station and Scott Base. Electrification, and subsequent major-scale industrialization, of Antarctica may be fostered by construction of a Bolonkin-Cathcart Antarctic Wind-Power Ring (AWPR), using a continuous Fabric Aerial Dam carrying air turbines, encircling Antarctica (Bolonkin and Cathcart, 2008). [Fabric Aerial Dams can be visualized as much like spinnaker boat sails or parakites.] The AWPR, in contrast with the many luxurious new scientific and sociological research centers (Princess Elisabeth, Concordia, Neumayer III, Amundsen-Scott South Pole Station and Sanae IV)—some even mobile (the UK's Halley VI Antarctic Research Station)—is fixed, attached to the ice near the shoreline bordering the Southern Ocean. Fishing off this shoreline and brief cruise-liner tourism during the summertime, both based abroad, are Antarctica's only economic activity currently. While the term "ecotourism" has existed since about 1988, only about 5% of world tourism is of this type. Suggestions have been proffered to designate Antarctica as a "World Park" (Tenenbaum, 1990). High-speed katabatic winds blow coastward off the higher elevation interior icecap (Lipzig, 2004); the winds blow with constancy in direction, often moving at 20-40 meters per second over the smoothest superficial icecap terrain but slow about 100 kilometers from the Southern Ocean when they flow above a rougher ice surface. Antarctica has a coastline—it varies with breakoffs of ice—that is about 17,000 kilometers in length. Deployment 100 kilometers from the coastline of an unbroken Fabric Aerial Dam, encircling the continent, would permit generation of approximately 450 Gigawatts of "Green" electricity by the Bolonkin-Cathcart Wind-Power Ring. Roughly speaking, that is the output of 225 nuclear fission power-plants. Excess electricity may even be exported to South America and/or southern Africa by utilizing a submarine High Voltage Direct Current (HVDC) transmission cable (Chatzivasileiadis et al., 2013). Such a linkage can be considered the final connection in R. Buckminster Fuller's 1969 macroproject proposal to create a HVDC "Global Energy Electric Grid" (Dekker, 1995). **Figure 7.** To properly service the Aerial Fabric Dam array regularly, a Hovercraft roadway—dubbed by Geographos "Antarctic-2"— will parallel the linear HVDC electric power-generation installation.

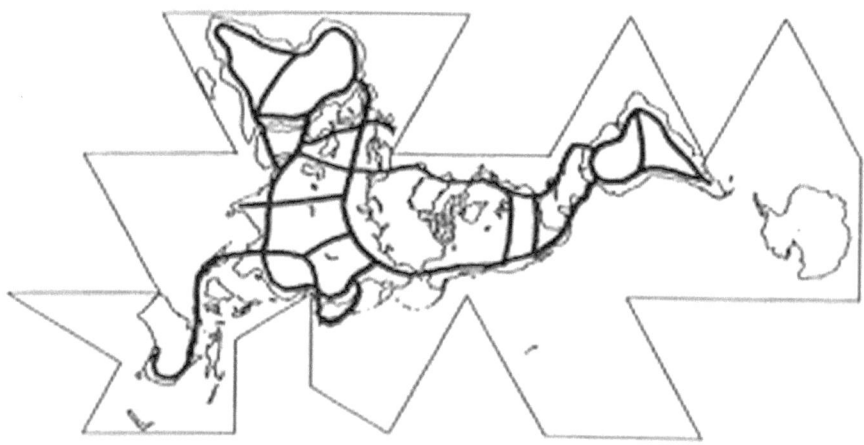

Figure 7. Our ocean-surrounded world of landscapes, the "Global Energy Electric Grid" as originally elaborated by Richard Buckminster Fuller (1895-1983). Fuller's power grid is a network of high-voltage transmission lines that provide long-distance transport of electric power within and between country-ecosystems; low-voltage lines provide local power delivery. (Image: Google Images.)

A complementary wind-energy extraction installation, Edward J. Schremp's "Macro-engineering process and system for all-weather at-sea wind-energy extraction", USA Patent 5,549,445, awarded 27 August 1996, to harvest the power of the Antarctic Circumpolar Current might also be considered for inclusion by Polar Zone-oriented Macro-Imagineers! By deploying 100 sea-worthy drifting generators, afloat in 10 separate wind-farms encircling Antarctica, a minimum of 5,000 Megawatts may be continually produced according to E.J. Schremp. It is just technically feasible that carbon dioxide gas deemed excessive might be abstracted from the Earth's air—of course, to thwart the alleged impending Anthropic Global Warming catastrophe—if 19,200 Megawatts of electricity can be transmitted by wire to freezers—employing over 446 closed-loop liquid nitrogen refrigerators and dry-ice deposition plants—installed at multiple locations on Antarctica (Agee et al., 2013). [Gaseous nitrogen was first industrially liquefied from the 1870s.] Such a spacially distributed, networked facility could remove one billion metric tonnes of gaseous carbon dioxide annually—a volumetric reduction of 0.5 parts per million—and store that undesired solidified material as a permanent unwanted material within a "landfill" that is well insulated to prevent any dry-ice sublimation. The lowest surface air temperature ever recorded for the Earth was measured at Vostok Station on 21 July 1983 at minus 89.2^0 C (Turner et al., 2009); the freeze-out of gaseous carbon dioxide withdrawn from the air should be expedited by such an energy-saving head-start in gas chilling! To become dry-ice, gaseous carbon dioxide must be chilled to minus 140^0 C. During 1876, the Russian geographer Peter Kropotkin (1842-1921) first suggested that freshwater lakes might underlie vast thick ice-sheets. Deliberate removal of the natural freshwater from such "lakes" beneath Antarctica may become the most practical and suitable sub-surface places, remote from the Earth-atmosphere, in which macro-engineers could inject artificial dry-ice for long-term storage with minimized future leakage possibilities. Confirmed to exist during 1993, by volume sub-surface "Lake Vostok" is the most capacious of all possible sub-glacial rift lake Antarctic landfill sites (estimated volume ranges from 3,800 to 7,000 cubic kilometers). Suppose mad macro-engineers extracted all carbon dioxide gas from Earth's air and converted it to storable dry-ice with a volume of approximately 2,023 cubic kilometers; anthropogenic carbon dioxide "snow" should all fit within the "cavernous" volume of "Lake Vostok" alone!

So far during our 21st Century, Antarctica is the only place in our Earth where national sovereignty is not exercised overtly and where there are as yet no skyscrapers. Antarctica's current geopolitical status is not likely to continue indefinitely, nor is it likely to remain a demilitarized Polar Zone. In other words, soon it probably will in many ways mirror what is happening now in the Arctic Zone (McCannon, 2012). The Antarctic Treaty, which entered legal force during 1961, revised in 1991, outlaws all mining and petroleum exploration in Antarctica until 2041. The Treaty governs the use of the landscape and the Southern Ocean south of the 60th degree of latitude in the Southern Hemisphere. Chile and Argentina are especially desirous in maintaining their "perpetual" territorial claims in Antarctica. At Expo '92—the universal exposition in Seville, Spain—Chile's pavilion housed a 60 metric tonne iceberg shipped from the "Chilean Antarctic". The Norwegian zoologist Nicolai Hanson (1870-1899) was the first human to be buried on the continent (at Cape Adare); the birth of the first person, Emilio Palma at Esperanza Base in January 1978, also made him the first Argentinian to be born on Antarctica whilst the first Chilean was not born there until 1984! Like C.A. Doxiadis, it is possible to envision Chile's Punta Arenas—currently Earth's southernmost major city—being in the same class as the Northern Hemisphere's Oslo, Stockholm and St. Petersburg. The presence of settlements near Antarctica should bring the legal question of Antarctica's ownership to the fore. Technological progress may embolden Chile and Argentina to press their territorial ambitions on Antarctica more vigorously in future. Antarctica's governance by the United Nations Organization seems rather preposterous because the fatal flaw in all proposals for global representational government—an authoritative body responsive to the average opinion of human beings—is that many of our world's more than seven billion living persons are ignorant (about themselves, others of their species and even planet Earth) and are truly provincial!

A hovercraft-based transportation system would supplement the Salterian "Planetran" and, at the same time, could counteract any noosystem governments that are unambiguously greedy and arrogant.

Ironically, the Antarctic Treaty was the first agreement by the immediate post-World War II Superpowers after the Cold War commenced; Article 1 specifies that use of the region "...shall be for peaceful purposes only" and clearly prohibits military bases, the carrying out of military training maneuvers, or weapons tests. [During World War II, William T.R. Fox (1912-1988) neologized the term "Superpower" in *The Super-Powers* (1944 at page 20): "There will be 'world powers' and 'regional powers'. These world powers we shall call 'superpowers' in order to distinguish them from the other powers...whose interests are great in only a single theatre of power conflict." However, the geopolitical concept of Great Powers, at least since the dueling Superpowers-centered Cold War's end, now "...extends well beyond its nineteenth century origins to current business in, for example, the European Union, the United Nations Organization Security Council, and the World Trade Organization" (Brenton, 2013).] In part, the USA approved of the Antarctic Treaty because it reduced fears that the continent might be used by another militarized ecosystem-state as a submarine and bomber-cruise missile staging theatre that could threaten commerce and military convoys, should the Suez Canal and/or Panama Canal become operationally compromised or destroyed. As McCannon (2012) has pointed out, the Arctic Ocean may soon be rendered nearly floe ice-free during summertime, while Antarctica's floe ice expanse will also be different than what is experienced by human traveler and shipping nowadays. A northern Polar Zone that is "without" a floe ice cover throughout most of the year means that naval submarines and civilian commercial surface shipping cannot operate therein with any more secrecy than in the Temperate Zone and Tropical Zone sections of the world-ocean. In other words, were it not lawfully prohibited by the Antarctic Treaty, the southern Polar Zone seasonally retaining all of its usual floe ice could serve as a hideout for stealthy warships, a strategic military stronghold. Unless, of course, the camouflaged warships emit a lot of heat [infrared] or are poorly disguised otherwise!

Global Nature—that is, Earth's biosphere—is constantly being re-built by, in part, the work of *Homo sapiens*; indeed, it might even be said that our species is creating a "Second Global Nature". The troposphere, located below the stratosphere, pulses to the heat emissions of humanity—that notion is ahistorical—and its Anthrome and Astythrome infrastructures, causing the air layer above our heads to restructure in response because of this newly-documented global ambient hot air "fingerprint" of *Homo sapiens* (Santer et al., 2013). If the Earth-biosphere is taken to be the cradle of our species, then we are now rocking that cradle all by ourselves; in other words, we have become so energy-using that mere urban "Heat Islands" have been surpassed—currently humans are generating a bulging "Planetary Heat Layering"! Speculatively, mobile ice-islands composed entirely of an icy anthropogenic material, pykrete, architecturally speaking, might even be considered as ARCHIGRAM group-fantasized plug-in cities (Sadler, 2005) consisting of mobile elements with architecture treated as a consumable or a vehicle—maybe replacing the naturally calved iceberg that shaped the now missing "Bay of Whales" (78° 30 South latitude by 164° West longitude)—which could be snuggled to Antarctica's mainland with hawsers braided of cold-proof artificial fibers tied off to bollards embedded in the mainland's ice; bollard emplacements in ice can be accomplished most elegantly by using, singly or in combination, blasting, melting and mechanical excavation (Colgan and Arenson, 2013). "Pykrete" is a namesake ice-wood pulp material concoction invented during World War II by Geoffrey N.J. Pyke (1893-1948) and proposed as a frozen construction material for ultra-large Allied aircraft-carriers called "Bergships" (Cross, 2012). More than one-half billion persons dwell on our world's many islands, islands which form slightly less than 6% of the planet's landscape. Can a C.A. Doxiadis "Ecumenopolis Ice Island" be anticipated by Macro-imagineers to become extant during the 21st Century? Yes, tentatively. The "...acts and thoughts of human beings are the final ground for judging quality" (Lynch, 1984, page 49). Forensic Macro-engineering stemming from all previously well-recorded macroprojects—completed or otherwise—should assist Macro-imagineers in correctly planning for future manipulations of everything within and, sometimes, exterior to, the Earth-biosphere (ICE, 2013).

Since we are Space Age macro-imagineers/macro-engineers, Earth-orbiting sunlight reflectors could be used to energize an expanding biosphere. It is feasible to bring materials to Earth's surface from inter-planetary space (asteroid mining) and other planet-places. Along with other Solar System objects, Earth's biosphere suffered and survived devastating blows inflicted by a chaotic Universe; undoubtedly, the Earth-biosphere will endure more similar catastrophes. Future comet and asteroid impacts are almost inevasible. Not even currently known types of nuclear explosives, stockpiled by the world's Superpowers and others can meaningfully compare to the mass destruction wreaked by our Solar System's development during its Astronomical Time. Earth's bombardment by comets and asteroids has eroded some of its air and seawater and created gigantic craters in its crust (some on land, some on the world-ocean's bed). Sometimes rich ore deposits are present in these shocked rock indentations. *Homo sapiens*, as Neil Paul Cummins opined in summary in Chapter 1, via quotation, must become proficiently prepared to defend the Earth-biosphere and a future Mars "Worldhouse" from rocky death-dealing Earth and Mars changers, or risk a possible extermination event-process. A damaged biosphere, disrupted or missing societies and markedly slowed advance of Science and Technology also may result from less-than-extermination collisions.

"Der Mensch als geologischer Faktor" an article authored by Ernst Fischer (????-1914) examined systematically various effects that humanity's everyday activities noticeably have on the Earth, including its atmosphere (Fischer, 1915; Steinman, 1915). He estimated that amount of rock and soil moved in mining, the handling of ores and non-metallic materials, and then continued to tally public infrastructure (roads, tunnels, dams, canals and the improvement of stream and river course and shores for navigation). Fischer judged from the obtainable pre-World War I database the volume of earth materials moved at human behest. Nowadays, people shift mechanically about twice as much as global Nature does annually (in the form of solid and dissolved products of landscape erosion by river, wind and glacial ice flows). In other words, in excess of 20 billion metric tonnes of earth materials are estimated moved during 2013. Depending on the economic state of humanity's nearly 200 ecosystem-countries, wars in various regions and other factors, this statistic is probably getting larger at a moderate, if unknown, rate; for other extant organisms, the transfer statistic is probably declining! Diethard E. Meyer (born 1938) thinks humans are effectively changing the "...upper third of the Earth's crust..." (Meyer, 1986, page 177). Ernst Fischer also considered the effects of agriculture in altering the chemical character of the soils and, to a limited extent, of relief, as in terracing and empoldering; then went on to the hydrosphere, starting, interestingly enough, with human effects on marine sediments. Fischer included the influence of all ships that have sunk with their cargoes, the total of which is very appreciable (2014 estimate: 300,000 wrecks). [Sunken ships and cargoes, of course, displace seawater. Earth's ocean masses 1.4 followed by 21 zeros kilograms.] Green environmental alarmists use the so-called "Titanic Analogy"—that is, negative criticism of pro-progress persons because of their Green-assumed mindset of "unsinkability" (Klinkenborg, 2012; Brown et al., 2013). The incomparable maiden voyage of the luxury liner **RMS Titanic** ended unexpectedly and tragically in April 1912 when the ship struck an iceberg in the North Atlantic Ocean and sank. Because of technology's progress, that once-lost ship is now visited by tourists since its 1985 discovery. "Der Mensch als geologischer Faktor" elucidates our species' encroachment upon the world-ocean by diking and reclamations, including coastal preservation by seawalls. Sometimes underwater sea-lane navigational hazards are artificially cured to promote commerce. [In the instance of the January 2012 mainland grounding of the largest Italian cruise-ship, **MS Costa Concordia,** that large vessel has been removed from the offending shoreline to a scrapping yard (Atkinson, 2013). The submerged rock ledge that punctured her moving, over-speeding hull, causing fatal flooding, has been left in place as a kind of scarred monument to fallible human judgment and inaction.] Since 1921, shipping interests had debated removal of Ripple Rock shoal at Seymour Narrows' south entrance in Canada's province of British Columbia; finally at 9:31:02.05 AM on 5 April 1958, that massive natural impediment to safe navigation in the Inside Passage was blown to bits by powerful modern explosives, the history of which is best described by Stephen R. Brown's *A Most Damnable Invention: Dynamite, Nitrates, and the Making of the Modern World* (2005). [The modern industry involved with the preparation of concrete, an Anthropic Rock, "...grew in parallel with the high explosives that made its raw materials affordable" (Kelly, 2006). Californian macro-imagineer Richard B. Cathcart's 115-page 2012 Thinker Media e-book, *Macro-Engineering the North Atlantic to Prevent Another Titanic*, recounts a proposal made by Carroll Livingston Riker (1854-1931) shortly after the 15 April 1912 sinking to install a huge jetty—a narrow backbone of rip-rap—extending 200 kilometers from Newfoundland, Canada into the North Atlantic Ocean for the purpose of diverting the southward flowing Labrador Current and the eastward flowing Gulf Stream, blocking the movement of icebergs pushed by the Labrador Current as well as the amount and density of foggy weather events. Ernst Fischer delved into history's register of swamp drainage and river channelization with the result that freshwater runoff is greatly hastened; lowering the groundwater table by drainage and pumping for farm use; and, changing the soils' texture and chemical composition through mixing of natural topsoil horizons via plowing, as well as destruction of original vegetations. Of course, there exists always the possibility of making new landscapes with materials shifted from the seabed. Coast defense public-works are today expertly built in many countries by The Netherlands' famous macro-engineers.

The year Henri Becquerel discovered radioactivity, 1896, Earth's anthropogenic atmospheric carbon dioxide gas buildup was first made public in an announcement issued by Svante A. Arrhenius (1859-1927). Ernst Fischer had little to say about the effect on of our existence on the air, only some speculation about a possible increase in the Earth-atmosphere's content of carbon dioxide gas and pollution of local air masses resulting from heavy industry's coal combustion. [Earth's major 21st Century air pollution, smog (nitrous oxide), regions are the USA, Europe, China and India. Climate control requires more cooperation amongst the Great Powers than currently is taking place: USA, European Union, China, Japan, Russia, Brazil, India and Canada emit about 70% of our world's anthropogenic carbon dioxide gas.] The retreat of valley glaciers in the European Alps after 1850, and until about 1930, has been tied to the precisely concurrent massive use of burnt coal in an industrializing Europe (Painter, 2013) and it is suspected that a similar soot-caused glacier mass loss occurrence is underway in the Himalayas because of an industrializing mainland Asia. Arrhenius, in **World in the Making** (1908), speculated that humanity's fossil fuel usage could forestall the onset of another Ice Age, like that of the Pleistocene. Living people are today observing the onset of the after-effect of anthropogenic atmosphere carbon dioxide gas buildup—but those specific climatic after-effects are not yet incontrovertibly obvious in Geographos' view, and Geographos does not issue this broad statement as a casual suggestion by an unthinking critic, as one of the so-called "climate change deniers". Fischer briefly explicated the effects on vegetations and animals of humankind's history, citing matters which are familiar to the well-informed person, giving particular attention to *Homo sapiens'* constant redistribution of biota in transporting domestic animals and cultivated plants throughout our biosphere, along with those pests and weeds which have uncontrollably accompanied our species' global migration. Organisms thriving in and on humans, dogs, apes, plants and insects—to name just a few creatures—have also traveled in interplanetary space and survived several years of exposure on the Moon's harsh surface! Molecular nanotechnologists intend taking precautions to prevent runaways from disorganizing Earth's precious biosphere.

Fischer's nearly century-old, but still profoundly useful, 1915 literary survey fits into the sequence of similar discussion since. He cited several items from the UK geoscientist Charles Lyell, but Fischer did not mention George P. Marsh, or Elisee Reclus (1830-1905), who paraphrased Marsh for Europeans. Traditional Natural History's central theme was "Balance of [Global] Nature"—this theme was discarded by Ecology after 1980—and radical Greens still subscribe to this notion-belief, vociferously touting that *Homo sapiens* must "return" to a prescribed geo-concept of Earth-biosphere constancy; radical Greens subscribe to the notion-belief of a perpetual State of Global Nature. In marked contrast, some radical macro-imagineers/macro-engineers subscribe to the notion-belief that they can plan a once-and-for-all-time technological fix! The working definition of "ecosystem" was settled generally circa 1901-1902. In Ernst Fischer's time, the idea of a planetary biophysical "Balance of life" was just beginning to be scientifically documented. Modern ecology's "Balance of Global Nature" propaganda storyline, formed during 1779 by Jan Ingen-Housz (1730-1799) when he discovered photosynthesis and conceptualized a harmonious Earth-biosphere balance of animal respiration and green plant transpiration. Geoscientists are yet to finish that formidable task of finding truth to the complete satisfaction of many legitimate 21 Century biologists.

Until the Space Age, the Anthropocene Earth biosphere was the planetary region of ultimate pollution dissipation and also the region of penultimate energy dissipation. Simple human Noosystems became technologically complex over Geologic Time (only a part of which was recorded Historic Time) as the amount of energy harnessed per capita annually increased, or as the efficiency of use increased—a scheme of thought borrowed from Leslie Alvin White (1900-1975) during World War II (White, 1943). Fischer evinced a strong belief that irreparable damage had been done—up to 1914—to some parts of our species' Earthly biosphere (the Mediterranean Sea Basin, for example, which will be broadly examined and summarized in Chapter 8) and that *Homo sapiens* was negatively affecting the world-ocean's wildlife. As long as some part of our Anthropo-cosmos progresses, the remained will not lapse into a stagnant state that could stall our advancement through economic growth. But, the major ecosystem-nations—those putative present-day [2014] Great Powers—find themselves in a simultaneous global economic slump or boom period of economic state because of "globalization" that is increasingly networked by the expanding and encompassing World Wide Web [Internet]. Doom-laden biosphere forecasts projected worldwide by the global media may prove harmful and certainly could make any bad economic situation worse than it is. Some propagandists, for ideological reasons chiefly, plus the intensity of our geographical studies, created a new scale of Earth-biosphere viewing, a new scale of opinion. C.A. Doxiadis' "Ekistics

Logarithmic Scale" is an operational scale designed for the classification of humankind's settlements, running from the human individual (Unit 1) as the basic, smallest unit of measurement to the whole Earth (Unit 5). Doxiadis' chart could become useful to future terraformers. Visually, these subdivisions are comparable to the entertaining movie (Jones, 2007) *Powers of Ten* (1962) as well as Gerard t' Hooft's *TIME in Powers of Ten* (2013) or even Mariam Thalos' *Without Hierarchy: The Scale of the Universe* (2013), a deeply metaphysical tome of considerable practical usefulness to unfettered Macro-Imagineering!

The ever-enlarging Anthropo-cosmos has replaced "Earth" as our known absolute region of human species containment. And, Terraforming has, in some respects, entered our everyday thinking. Data gathered and divulged during the 20th Century has formed and squarely set a viewing framework for our Solar System's planets and other celestial bodies (moon, asteroids et cetera). [The NASA's Deep Impact probing spacecraft actually blasted a crater into the fluffy surface material of comet Tempel-1 on 4 July 2005.] Future humans may make it much less essential that people procreate to dominate additional, artificial biospheres—in other words, bluntly said, a human "reproductive imperative" (and its co-instantaneous "territorial imperative") is not a constraining commandment.

Should our Earth be made more massive? How might *Homo sapiens* deposit additional materials, applying extra-terrestrially garnered matter to our global Nature? Launchings of fabricated objects into interplanetary and interstellar space must be considered an entirely new kind of erosion process-event, which thereby reduces artificially our cherished world's mass by newsworthy subtraction. Since 1957, spacefaring ecosystem-countries have managed to eject thousands of metric tonnes of materials (both natural and anthropogenic) into outer space; since 1957, for instance, over 5,000,000 kilograms of stuff has been ascended into planetary orbit, probed other celestial bodies (planets, comets et cetera) and successfully propelled entirely out of the Solar System. Meanwhile, small-sized bits of cosmic debris, which naturally falls into the Earth, amounts to over 30,000 kilograms daily! Most of these natural debris pieces reach our ocean and landscape as a fine dust-fall, or like the scary Chelyabinsk, Russia, super-bolide of 15 February 2013, as enormous chunks of descending material. *Homo sapiens* have, so far, been unable or incapable of tipping decisively global Nature's so-called balance in terms of material input/output trends. The NASA's 1977 *Special Publication 413: Space Settlements: A Design Study*, suggested the manufacture of capacious lifting-body non-crewed, remote-controlled flight vehicles to convey to needy Earthlings bulky cargoes of processed ores originating at open-pit mines on the Moon. Such robotic, standardized atmospheric entry vehicle-containers—the outer space version of the standard shipping container use in the maritime and trucking industries—would then be retrieved (from their Southern Hemisphere splashdown sites) and towed by ocean-going powerful super-tugboats to seaport-served industrial complexes near coastlines. The NASA's 1979 *Special Publication 428: Space Resources and Space Settlements* ignored this exciting resources acquisition and delivery macroproject idea! Since there is the real possibility for extracting valuable metals from near-Earth resources such as metalliferous asteroids and bringing the obtained materials to Earth for further industrial processing and use, R.B. Cathcart, A. Bolonkin, V. Badescu and D. Stanciu) elaborated and mathematized the dormant NASA concept by 2013 [in Chapter 22, "Shaped Metal Earth-Delivery Systems", pages 508 to 537 of Viorel Badescu's edited *Asteroids: Prospective Energy and Material Resources* (The Netherlands: Springer)]. During 1980, Cathcart had suggested, following an unpopular Macro-Imagineering proposal by Samuel Herrick (1911-1974), **Figure 8**, in March 1971 the potential un-trammeled excavation function of untransformed parts of the Asteroid

1620 Geographos as a Directed Meteorite Excavators (DMEs). Herrick intended his selected DME to plow an Interoceanic Crater-Canal in northwestern Columbia's Atrato River Valley of South America!

Figure 8. A formal circa 1960s photograph of the imaginative and exceptionally daring American astronomer Samuel Herrick. (Image: Wikipedia Images.)

Using a DME peeled from the still insufficiently defined Asteroid 1620 Geographos, a very rapidly dug new sea-level canal traversing the northern part of South America, to complement, supplement or supplant the recently updated Central America canal located in Panama. Because it crosses the orbit of Mars along with Earth, perhaps a more practical use of Asteroid 1620 Geographos might be found by a coterie of future Mars terraformers! DMEs hits on vacant wastelands (like the most aridic parts of Chad, in northern Africa's

Sahara) would be a cheap means of bringing construction and industrial materials to isolated Noosystems tied to marginal landscapes that are coast-less; locally collected solar energy, especially in the Sahara and its southern bounding slightly-less-arid border zone, the Sahel, could become the main source of transformational power used by humans. Macro-engineers would never plan a simultaneous arrival of Directed Meteorite Excavators such that their dust-generating and dispersed explosive touchdowns would come close to equaling the 100 Megatonnes calculated by Nuclear Winter experts to be a threshold level. "Nuclear Winterists", circa 1982-1983, came to utterly chilling conclusions about general thermo-nuclear warfare (also known as "macro-war"): that post-World War III survivors would witness severe air temperature decreases (as low, perhaps, as minus 25^0 C) in the Northern Hemisphere, and that dwellers in Earth's Southern Hemisphere would likely have to endure sub-freezing temperatures! In other words, Nevel Shute Norway (1899-1960), in his 1957 novel and 1959 film **On the Beach**, got only half the post-World War III story! Geographos foresees that Samuel Herrick and the NASA-style earth-moving will, inevitably, foster Rodoman-ALPS, the Geographos Earth-defense scheme reviewed in Chapter 3; Rodoman-ALPS' functionality merely needs to be slightly augmented. Mining of asteroids and comets that pose a known natural geo-hazard to our Earth-biosphere, sending riches piecemeal to a needy mankind, could be accomplished—in other words, humans might simply reduce the mass of asteroids and comets by mining. Imagine if a swarm of space-faring molecular Nanotechnology miners consumed ("Ecophagy"!) any biosphere-threatening space debris! Geographos adds its "voice" to the chorus of other investigators who suggest that the trajectory in outer space of the Asteroid 1620 Geographos be carefully altered by capitalizing on the Yarkovsky Effect—the radiation thrust due to the anisotropic radiation of heat from a sunlit rotating solid body in space

(Spitale, 2002). The asteroid's deliberately altered course through space could be done most simply by stationing an umbrageous robotic close-orbiting anthropogenic sunshade-monitor space probe.

Although people cannot as yet control the amount of energy that our Sun gives forth, macro-imagineers think that humanity will someday be capable of controlling how much sunlight is captured by Earth. Herman Oberth's 1923 pamphlet *Die Rakete zu den Planetenraum* proposed construction of Earth-orbiting circular sunlight reflectors with diameters of 100 kilometers. Oberth (1894-1989) claimed that focused sunlight from a single mirrored satellite "...could make large...[Northern Hemisphere regions] of the North [Arctic Zone] inhabitable; in the middle latitudes [Temperate Zone] it could prevent sudden drops of [air] temperature...prevent freezing temperatures at night". Such a satellite could melt snow cover, which is complex and short-lived sediment with many effects on the natural and anthropogenic surfaces upon which it comes to rest. He might easily have also claimed that orbital mirrors could replace artificial street lighting reducing electricity use in outdoor lamps. Use of these devices might resemble the trends in the northern Polar Zone toward warmer and unstabale Winters which could both alleviate and exacerbate the current technical challenges for cold climate regime hydropower operations (Gebre et al., 2013). Incautious use of space mirrors at night could cause damage to human eyesight (Laframboise and Chou, 2000). Possibly 220,000 square kilometers of Earth's Northern Hemisphere—mainly the Eastern Mediterranean Sea Basin—were affected by a regionalized landscape-smothering ash fallout resulting from a violent volcanic eruption that occurred about 3,700 years ago (Friedrich et al., 2006). A prehistoric, abrupt colder and dryer-than-previous-period of the immediately preceding time, global climate regime change occurred about 12,900 years ago [the Younger Dryas stadial]. In the New World the Clovis people changed their living habits from mostly big-game hunting to a hunter-gatherer subsistence diet of small-game, roots and berries; in the Old World [Middle East] the Natufians for the first time settled, pursuing agriculture almost exclusively. Why? Perhaps a meteor or comet struck the thick continental ice-sheet covering North America, the Pleistocene Laurentide Ice Sheet (**Figure 9**) then resting upon northeastern Canada's Quebecia Terrane of North America's Grenville Province near the Gulf of Saint Lawrence (Wu et al, 2013). [Geographos here wonders aloud if the proposed bolide impact's fireball could have sublimated so much ice so quickly that the moisture-carrying clouds of the Genesis Flood might be unclearly stating multiple person-made observation(s) eventually described in a primitive, allegorical format many years after the fact of the terrifying and widespread event-process of *The Holy Bible* story.]

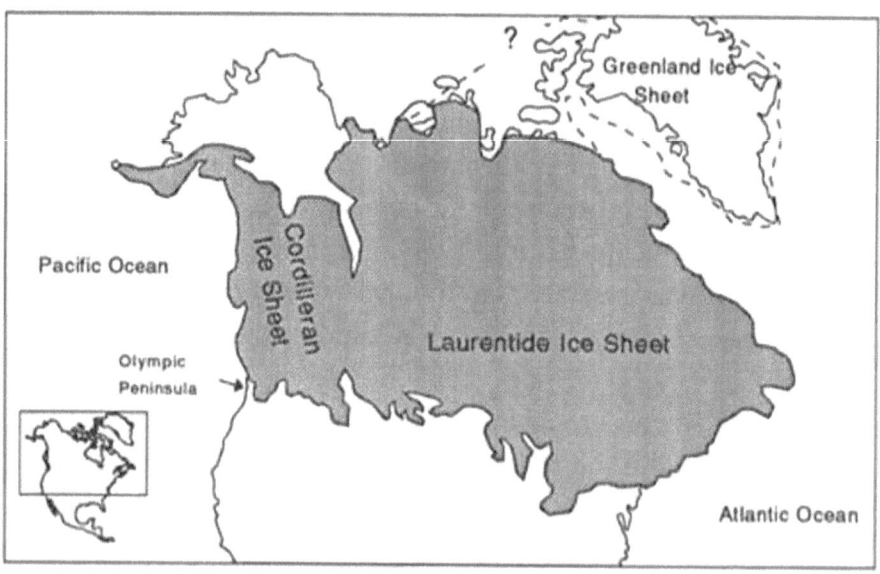

Figure 9. The Laurentide Ice Sheet was possibly two kilometers thick over Quebec, Canada, during the Pleistocene. Receding glacial ice in Greenland has revealed long-ago human settlements. More may become visible and, as well, new ore deposits too. Currently, Greenland has six known major mineral deposits. (Image: Google Images.)

Barring similar—genetically related?—Aliens living in another Solar System or interstellar space Ark somewhere in the Milky Way Galaxy, our unique species has been extending its domain in Earth on both sides of a systematic interface, situated between the crust-ocean-atmosphere) since *Homo sapiens* first became a builder and digger. Crustal excavation is a special aspect of human intra-terrestrial exploration activities and imaginative descriptions of trips to Earth's interior are often loaded with deep psychological implications (Meakin, 1971; Debus, 2006). The mass of the Earth's crust is about 0.71% of the planet's mantle. In International Law, ownership of the land and sea-bottom is held to extend ad inferos. If Earth's sub-crust mass (that is, mantle and core) is a kind of very hot, circulating "fluid", then ownership has no significance below the crust until some amazing future technology becomes available to extract energy or materials (methane gas) from far below. [See: Cathcart, 2006.] In that event-process, humans would be tapping global Nature's basement! Humans can find little geoscience literature dealing with ownership issues of bounded volumes of the world-ocean, except what is presented in the text of the 1982 United Nations Law of the Sea Treaty. "Outer Space" is internationally regulated, but the boundary between air and interplanetary space remains legally confused. Clearly, and international legal regime is needed if Space Elevators, as technically presented by many publications are ever to be constructed to serve Earthlings' purposes (van Pelt, 2009; Howe and Sherwood, 2009; Lee, 2012). Both the USA's NASA and Europe's ESA leaderships see as desirable the certain development of an international strategy to coordinate, monitor and control interactions of spacecraft in Earth-orbit.

SYNCOM II became humanity's first geosynchronous artificial Earth satellite during 1963. A geosynchronous satellite has an orbital period synchronized with our planet's axial rotation. A 266,000 kilometer-long geosynchronous orbit zone is roughly circular, about 35,900 kilometers above the Earth-surface. A carbon nanotube cable-anchored satellite of this type would eliminate the need for fiery rockets taking payloads and personnel into Earth orbit and would also reduce the need for rather spectacular, and potentially fatal, heat-shielded atmospheric re-entries of homeward-bound spacecraft launched from Earth. Future aerospace planes, such as the UK's SKYLON, might become obsolete too. In addition, scientific instruments and tourists could be made immobile at any altitude above Earth's Equator. Geographos pities those future hearty souls who could endure several days of piped melodious Muzak (Lanza, 1994) inside a capsule that is little more interesting architecturally than Seattle, State of Washington, Space Needle (Spector, 2002)! The Base Station for a Space Elevator can be installed on a landscape or on a self-positioned, wave-stabilized floating sea platform. Practitioners of Macro-Imagineering envision land terminals at central Africa, northern South America and sea terminals at various locales atop the world-ocean. Near Jarvis Island (0^0 22 South latitude by 160^0 03 West longitude), **Figure 10**, a large surface Base Station could be kept operational without much difficulty and it could prove to be an ideal tourist and worker quarantine facility since it is landscape, solid ground (Gove et al., 2006). Jarvis Island is central to the Pacific Ocean region often referred to as "The Pacific Rim" (Biribo and Woodroffe, 2013). A formal content-analysis of all issues of the periodical *Zeitschrift fur Geopolitik* from 1924 until 1965 showed that "...the most thought-provoking contribution of the geopoliticians centers on their belief that, in spite of its European origins, the transcending long-range interests of the United States [of America] lie in the Pacific [Ocean] rather than the Atlantic [Ocean] world" (Blumenthal, 1967).

Pacific Remote Islands
Marine National Monument

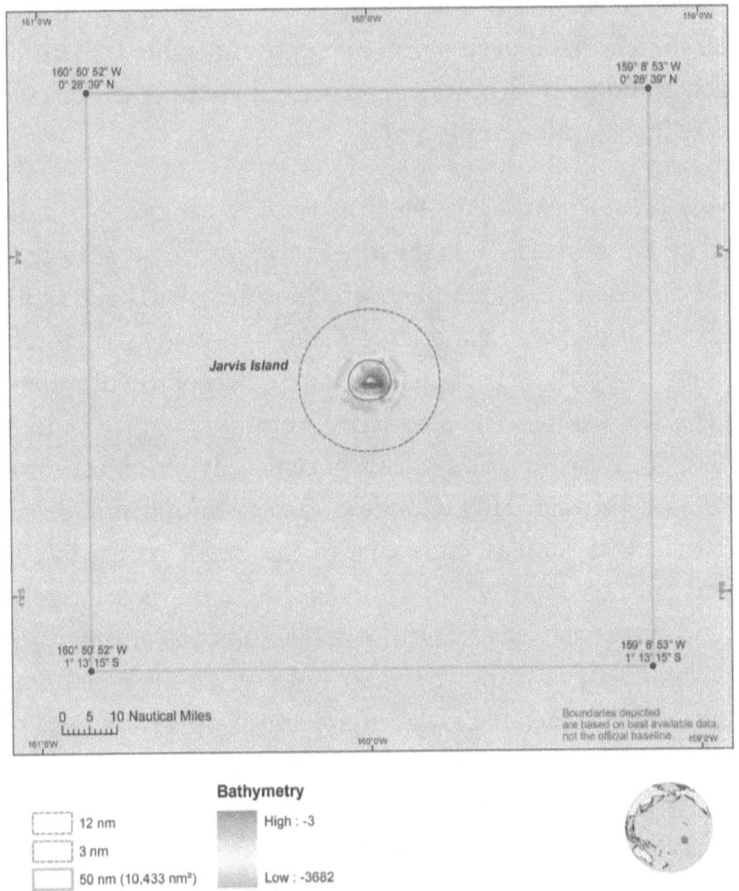

Figure 10. Jarvis Island, a territory that is part of the United States of America. The island is very arid, a climate regime situation caused primarily by its position south of the Equator (Hwang et al., 2013; Dickinson, 2009). Since the US Government ceased print lithographic nautical chart production on 13 April 2014, only map-library archived traditional paper marine charts are available. (Image: Google Images.)

Space Elevators are being made possible by the fact that Science and Technology R&D has brought humans to an Anthropocene moment when humans can design materials with predictable properties. Space Elevators would constitute a new kind of urban development, a vertical rather than horizontal city form (of almost unlimited extensibility) on a much grander physical scale than mere skyscraper clumps such as New York City's Manhattan Island.

In *Voprosy filosofii No. 12* of 1962, the Russian philosopher-biologist G.F. Khil'mi announced: "…in the foreseeable future the surface of the Earth, the atmosphere, hydrosphere and biosphere will be so saturated with technological devices and large-scale Man-made structures that the outer envelope of the Earth will become a new object of reality that will develop according to its distinctive, yet unknown laws". In other words, like Timothy Morton's Anthropocenic "Hyper-object" brought into existence through H.D. Goode's "Geo-evolutionism" discussed in Chapter 3. Was Khil'mi' speaking of the Earth's crust totally restructured by molecular Nanotechnology? Khil'mi' is a foreseer. He warns, and vigorously advocates the activation of supra-mantle/sub-geosynchronous orbit things by means of human-managed dynamic-constructive systems of Earth bio-apparatus operation. Molecular Nanotechnology's future controlled techno-wizardry promises to realize Khil'mi's shell-shaped mantle-enveloping zone a place of only two kinds of nano-machines, the operational and the non-functioning. *Homo sapiens'* immediate planetary spatial context won't be global Nature (the Earth-biosphere), as formerly, but organization of (self-controlled matter). Even persons living could be constantly enduring repair work within their bodies and, therefore, be considered as (120 kilogram) masses of biological material separating nano-sized robots. Earth's future robosphere could be penetrated by Space Elevators and, in addition, be a field of activity for molecular Nanotechnology. Eventually, Geographos suspects, all human industrial production and all human consumption would possess essentially the same ideology—molecular Nanotechnology's programming governed by humankind's artistic sensibilities. Rodoman-ALPS seems to fit into G.F. Khil'mi's concept in that unarmed automatons able to select and destroy their targets with non-explosive inert masses of dense matter (not warheads) using guidance or direct assistance from other spacecraft or ground-based controllers. Rodoman-ALPS will, perhaps, first participate in the clearance from Earth orbit all the non-working satellites and rubbish posing a partly calculable hazard to aerospace planes, crewed space-craft and other useful space probes. Unmanageable space debris poses a significant supra-stratospheric collision threat to extra-atmospheric parts of all Space Elevators projecting above 50 kilometers altitude. The Chelyabinsk incident revealed that, even below 50 kilometers, there is still much danger to living persons and cherished property! Rodoman-ALPS is but one excellent possible Macro-engineering solution to the macro-problem of how to destroy this "cloud" of useless, misplaced junk, which is beginning to enshroud Earthly global Nature already polluted by street-lighting and a huge assortment of chemicals and aerosols. It is fascinating to Geographos that cartographic Earth control facilities, much like those intended for Rodoman-ALPS, were first projected (onto big theater screens) in Hollywood movies (Caquard, 2011)! Very litter-wary spationauts appreciate a clean vacuum during their danger-filled travels in outer space near the Earth!

How might geoscientists examine any proposed Base Station grounding site necessary for our planet's first operational Space Elevators? In the field of seismic studies, computers and remote sensors may allow an ever more; actually visible, imaginary worldscape of seeing (via Virtual Reality) the Earth's crust, in part or whole. Someday there will be very detailed geological mapping from the core of the Earth outward. The center of the Earth is a geographical place (Harrington, 1965). Perhaps the most plausible macroproject to reveal Earth's innards to humanity is via the remote sensing of the sub-surface leading to maps illustrating its deposits, including cities, via neutrino tomography (Millhouse and Latimer, 2013; Witze, 2013; Tang and Winter, 2012); a true determination of our home-world's global density profile may result from such scrupulous scientific examination. The propagation of neutrinos is affected by the presence of Earth matter through neutrino absorption and attenuation. Perhaps such exploration, perhaps with an accelerator/receiver system dubbed "Geotome" with its graphic results displayed via a Jumbotron, will confirm the prediction of Thomas Gold (1920-2004), made most fully in *Power from the Earth* (1987), that voluminous reserves of methane locked in deep crustal strata could fuel our planetary civilization for centuries. Such a fully globalized upper Earth-mantle methane gas energy supply system naturally would make world petroleum and world natural gas pricing/natural gas production no longer assuredly sustainable. If wholesale price determination were wrested from the control of a few ecosystem-nations and a few international corporations and organization, the very raison d'etre for their current existence would be rapidly dissolved. Methane gas extraction, unlike shallower fluid and gas withdrawals from the Earth-crust or surface rock quarrying, is unlikely to cause earthquakes.

Geographos' macro-imagined Geotome conceptual design might be floated and moored anywhere within, submerged, or on the world-ocean and subject to bad weather and rough seas, and like David Noel's freshwater collectors, "covering" a circular region of possibly 700 square kilometers and be tiltable up to 90^0 from the vertical. If a Geotome were located in the Indian Ocean, at 106^0 East longitude by 40^0 45 South latitude, then it could investigate an antipodal place—the point on the exact opposite side of the Earth—such as New York City's famous Times Square. Times Square has multiple Jumbotrons, according to Marshall Berman's *On the Town: One Hundred Years of*

Spectacle in Times Square (2006). Besides "x-raying" Earth, experimental physicists had first predicted in 1979 that the neutrino could be the basis for a revolution in telecommunications in which messages, perhaps even picture signals, could be sent and received via collimated neutrino beams shooting through the whole planet (Uberall et al., 1979). Neutrino beams would not be a human health hazard and could be sent with no loss of power through thousands of kilometers of planetary material (solid, liquid, hot, living, or whatever) mass as though it were non-existent to precisely picked receiver target arrays where the cycle of transubstantiation is completed; a decided advantage over laser beams, microwave beams, or other electromagnetic transmissions. Communications, therefore, could be totally secure and very flexible in routing—certainly more so than fiber-optic cables strung across the landmasses and draped on the variable relief ocean floor. Undersea earthquakes south of Newfoundland, Canada generated huge masses of sliding sea-bottom material in 1929, severing many telegraph cables.

James Graham Ballard's pithy anti-utopian "New Wave" science fictions are, even after his death, still remarkably popular. The spirit of that genre is iconoclastic, nihilistic and pessimistic and his short-story, "Build-up" [**New Worlds** 19: 52-70, January 1957], set three million years in Earth's future Geologic Time, plays with a future exacerbation of our post-World War II macro-problems, such as concretized bureaucracies, uncontrolled urbanization and over-population. Ballard's truly sad tale—"sad" because it is a symbol of disunion between humans and the Divine—centerpieces a single monstrous building, a virtual Megastructural Hive, unaffected by Earth-crust dynamics (tectonic stresses, earthquakes) entirely enclosing our planet, which is many levels high and populated by multitudinous humans. Ballard's storied building, which Geographos dubs "Ballardia", covers not only all landscapes—that is, 29% of Earth's surface—but the world-ocean too. Conceptually, it markedly outdoes Richard L.S.

Taylor's proposed "Mars Worldhouse". Ballard's anthropogenic material mega-formation—a term borrowed from Geoscience—is entirely devoid of geotechnical macro-problems. Apparently, people first entered Ballardia to escape the non-functioning eco-systemic mess they had made of global Nature! Occupants of this unitary building would no longer experience the 0.5% personal weight change present-day persons "feel" as they move about from place to place on our planet's surface owing to the different values of gravity; agoraphobic humans housed within Ballardia would cease experiencing the 0.3% weight change Terrans feel which is caused by moving air masses that are weather. The ultimate geopolis, Ballard gives new meaning to the ancient Greek phrase, "he oikomene ge" ["the inhabited earth']! Were J.G. Ballard's immense multi-layered apocalyptic skyscraper extant, a special Papal Benediction—"urbi et orbi" ["to the city and the world"]—might require another title to reduce popular confusion. Manhattan Island, a 41 cubic kilometer section of New York City, for 2014 three-dimensional property taxation purposes, is valued at about USD1,000,000,000; if macro-economists assume that Manhattan Island is a very small fraction—it would be in the billionths as a property parcel—of Ballardia, then a built, monetized, actualized Ballardia would be entirely urbanized real estate grossly valued at USA$50,000,000,000,000,000,000,000,000. James Graham Ballard's future humans have become trapped within themselves, inside a stark cellular worldscape, an unhealthy state of affairs. The inhabitants and combatants of his huge tenement become entirely ignorant of the existence of what remained of a poetized global Nature technologically un-interfered with by humans. Like all people without their feet on the ground, Ballard's unhappy populace lacks commonsense. Ballardia might be named the equivalent of a womb except that it is the walls, floors, ceilings and roof that really matters in his presumptuous, far-from-junior-grade myth.

Any building that excludes Earth's global Nature enlarges the many nameless persons housed within and emphasizes that the very smallness of the living space allotted to each peripatetic human is the ultimate guarantor of security, a completely governable and manageable "bio-apparatus", a multi-storey graveyard or tomb! In essence, Ballardia is equivalent to the Club of Rome's idea of our Earth-world as a single finite geo-system, which it most certainly is not. In his narration, J.G. Ballard, who spent much of his early youth incarcerated in Japanese-run wartime prison camp—goes beyond mere lampooning; Ballard shocks his readers with a potential reality. Ballardia should be seen as a second choice in a logical dichotomy of technological methods for maintaining our Earth-biosphere: continuing our futile, costly labors to artificially maintain a complex genetic biodiversity or, in future, devising a single technological fix via a huge dwelling's construction (Cathcart, 2002; Cathcart et al., 2006). Ballardia's timid people never visit its roof and, as a consequence, they are incognizant of the Solar System's existence. [Rather worse than today's Astythrome populations unable to see the Milky Way Galaxy at night because of artificial light pollution!] Ballardia's basement would have to be a "15" on C.A. Doxiadis' "Ekistics Logarithmic Scale"! A cowed population's total lack of data and framework about Ballardia's sub-basement level—the Earth's mantle and core—astounds the science-fictionist novelist's enraptured readers. All future construction, truly in a Macro-engineering matrix, ever attempted would use only recycled materials except if molecular Nanotechnology permitted reconstitution of matter caused by stacked nano-robots since even unstoppable natural space debris (dust et cetera) is naturally added to Ballardia without the knowledge or understanding of Ballardia's inhabitants. Such nano-robots exceed the construction uses of toy LEGOs, both in the artistic and scientific realms (Herman, 2012; Robertson, 2013; Doyle, 2013). **Figure 11**. Interestingly, a highway overpass located at

Schwesterstrasse, Wuppertal, Germany was painted during 2011 to look like a LEGO bridge (Herman, 2013, page234-236)!

Figure 11. An assortment of LEGO build-it-yourself plastic bricks. Children as well as some adults develop a pride of creation by using these simply connected objects. During February 2014, Hollywood film-maker Warner Brothers released a three-dimensional computer-animated, heartless, February 2, 2014-released movie, *The LEGO Movie*, featuring the LEGO mini-figure adventures of construction worker "Emmet" as well as a plethora of product placements, making the film a TV commercial for toys! (Image: Google Images.)

Unlike the urban fabric of extant large cities such as Manhattan Island with its canyon-like streets, Ballardia has not such outdoor activity spaces and so the sky is never viewable. J.G. Ballard's lethargic fictional people are unaware they are residing inside anything, the building has no discernible architecture, merely interior decoration, and their geopolitics would take on the character of domestic relations. Unprotected from colliding asteroids and comets, Ballardia is a bunker, a homeland for passive persons. Lewis Mumford (1895-1990) warned that Manhattan Island would become so crowded with buildings that Architecture would someday cease to matter. In *The Architecture of the Well-Tempered Environment*, Peter Reyner Banham (1922-1988) expressed his projection that global climate control would eliminate the need for massive buildings (Langevin, 2011).

Better than any other imagery that Geographos can recall, Ballardia symbolizes R. Buckminster Fuller's "Spaceship Earth" from the 1960s. Buildings function as second skins, protecting people from weather, storing food, and regulating internal body temperature. Because of buildings, *Homo sapiens* became a super-organism, capable of adapting to conditions of other planets such as Mars. Wyville Thomson (1830-1882), author of *The Depths of the Sea* (1873), is the most successful claimant to the honor of being the first published writer of a modern oceanographic textbook. Who will be the last?

Cargoes can be containerized for transport. Can a planet be "containerized"? That is the key question Viorel Badescu and R.B. Cathcart, both members of the prestigious Candida Oancea Institute (Bucharest, Romania), answered affirmatively! Earth's human 2014 human population exceeds seven billions, well more than twice what is was during 1964 when the UK physicist John Heaver Fremlin (1913-1995) speculated on the ultimate technical limits to human population increase in *New Scientist*, a UK-published weekly Science and Technology round-up periodical. On 29 October 1964, Fremlin asserted public-ly his back-of-the-envelope calculation for his figment of imagina-tion—perhaps best termed a "phantasmagoria"—that proved that, in less than1,000 years, 20,000,000 times as many persons could be alive than were alive in 1964. He further stated that all people alive near the end of the Third Millennia would be adequately fed and housed. Re-examining Fremlin's preliminary mathematics, Badescu and Cathcart (2006) were intrigued that J.H. Fremlin was, in most particulars, essen-tially correct: the "...maximum world population ranges from 1,600 to 4,000 millions for various cases detailed". Humans of the far-future may dwell inside a 2,000 storey eco-house obscuring the entire Earth-surface. As a consequence, Earth will give off an entirely anthropogen-ic radiation signature owing to civilization's encasement. Aliens with SETI programs will take notice! This building, as yet unnamed, would be a true solar-powered machine for living. Interestingly, Badescu and Cathcart suspect that J.H. Fremlin may have been influenced to ad-vance his big architectural idea by the 1957 science-fiction short-story, "Build-up", penned by James Graham Ballard!

Advocates of "ecological foot-printing"—that is, the calculation of carrying capacity to provide Earthlings with a near-mandatory final target population—usually exclude the world-ocean and few such ad-vocates even bother to guess or estimate what global Nature's toler-ance for *Homo sapiens* might actually be (Valero et al., 2011). However, most settle on "about two billions" semi-naturists. Still, Macro-Imagineering/Macro-Engineering professionals like Cesare Marchetti see a far bigger total of well-dressed humans; Marchetti says "...from a technological point of view, a trillion people can live beautifully...[in] the Earth, for an unlimited time and without exhausting any prima-ry resources and without overloading the environment" (Marchetti, 1979). Chapter 4 has dealt only with macroprojects of such geographi-cal grand-scale, and usually promoted by the leading personalities of

Macro-Imagineering/Macro-Engineering, that they would encompass our planet and nearly every living person therein. Chapter 5 hereafter focuses on local societal core macroprojects here in Earth's biosphere as well as elsewhere our Solar System. It does not yet appear as if humans will have to compete, in this Solar System, with pestilent or benign Aliens.

References Cited

Agee, E. et al. (2013) *CO2 Snow Deposition in Antarctica to Curtail Anthropogenic Global Warming.* Journal of Applied Meteorology and Climatology 52: 281-288.

Allen, T. (2011) *Virtual Water: Tackling the Threat to Our Planet's Most Precious Resource.* (The Netherlands: I.B. Tauris) 384 pages.

Anon. (May-June 2010) *The "Greening" of Antarctica.* The Futurist 44: 2.

Ananthaswamy, A. (20 July 2013) *Making Waves.* New Scientist 219: 34-37.

Atkinson, K. (December 2013) *Salvaging Costa Concordia.* Ships Monthly 49: 18-21.

Badescu, V. and Cathcart, R.B. (2006) *Environmental thermodynamic limitations on global human population.* International Journal of Global Energy Issues 25: 129-140.

Badescu et al. (2013) *Macro-engineering Australia's Lake Eyre with imported seawater.* International Journal of Environment and Sustainable Development 12: 264-284.

Biribo, N. and Woodroffe, C.D. (2013) *Historical area and shoreline change of reef islands around Tarawa Atoll, Kiribati.* Sustainability Science 8: 345-362.

Blumenthal, H. (June 1967) *The Image of the United States in the Zeitschrift fur Geopolitik.* Southwestern Social Science Quarterly 48: 44-52.

Bolonkin, A.A. and Cathcart, R.B. (2008) *Antarctica: a southern hemisphere wind power station?* International Journal of Global Environmental Issues 8: 262-273.

Bordi, I. et al. (2012) *On the climate response to zero ozone.* Theoretical Applied Climatology 109: 253-259.

Brenton, A. (2013) *"Great Powers" in climate politics.* Climate Policy 13: 541-546.

Brown, S. et al. (2013) *Titanic: Consuming the Myths and Meanings of an Ambiguous Brand.* Journal of Consumer Research 40: 1-21.

Brunner, B. et al. (2013) *Nitrogen isotope effects induced by anammox bacteria.* Proceedings of the National Academy of Sciences 110: 18994-18999.

Brunt, K.M. et al. (2011) *Antarctic ice-shelf calving triggered by the Honshu (Japan) earthquake and tsunami, March 2011.* Journal of Glaciology 57: 785-788.

Cathcart, R.B. (October 1980) *How a Global Warming Could Change the Geographical Future.* The Futurist 14: 28.

Cathcart, R.B. (1997) *Greenhouse atmospherics: Mega-death or Macro-engineering?* Speculations in Science and Technology 20: 17-20.

Cathcart, R.B. (2002) *Unnatural Envelopment: Fieldwork on the Active Tectonics of J.G. Ballard's "Build-up" (1957).* Journal of Geoscience Education 50: 176-181.

Cathcart, R.B. (2006) *Extreme Climate-Control MembraneStructures: Nth Degree Macro-Engineering.* Chapter 9, pages 151-174 **IN** Badescu,

V., Cathcart, R.G. and Schuiling, R.D. (Eds.) *Macro-Engineering: A Challenge for the Future*. (The Netherlands: Springer) 316 pages.

Cathcart, R.B. (25 April 2005) *Nautical jugs or not?* Current Science 88: 1211-1212.

Cathcart, R.B. (2007) *Tapping Earth's upper-mantle methane gas resource at an Independent Nuclear Drilling Initiative Area (INDIA) Sited in Palk Bay, India/Sri Lanka*. Current Science 92: 729-732.

Cathcart, R.B. (2011) *Freshwater Supplies Necklace Super-Project: Floating Bags and Rolling Freshwater Tires Facilitating Future India-China-Bangladesh Life Necessities Trade*, Chapter 87, pages 1531-1540, **IN** Brunn, S.D. (2011) *Engineering the Earth: The Impacts of Megaengineering Projects*. (The Netherlands: Springer) 2266 pages.

Caquard, S. (November 2011) *Cartographies of Fictional Worlds: Conclusive Remarks*. Cartographic Journal 48: 224-225.

Chatzivasileiadis, S. et al. (2013) *The Global Grid*. Renewable Energy 57: 372-383.

Cohen, J.J. (Ed.) (2013) *Prismatic Ecology: Ecotheory Beyond Green*. (Minneapolis: University of Minnesota Press), 349 pages.

Colgan, W. and Arenson, L.U. (January 2013) *Open Pit Glacier Excavation: A Brief Review*. Journal of Cold Regions Engineering. DOI: 10.1061/(ASCE) CR.1943-5495.0000057.

Convey, P. (2011) *Antarctic terrestrial biodiversity in a changing world*. Polar Biology 34: 1629-1641.

Crescenti, U. and Mariani, L. (2010) *Carbon dioxide and global temperatures: A causal and historical perspective.* Italian Journal of Engineering Geology and Environment 2: 51-62.

Cross, L.D. (2012) *Code Name HABBAKUK: A Secret Ship Made of Ice.* (Victoria, Canada: Heritage House)137 pages.

Dalziel, I.W.D. et al. (2013) *A potential barrier to deep Antarctic circumpolar flow until the late Miocene?* Geology 41: 947-950.

Davidson, F.P. (1989) *A School of Engineering and Diplomacy.* Interdisciplinary Science Reviews 13: 9-11.

Debus, A.A. (November 2006) *Re-Framing the Science in Jules Verne's Journey to the Center of the Earth.* Science Fiction Studies 33: 405-420.

De Jong, R. et al. (2012) *Trend changes in global greening and browning: contribution of short-term trends to longer-term change.* Global Change Biology 18: 642-655.

Dekker, P.M. (1995) *The GENI Model: The Interconnection of Global Power Resources to Obtain an Optimal Global Sustainable Energy Solution.* Simulation 64: 244-252.

Dickinson, W.R. (March 2009) *Pacific Atoll Living: How Long Already and Until When?* GSA Today 19: 4-10

Doxiadis, C.A. (1974) *Ecumenopolis: The Inevitable City of the Future.* Page 345.

Doyle, M. (2013) *Beautiful LEGO.* (USA: No Starch Press) 280 pages.

Ebbesmeyer, C. and Scigliano, E. (2009) *Flotsametrics and the Floating World*. (New York: HarperCollins) 286 pages.

Edwards, L.K. (1965) *High-Speed Tube Transportation*. Scientific American 213: 30-40.

Fasullo, J.T. et al. (2013) *Australia's Unique influence on global sea level in 2010-2011*. Geophysical Research Letters 40: 4368-4373.

Firing, Y.L. et al. (2011) *Vertical structure and transport of the Antarctic Circumpolar Current in Drake Passage from direct velocity observations*. Journal of Geophysical Research 116: C08015.

Fischer, E. (1915) *Der Mensch als geologischer Faktor*. Zeitschrift der Deutschen Geologischen Gesellschaft 67: 106-148.

Forgacs, R.L. (1973) *Evacuated Tube Vehicles Versus Jet Aircraft for High-Speed Transportation*. Proceeding of the IEEE 61: 604-616.

Friedmann, J. (awarded 3 October 2006) *Aqua-Terra Planetary Transport System and Development Pneumatic and Electro-Magnetic Underwater Tube-Link Transportation System*. US Patent 7,114,882 B1.

Friedrich, W.L. et al. (2006) *Santorini Eruption Radiocarbon Dated to 1627-1660 B.C.* Science 312: 548.

Fuchs, V. (January 1966) *Hovercraft in Polar Regions*. Polar Record 13: 3-5.

Fyfe, J.C. and Saenko, O.A. (2005) *Human-Induced Change in the Antarctic Circumpolar Current*. Journal of Climate 18: 3068-3073.

Fyfe, J.C. et al. (2013) *Overestimated global warming over the past 20 years*. Nature Climate Change 3: 767-769.

Gebre, S. et al. (2013) *Review of Ice Effects on Hydropower Systems.* Journal of Cold Regions Engineering 27: 196-222.

Gove, J.M. et al. (2006) *Temporal variability of current-driven upwelling at Jarvis Island.* Journal of Geophysical Research 111: C12011.

Harrington, H.J. (October 1965) *Space, Things, Time and Events— An Essay on Stratigraphy.* Bulletin of the American Association of Petroleum Geologists 49: 1606.

Herman, S. (2012) *A Million Little Bricks: The Unofficial Illustrated History of the LEGO phenomenon.* (New York: Skyhorse Publishing) 303 pages.

Herman, S. (2013) *EXTREME BRICKS: Spectacular, Record-Breaking, and Astounding LEGO Projects from Around the World.* (Delaware: Skyhorse Publishging) 242 pages.

Hoekstra, A.Y. (2013) *The Water Footprint of Modern Consumer Society.* (New York: Routledge) 224 pages.

Higuichi, K. (1971) *A Possibility of Constructing a Dam to Change the General Oceanic Circulation.* Collected Papers on Science of Atmosphere and Hydrosphere 9: 355-363.

Hohn, D. (2011) *Moby-Duck.* (New York: Viking) 402 pages.

Howe, A.S. and Sherwood, B. (Eds.) (2009) *Out of This World: The New Field of Space Architecture.* (USA: American Institute of Aeronautics & Astronautics) 400 pages.

Hutchinson, D.K. et al. (2013) *Interhemispheric asymmetry in transient global warming: The role of Drake Passage.* Geophysical Research Letters 40: 107.

Hwang, Y-T. et al. (2013) *A new look at the double ITCZ problem: Connections to cloud bias over the Southern Ocean.* Proceedings of the National Academy of Sciences 110: 4935-4940.

ICE (2013) *Forensic Engineering: Informing the Future with Lessons from the Past.* (London: ICE) 470 pages.

Jones, M.G. et at. (2007) *Understanding Scale: Powers of Ten.* Journal of Science Education and Technology 16: 191-202.

Karl, D.M. (January 2014) *Solar energy capture and transformation in the sea.* ELEMENTA: Science of the Anthropocene 2: 1-6.

Kay, L. (2013) *Technological Innovation and Prize Incentives.* (New York: Elgar) 256 pages.

Kelly, J. (Summer 2006) *Big Bang: The Deadly Business of Inventing the Modern Explosives Industry.* Invention & Technology 22: 40-51.

Klinkenborg, R. (April 2012) *Titanic Repercussions.* American History. Pages 34-41.

Koepple, G. (2009) *Bond of Union: Building the Erie Canal and the American Empire.* (New York: Da Capo Press) 464 pages.

Kokusho, T. et al., (March 2013) *Sailing Solar-Cell Raft Project and Weather and Marine Conditions in Low-Latitude Pacific Ocean.* ASCE Journal of Energy Engineering 139: 2-7.

Laframboise, J.G. and Chou, B.R. (December 2000) *Space Mirror Experiments: A Potential Threat to Human Eyes.* Journal of the Royal Astronomical Society of Canada 94: 237-240.

Langevin, J. (2011) *Reyner Banham: In Search of an Imaginable, Invisible Architecture.* Architectural Theory Review 16: 2-21.

Lanza, J. (1994) *Elevator Music: A Surreal History of Muzak, Easy-Listening, and Other Moodsong.* (NY: St. Martins Press) 280 pages.

Leane, E. (2005) *Locating the Thing: The Antarctic as Alien Space in John W. Campbell's "Who Goes There?".* Science Fiction Studies 32: 225-239.

Lee, R. (2012) *Law and Regulation of Commercial Mining of Minerals in Outer Space.* (The Netherlands: Springer) 365 pages.

Li, K-F. et al. (2009) *Atmospheric pressure as a natural climate regulator for a terrestrial plant with a biosphere.* Proceeding of the National Academy of Sciences 106: 9576-9579.

Lipzig, N.P.M.V. (2004) *The Near-Surface Wind Field over the Antarctic Continent.* International Journal of Climatology 24: 1973-1982.

Lynch, K. (1984) *Good City Form* (Cambridge: MIT Press).

MacDonald, G.K. (2013) *Eating on an interconnected planet.* Environmental Research Letters 8: 021002.

Maeda, H. (2003) *Ultra-Ultra Large Scale Floating Structures for the Stable Energy Supply.* Proceedings of the International Symposium Ocean Space Utilization Technology, pages 387-389.

Martinson, D.G. (June 2012) *Antarctic circumpolar current's role in the Antarctic ice system: An overview.* Palaeogeography, Palaeoclimatology, Palaeoecology 335-336: 71-71.

Marchetti, C. (December 1979) 10^{12}: A Check on the Earth-Carrying Capacity for Man. Energy 4: 1107-1117,

McCannon, J. (2012) A History of the Arctic: Nature, Exploration and Exploitation. (London: Reaktion Books) 349 pages.

McGreevy, P.V. (2009) Stairway to Empire: Lockport, the Erie Canal, and the Shaping of America. (New York: State University of New York Press) 309 pages.

Meakin, D. (April 1971) Jules Verne's Alchemical Journey Short-Circuited. French Studies: A Quarterly Review XLV: 152-165.

Meredith, M.P. et al. (2011) Sustained Monitoring of the Southern Ocean at Drake Passage: Past Achievements and Future Priorities. Reviews of Geophysics 49: RG4005.

Meyer, D.E. (January 1986) Mass displacement by mineral exploitation and it impact on the geologic environment. Zeitschrift der Deutschen Geologischen Gesellschaft 137: 177-193.

Millhouse, M.A. and Latimer, D.C. (2013) Neutrino Tomography. American Journal of Physics 81: 646.

Mirza, M.M.Q. et al. (2008) Interlinking of Rivers in India: Issues and Concerns. (London: CRC Press)298 pages.

Moody, S. (2006) Washed Up: The Curious Journeys of FLOTSAM and JETSAM. (Seattle: Sasquatch Books) 232 pages.

Mulder, E.F.J. de et al. (2014) Underground Cities, Chapter 3, pages 25-32 IN Kraas, F. et al. (Eds.) Megacities: Our Global Urban Future. (The Netherlands: Springer).

Murtinho, F. et al. (2013) *Water Scarcity in the Andes: A Comparison of Local Perceptions and Observed Climate, Land Use and Socioeconomic Changes.* Human Ecology 41: 667-681.

Noel, D. (June 1980) *Fresh Water from the Sea.* Speculations in Science and Technology 3: 222-223.

Oster, D. et al. (2011) *Evacuated tube transport technologies (ET3): a maximum value global transportation network for passengers and cargo.* Journal of Modern Transportation 19: 42-50.

Painter, T. (2013) *End of the Little Ice Age in the Alps forced by industrial black carbon.* Proceedings of the National Academy of Sciences 110: 15216-15221.

Passerini, P. (1984) *The Ascent of the Anthropostrome: A Point of View on the Man-Made Environment.* Environmental Geology and Water Science 6: 211-221.

Post, V.E.A. et al. (2013) *Offshore fresh groundwater reserves as a global phenomenon.* Nature 504: 71-78.

Pyne, S.J. (2007) *The extraterrestrial Earth: Antarctica as analogue for space exploration.* Space Policy 23: 147-149.

Qian, W-H. and Lu, B. (2010) *Periodic oscillations in millennial global-men temperature and their causes.* Chinese Science Bulletin 55: 4052-4057.

Robertson, D.C. (2013) *Brick by Brick: How LEGO rewrote the rules of innovation and conquered the global toy industry.* (New York: Crown Publishing) 305 pages.

Ross, M. et al., (2009) *Limits on the Space Launch Market Related to Stratospheric Ozone Depletion.* Astropolitics 7: 50-82.

Sadler, S. (2005) *ARCHIGRAM: Architecture Without Architecture.* (Cambridge: MIT Press) 242 pages.

Salter, R.M. (February 1978) *Trans-Planetary Subway Systems—A Burgeoning Capability.* RAND P-6092. 35 pages.

Santer, B.D. et al. (2013) *Human and natural influences on the changing thermal structure of the atmosphere.* Proceedings of the National Academy of Sciences 110: 17235-17240.

Sawyer, J.E. (1952) *Entrepreneurial Error and Economic Growth.* Explorations in Entrepreneurial History 4: 199-204.

Scambos, T. (2005) *On the current location of the Byrd "Snow Cruiser" and other artifacts from Little America I, II, III and Framheim.* Polar Geography 29: 252-267.

Scholz, F. et al. (2014) *Beyond the Black Sea Paradigm: The sedimentary fingerprint of an open-marine iron shuttle.* Geochimica et Cosmochimica Acta 127: 268-380.

Schwendtner, B. et al. (2013) *Risk evolution: how can changes in the built environment influence the potential loss of natural hazards?* Natural Hazards and Earth System Science 13: 2195-2207.

Sherwood, S.C. and Huber, M. (2010) *An adaptability limit to climate change due to heat stress.* Proceedings of the National Academy of Science 107: 9552-9555.

Solomon, S. et al. (2007) *Contrasts between Antarctic and Arctic ozone depletion.* Proceedings of the National Academy of Sciences 104: 445-449.

Spector, R. (2002) *Space Needle: Symbol of Seattle.* (UK: Documentary Media Llc) 96 pages.

Spitale, J.N. (2002) *Asteroid Hazard Mitigation Using the Yarkovsky Effect.* Science 296: 77.

Steinman, ? (15 December 1915) *Ernst Fischer.* Geologische Rundschau 6: 325.

Stephens, B. (25 October 2011) *How Many Nukes Does China Have?* The Wall Street Journal CCLVIII: A17.

Sun, Y. et al., (2012) *Lethally Hot Temperatures During the Early Triassic Greenhouse.* Science 338: 366-370.

Tang, J. and Winter, W. (February 2012) *Requirements for a new detector at the South Pole receiving an accelerator neutrino beam.* Journal of High Energy Physics, Volume 2012, Issue 2, Article 28.

Tenenbaum, E.S. (Fall 1990) *A World Park in Antarctica: The Common Heritage of Mankind.* Virginia Environmental Law 10: 109-136.

Torgersen, T. (2006) *Observatories, think tanks, and community models in the hydrologic and environmental sciences: How does it affect me?* Water Resources Research 42: W06301.

Tournadre, J. et al. (2012) *Antarctic Icebergs Distribution, 2002-2010.* Journal of Geophysical Research 117: C05004.

Uberall, H. et al. (1979) *Neutrino Beams: A New Concept in Telecommunications.* Journal of the Washington Academy of Sciences 69: 48-54.

Valero, A. et al. (2011) *The crepuscular planet: A model for the exhausted atmosphere and hydrosphere.* Energy 36: 3745-3753.

Van Pelt, M. (2009) *Space Tethers and Space Elevators.* (The Netherlands: Copernicus) 225 pages.

Watson, J.E.M. et al. (2013) *Mapping vulnerability and conservation adaptation strategies under climate change.* Nature Climate Change. DOI:10.1038/nclimate2007.

White, L.A. (1943) *Energy and the Evolution of Culture.* American Anthropologist 45: 335-356.

Witze, A. (2013) *Detectors zero in on Earth's heat.* Nature 496: 17.

Wu, Y. et al. (2013) *Origin and provenance of spherules and magnetic grains at the Younger Dryas boundary.* Proceedings of the National Academy of Sciences 110: E3557-E3567.

CHAPTER 5

21ˢᵀ CENTURY MULTI-PLANETARY REFORMATION

Our pre-human and human ancestors freed our species from Earth's presumed slow pace of biological change, surviving by gaining control of ever-larger units of an Earthly global Nature. *Homo sapiens* needed technology as a shield between itself and every single bit of our biosphere—an Earth-biosphere that is ever more frequently described in untold computer bytes resulting in a globalized "datascape" and regionalized "datascapes". Humanity's technological protections formed by Science may be underpinned by religions and society-churning geopolitical ideologies. Science seeks to manipulate Earthly global Nature's forces according to our will in the interests of species comfort, health, longevity and full satisfaction of inborn curiosity. 21ˢᵗ Century humans perform in a Noosphere that ranges from Earth's geosynchronous orbit to a maximum depth in the crust of less than 15 kilometers, so that living contents of the human-expanded Earth-biosphere operate through a vertical cross-section almost 35,915 kilometers thick. [Until Voyager-1 ceases to report, we could consider human influence as ever-extending our Noosphere and when it becomes permanently inoperative, becoming "trash", it will then commence to demarcate our "sphere of influence" as a relic with only minimal human civilization context appreciable to Alien finders; New Horizons 2015 fly-by mission to Pluto eventually will exit our Solar System. It was loaded with a container for the ashes of Clyde Tombaugh, the astronomer who discovered Pluto in 1930.] Human alterations of rainfall patterns that finally impact the landscape, as well as the consequent landscape erosions and depositions due to precipitation, ultimately influences all the tectonic plates atop the Earth's mantle and our planet's core (Pysklywee, 2006). [Such would also be the case, probably, for terraformation, even on planets without known tectonic plate geologies. For example, the artificial creation of a modern ocean on Mars could cause massive readjustments of that planet's crust. Terraforming of Venus, for instance, by solidifying most of its oppressive carbon dioxide gas atmosphere also would cause similar crustal landscape shifting.] Our world's Earth-biosphere, practically speaking, can only be enlarged by *Homo sapiens*' downward extension management capabilities. So far, humans have not penetrated deeper than jet airliners fly above us! In all Earth's existence, a single species has achieved global dominance of its biosphere, creating a world from a given globe for the first time in our Solar System's Astrologic Time. As of 2014, Astrologic Time [also known as Astronomical Time] runs concurrently with Earth's Geologic Time.

During the late-20th Century, however, geographers came to realize that technology, by itself, was a reorganizer of the Earth-biosphere's landscapes: "...it is not premature to investigate the idea that the habitable area of the world itself may be in the process of being reduced quantitatively and qualitatively through impact of modern technology. Machine space, or territory devoted primarily to the use of machines, shall be so designated when machines have priority over people in the use of territory" (Horvath, 1974). Ronald Horvath's quoted comment converges on North America's urban carnage created by automobiles and heavy-duty trucks, not on future autonomous machines and the potential LEGO-like creations of future perfected molecular Nanotechnology, namely a plethora of nano-robots. During 2013, a new professional journal started publication, **Unmanned Systems,** to explicate the advancements being made in the field of unscrewed vehicles capable of performing missions autonomously in the air, on landscapes, underwater, anywhere in outer space and even inside conventional buildings and unconventional structures. Another, previously referenced, geographer, William Bunge, in 1988, offered the extreme admonition that Technology and Science's "progress" is machinekind-evolving, not mankind-evolving. Will there ever come a time when robots replace people? Humans do have one important—perhaps a great advantage—over all extant and future robots: while humans are liberated by biological "Evolution" and technical "Progress", robots are unrestrained by "Evolution" and responding only to technological "Progress".

Amongst structure-oriented macro-engineers specifically, it is an old drollery to wish for a wonderful construction material called "Unobtainium", a material that is weightless, costs nothing, and is capable of shaping by childish technicians. It might, just conceivably, exist but, of course, it is unprocurable, except in maligned and satirized Hollywood science-fiction films such as **The Core** (2003) where it was invoked to allow a manned vehicle to travel in the Earth's hot, molten mantle-material. "Unobtainium" is also an excuse for non-delivery. However, 21st Century molecular Nanotechnology promises *Homo sapiens* the capability to put together molecules atom by atom—"Unobtainium" could, indeed, one day be manufactured! In other words, professionals of all scientific persuasions must learn to appraise, predict and direct all future planet-based macroprojects attempted by *Homo sapiens*-made nano-robots—macroprojects reorganizing Earth and other suitable planets and celestial bodies within our Solar

System. Even if large-scale human migration to unterraformed planets and moons never takes place, it could well be that Mars and Venus are converted from people-less orbs into phytotrons feeding food, fiber, other substances and energy to Earth's life and its robotized regions as implied by Horvath so many years ago. Such macroprojects might justify humankind's (and/or mindkind's) pharaonic investments of money (or merely time and energy) in extra-Earth macroprojects.

"Infrastructure" is a vogue word, yet it serves a very useful function; its retention is vital to any discussion of Geopolitics, Macro-Imagineering/Macro-Engineering and Terraforming. More than sixty years ago, circa 1950, "infrastructure" meant fixed facilities supporting military plans and operations. In 2014, the term is defined as referring to geographically fixed facilities on which a community's continuance and future spacial enlargement depends. As hinted by Ronald J. Horvath, molecular Nanotechnology's creations will not be static or geographically affixed to Earth's landscape; yet they too will become a significant part of the 21st Century's *Homo sapiens* civilization infrastructure. While William Bunge surmised this kind of technological progress as probable, but undesirable, others have often and rather emotionally too claimed mindkind as "unnatural" its probable geographical impact is estimated at close to "nil". Why "unnatural" should always be, as "un-Earthly" is not, a reprobative term is difficult to comprehend. Until mindkind begins directing its own non-biological evolution via technological progress, these nano-robots creatures cannot be considered as much more than extensions of our bodies and present-day machinery. Bodies, buildings, succeeded by mindkind embodiments—that seem to be a reasonable natural progression! Planet-altering mindkind—the sturdier offshoots of humans, or possibly as a sub-species of *Homo sapiens* is reportedly well along towards full realization. As long ago as the 17th Century, Robert Boyle (1627-1691) perceived Earthly global Nature as a

"...great...pregnant automaton" (Easley, 1980, page 214). K. Eric Drexler, during 1986, first voiced the idea that, over a period of single decade, molecular Nanotechnology might reverse the carbon dioxide buildup in our air, thereby eliminating anthropogenic "global warming", but not natural global climate change, as a macro-problem vexing people! Such an event-process surely would be 21st Century molecular Nanotechnology's answer to a misguided international, "Green", 20th Century sententious cult slogan: "think global, act local", derived from E.F. Schumacher's **Small is Beautiful: Economics as if People Mattered** (1973). "Think global, act local" is an excellent motto for molecular Nanotechnogists and Macro-Engineers. The Earth's atmosphere can be scrubbed clean of anthropogenic carbon dioxide gas employing correctly applied chemistry harnessed in modified biota alone (Keasling, 2010). Macro-Imagineering presages a world with no future fossil fuel use restrictions and no anthropogenic global warming.

Antarctica—our planet's largest contiguous icescape—contains 91% of Earth's glacier ice. Melting of mountain glaciers, mostly caused by the deposition of industry-emitted black carbon [soot], and non-floe ice located at the Polar Zones might raise the global sea-level; future sea-level change could hasten the melting of Earth's marine glaciers, those land glaciers that are partly afloat on the ocean. Soot is emitted by airplanes and ships so that increased use of the Polar Zones by cruise and commercial ships or overflying aircraft will amplify the deposition of soot and, obviously, increase the sublimation of blackened snow and ice covers (Stettler et al, 2013). Any rise of sea-level will affect the Suez Canal's operation, but not adversely, except that the entrances/exists of the Canal, as well as the Canal proper, will become significantly more vulnerable to devastating tsunamis propagating throughout the Canal's channel (Finkl et al., 2012). However, a higher sea-level could cause radical reorganization of the Panama Canal, which since 1 January 2000 is Panama's military and commercial liability.

Post-Panamax container ships cannot use the Canal and new, even larger, ship types will assuredly emerge from the world's shipyards, most of which are located across the Pacific Ocean in Asia. Panama Canal modernization, completed by 2014-2015, resulted in a new set of sea-locks and a new navigation channel. [During 2014, China and Nicaragua announced their joint construction of a USA$60 billion, 290 kilometer-long route, possibly finished by 2019, through Central America, connecting the Atlantic and Pacific Oceans.] Completed in 1914, the Panama Canal is located nearly at 9^0 North Latitude where a rise of global sea-level would be exaggerated by Earth's rotational dynamics that close to the Equator. The local tidal range for the Pacific Ocean is quite remarkably big—so great in fact that even a new, sea-level Canal [such as the one proposed by Samuel Herrick, see Chapter 4] would require some kind of tidal-lock system—and this geographical fact could become of critical importance should our globe's ocean become significantly more voluminous! The existing Panama Canal's infrastructure is obsolete; true, the Canal's lock miter-gates—designed during 1906-1914 by Henry Goldmark (1857-1914)—are a century old and possibly mechanically unsound owing to numerous loading cycles; gate failure at Pedro Miguel Locks, or Gatun Locks, could drain Gatun Lake, catastrophically terminating all trans-Panama ship movement for an long time—some intact, almost upright ships will be carrying time-sensitive cargoes that would rot, become outdated or even dangerously unstable before necessary sea-lock repair and expedited lake refilling could be accomplished by jittery macro-engineers. Some preparations have already been done to ensure the watershed for Gatun Lake retains enough freshwater to operate the new and improved Panama Canal normally (Simonit and Perrings, 2013). To speed recovery from such an operational setback as Gatun Lake evacuation, cloud seeding could be carried out to enhance local rainfall although smog generated by idling trapped ships could interfere chemically with rain-making efforts (Hagler et al., 2013). Freshwater-conveying plastic pods should be used to import enough fluid from some reasonably nearby major flowing river source(s) to fill Gatun Lake (Cathcart et al., 2011). Realistically, since prolonged lake draining kills vegetations and fish, it may be feasible to simply fill the supra-Panama Canal service lake with pumped seawater. And, seawater would kill off the clogging aquatic weed *hydrilla verticillata*, which is a macro-problem in the Panama Canal Zone. Accomplishment of this effort would affect infra-service lake groundwater paths, biodiversity and watershed soil erosion. A worldwide rise of sea-level would reduce sea-lock cycling periods (at Miraflores Locks and at the lowest section of Gatun Locks), since a reduced volume of freshwater leaving Gatun Lake would then be necessary to fill lock-chambers. The most basic fact is that because large ships generate economies of scale, and containerization reduced labor costs, whilst Information Technology improvements improved shipping resource allocation immensely a significant amount of shipping has been diverted, undercutting the commercial value of this strategic waterway. And, in the Information Economy—a somewhat debatable economic concept—products of the data industries cost very little to transport and, sometimes, can be exported via the Internet!

Our world-ocean intercepts more than 75% of Earth's precipitation. Luckily for Australians, an effective Noelian "lily-pad" sea-floating rain collector could be laid out on the oceanic surface south of that continent. The only practical close-by collection region for North America is in the stormy mid-North Pacific Ocean and a "lily-pad" installed there could be shared by Japan and, very likely, would have to be installed with Russia's acquiescence. However, there are several appropriate sites south of the Hawaiian Islands which could serve the USA alone. Operational Noelian "lily-pads" there would risk an assortment of ultra-hazardous human-instigated threats: ship collisions, airplanes ditching and the violent crashes of wayward aerospace planes and errant in-falling spacecraft. Since humans have been stoned by small meteorites there is a chance—approximately a 1% risk yearly— that Noelian freshwater collectors could be damaged or destroyed by an asteroid impact in the Pacific Ocean during the 21ˢᵗ Century. Whatever suitable, freshwater collected by David Noel's temporary freshwater reservoirs may be delivered to consumers via floating, su- per-tugboat-towed plastic pod-like Stauber Bags (Cathcart et al., 2011) or, via appropriately directed submarine pipelines floating just below the maximum known navigational depth of surface shipping of the world-ocean, extending from the "lily-pad" to nearby shores where recipients of the freshwater can do with it as they please, but at a cost of course.

In 1980 the Massachusetts Institute of Technology's Energy Laboratory (organized 1972) convened a committee of experts to review all ex- tractive technologies related to oil shale mining. The Committee's rec- ommendation involved a solution only slightly more high-technology than the continuous coal-mining machine invented, in 1947, by Harold Farnes Silver (1904-1974) (Arrington and Alley, 1992). [Nearly a quar- ter of a century later little further improvements in shale mining have occurred.] The group in 1980 proposed an almost Indian Juggernaut- like 1,000 metric tonne mechanical mole (with a spinning 10 square meter gnawing face) driving itself forward through oil shale forma- tions at one meter per second. The excavation device would break the rock, remove the oil by in situ rubble heating, and then dispose of the unused pulverized rocky residue (called "muck") behind the perpetually roving mole-like vehicle. Shields on modern tunnel bor- ing machines are equipped with movable fins, which help operators to control the direction of forward movement. In effect, this changes the concept of a tunnel-boring machine from an over-sized drill bit working in a horizontal mode, to a sub-surface ground-craft. Science has already elaborated data on aerodynamics (from 1866) and hydro-dynamics (from 1738). Will 21st Century macro-engineers organize a special study named "lithodynamics", a geo-scientific lore dealing with controlled maneuvering of planet crust penetrators functioning with robotic intelligence? ["Subterrene" is one type of unmanned underground cruiser that has been proposed.] Should they come into existence, such ground-craft (on the analogy with spacecraft, aerospace planes, and watercraft) could be guided to their intra-crust goals by geological maps resulting from Geographos' imagined Geotome design. A major unexamined macro-problem associated with Robert M. Salter's Planetran is what inevitably might be encountered underground by diggers—fluids and gases that pose great risks to machines and people. Planetran and John H. Fremlin's unsurpassable edifice were both, remarkably, prefigured conceptually in print during by Edward Morgan Foster (1879-1970) who penned "The Machine Stops" published in the November 1909 issue of *The Oxford and Cambridge Review*; therein, Foster imagined a world where many humans live inside a failing global machine for living linked by an underground subway network extending from England to Indonesia: "...the rail-network stretches from Somerset to Sumatra" (Ashford, 2013, page 1)!

Such robotically brilliant ground-craft could then be given updated data, not instructions as in increasingly old-fashion human-formulated computer programming, via collimated neutrino telecommunications transmitted from a Ground Control Center, set up to function in a way similar to any relevant Rodoman-ALP facility, in its daily operational milieu. MIT's smart robotic moles were to be tele-controlled, and the valuable oil extracted was to be pumped upwards to the surface through trailing hoses at a rate of five million US barrels [794,936,500 liters] daily. Machines of this kind could well become, also, the very first adaptable to the self-repairing mode so much desired by robotics experts as well as molecular Nanotechnogists. At first, robot manufacturers could create a "breed" of intelligent planetary crust moles capable of commanded crust burrowing—even in the seabed—and delivery whatever is extricated that is valuable or is produced on the spot to the human masters. Then, later, such robots might be powered by some of the very materials mined, motivated by solar power (collected and stored for 24-hour-a-day work shifts and delivered downwards in exchange for products sent upwards), or fueled by gases extracted from deep geological strata, where it is thought methane exists naturally. If mole-machines could be powered by heat given off by magma, melted rock that has not reached the surface to become lava, then the topmost part of our Earth's mantle, just below the Earth-crust, could become a valuable resource (Haraden, 1989).

With the approval of archaeologists and anthropologists solicited, independent moles could process urban ores (landfills), global Nature-made ore bodies, and even toxic waste dumps. Such anthropogenic places exist because land-fillers wish to entomb civilization's castoffs. Humans have always made some additions to Earth's paleontological record; contributions by robots may be nil—that is, their "lifestyles" may be entirely waste-free. Assuredly, that would be the case with molecular Nanotechnology.) If 21st Century people fail to acquire off-planet resources bases on the Moon, Mars and Venus, then such "desecrations" may become necessary as well as desirable. One of the greatest advantages that will accrue from a future use of field robots in Macro-engineering's endeavors is that there is no instant underdevelopment following a macroprojects completion as too often happens when big things are built in not very well developed Earthly ecosystem-countries. There, poverty often succeeds the disappearance of jobs and public services after a particular macroproject construction phase terminates. In other words, macroprojects can be placed just about anywhere in the Earth-biosphere—anywhere that its approvers deem okay—built without regard for human labor supply centers, human health constraints, sans people costs (insurance, housing, social welfare, recreational needs, crime)! Macroprojects built almost entirely by robots would, therefore, entail no loss (to humans) of democratic freedoms through a regionalized bureaucracy buildup—hence, definitely no "metaphysical pathos" (Goulder, 1953) in the ranks of busy local macro-engineers!

Henry Alexander Murray (1893-1988), co-inventor of the Thematic Apperception Test used in psychiatric diagnosis, though all humans has a "construction need", an imperative to organize and to build things. American psychologist William James (1842-1910) recognized an expressed "construction instinct" in *Psychology*, a human instinct he described as "...genuine and irresistible in man as in the bee or the beaver. Whatever things are plastic to his hands, those he must remodel into shapes of his own [desire], and the remodeling, however useless it may be, gives him more pleasure than the original thing" (James, 1890, Volume II, page 426). There many observed and noted connections linking LEGO-like children's model construction sets to various architectural social movements, myths, civilization history and innumerable ecosystem-nation human identities (Vale and Vale, 2013). Because global Nature is expanded in volume by *Homo sapiens* the psychological insights of both Murray and James amply justifies a poetic remark to the effect that "Nature imitates Art" made long ago by Oscar Wilde (1854-1900)! Techno-Art and Science (Edwards and Bailey, 2012)—each taken in several senses of meaning—do enlarge humankind's commonly apprehended reality! Possibly our penchant as a species for building small and large things is a genetic effect.

Most geoscientists are obsessed with the desire to know, to learn something, to create new about planets today that no one knew yesterday; macro-engineers are obsessed with the desire to build something, to create new, functioning structures, buildings and devices. More often than not, both professions embrace a macroproject because it serves their separate needs. Indeed, civil and military macro-engineers never encounter a biosphere macroproject they cannot be successfully resolved, given enough money or free labor; geoscientists never envision a question about global Nature they cannot as least try to answer, given the sensors and other necessary tools needed to do so. Psychologists who are proponents of "carbon dioxide therapy" have related that human inhalation of a gas mixture of 30% carbon dioxide and 70% oxygen—normal air composition is 78% nitrogen, 21% oxygen and just 0.03% carbon dioxide—several times a week to a moment of near-unconsciousness alleviates psychotic symptoms in some patients so treated. Is it possible the human-augmented Greenhouse Effect—a 21st Century ecological "crisis" overriding all other macro-problems, some strongly allege—may elicit Murray's "construction need" via a carbon dioxide-induced cenotrope? Could Macro-Imagineering/Macro-Engineering be the main means by which the need is manifested? To be sure, there are persons who actually fear humans ever acquiring "control" over the aerial carbon dioxide gas ratio (Matsumoto, 2006). Macro-engineers need a comprehensive overview of the problems they face, solve and create through their existence! A person feels good when that person is happy! A person is generally happy when doing Id and Ego fulfilling task(s)—that is, any project(s) of any size challenging enough to leapfrog boredom and yet feasible enough to prevent anxiety (Tay and Diener, 2011). When a person's self-consciousness is temporarily controlled or mentally disentangled from Earthly geophysical reality, a personal state or sense of purposefulness and harmony is thought to happen. Electronics is very progressive technologically, so much so that before or by 2050 humanity may be introduced to environments created by tele-present macro-engineers and automatic terraformers! Wholly changed Solar System planets are not going to be escapist enclaves. Not to be overlooked is Art-as-reclamation ("Earth

Art"), as practiced by Robert Smithson (1938-1973) and many others, and which could be one expression of robots. "A visually striking reminder that art is central to the earth can be shown dramatically by covering the first and last letters of the word 'Earth'" (Pestrong, 1994). Tele-present macro-engineers and tele-present terraformers might be said to practice a unique form of Art, "Arte Povera"!

George Perkins Marsh would have found little that was unfamiliar in Ernst Fischer's 1915-published essay. Marsh, born during 1801, served the USA in several delicate overseas appointments, most notably to the Turkey legation in 1848 as well as a special mission to Greece. These onerous diplomatic tasks enabled him to travel in Europe and North Africa. On these trips, Marsh collected flora and fauna samples for the Smithsonian Institution in Washington, D.C. Marsh then returned to the USA in 1854; was appointed by President Abraham Lincoln as representative to Italy, where he served the USA from 1861 until his death and burial in Rome during 1882. His compendium of personal observations, readings, and deductions, *Man and Nature*, was first published in 1864, eventually was revised in several larger editions. With *Man and Nature*, Marsh set out to depict accurately the character and area of Earth's anthropogenic changes with an early kind of Environmental Impact Statement. Quite evidently, Marsh was overly impressed— particularly during his Mediterranean Sea Basin excursions—by regional-scale landscape destruction. Geo-philosophical bent as well as in-persons visits impelled Marsh to emphasize human freedom of will to impart motion to matter. Human reaction to a region's deterioration can be either remedial (reconstruction, restoration, redesign) or removal (emigration), depending upon available ecosystem-nation technologies involved with operative geopolitical conditions. Since G.P. Marsh lived more than 75 years before the Space Age's onset, obviously he dealt with only actual, not potential emigration options and he chiefly sought to foster local reharmonizations of people with local Nature. "The Environment", a noun that denotes rather a lot (the "Universe"!), and "The Ecology" after the first Earth Day period would probably have baffled Marsh. 2014's popular "Environmentalists" can be outlandish gurus, tenured university professors, dedicated preachers, lawbreakers, glib politicians and geopoliticians, and cyberpunks; today's reasonable Environmentalists, as individuals, can be members of the same groups! If such persons are the defining counter-culture of our Space Age, the most recent period of humanity's Anthropocene, then the planning skills of such persons ought to be enlisted by 21st Century Macro-Imagineering/Macro-Engineering remaking our Solar System.

Marsh stipulated that he was especially concerned about Europeans along with their colonial offshoots, because that geographically far-flung group was technologically ahead of other peoples. It is rather disconcerting that none of the participants at the post-1977 Macro-Engineering seminars have drawn their audience's attention (or their article and book readers) to a still-pertinent Chapter VI, "Great Projects of Physical Change Accomplished or Proposed by Man" found in Marsh's **Man and Nature.** Perhaps this is one of the reasons Macro-Engineering is not widely understood by Americans or others [in China, Europe, India and Brazil, for example], even though the term "macro-engineering" first appeared in the popular British news press during 1964 and America's popular press during the 1970s—for instance, see **The New York Times,** Section 4, page 7 in its 19 February 1978 issue. Some daily American newspapers, weekly and monthly published news magazines, Internet bloggers as well as a myriad of World Wide Web sites that sensationally bemoan publicly the so-called apparent decline of USA leadership in high-technology have influenced legislative decisions to support costly mega-Science and Macro-Engineering R&D. [**The New York Times'** management cadre waited until 1978 before establishing that newspaper's first "Science

Times" section, devoted to matters centered upon modern Science and Technology.] Macro-Engineering has a wonderful and inspiring history. A baseless research and development—R&D is supposed to be the cause of technical progress, product innovation and ecosystem-nation economic growth—adverse to vigorous commerce and national security. A rebuilding of the USA's infrastructures might be the means by which barriers between academic disciplines, between Technology and Science, and between universities, national and regional government, and private-sector industry are broken. Certain macro-engineers never fail to hype silly ideas in the national public media while ignoring practical macroprojects plans. In short, some macro-engineers sometimes short-circuit their own profession's expansion. The idea that a new thing or concept sometimes sparks another is, of course, commonsensical although undocumented before the advent of the Internet where encounters with mundane novelties of thought, real-world action and actual fabrication can now be studied by its recordation in bits and bytes.

The English language word "Earth" comes from Old Norse and Old German. Put together from many linguistic elements, English used a northern European word to identify the planet; so, it is called Earth, with the "E" properly capitalized to name our planet. [Small "e", as in "earth" is the planet's variegated soils, or at least broken rock, fragments.] After describing humankind's depredations on Earth's biosphere, G.P. Marsh then hinted that *Homo sapiens* may one day transform the Solar System into an Anthropo-cosmos: "Yet among the mysteries which science is hereafter to reveal, there may still be undiscovered methods of accomplishing even grander wonders than these" (Marsh, 1974, page 41). Circa 2014, Voyager-1 physically demarcates the Anthropo-cosmos for every member of the species *Homo sapiens*; it is, approximately, a human-populated spacial bubble, of mostly hard vacuum seeded with natural particles and energies of many kinds immersed in interstellar space, with a diameter greater than 200 Astronomical Units, each AU equals about 150,000,000 kilometers. The AU's value—the average distance between the Sun and our Earth's orbit—was first determined accurately by electrical macro-engineers circa 1961, when Planetary Radar Astronomy became established as a profession. Members of an Alien civilization, scouting the spacey fringe of our Anthropo-cosmos, and meeting some of our Solar System-exiting trash, might develop a weird (to humans) concept of *Homo sapiens'* Cosmology of Consumption!

In 1969, stimulated by J.H. Fremlin's 1964 insights on Earth's "carrying capacity" (Sayre, 2008)—or Ballardia's holding capacity—Lamont Cook Cole (1916-1978) concluded that in less than one thousand years the mean air temperature of our planet would be doubled (from 15^0 to 30^0 C) and Earth's biosphere would be made effectively deadly (Cole, 1969). Like Marsh before him, Cole was too subjective and overly impressed by examples of then-prevalent energy macro-mismanagement, rather than creative and successful energy macro-management. Sagely, Lamont C. Cole warned humanity that any use of fusion nuclear reactors would produce a byproduct—tritium, a radioactive form of hydrogen—that would become a constituent of Earth's water and, therefore, would likely contaminate all of the biosphere and all living organisms. Disruption of the Anthropocene by chemical pollution is already well underway (Persson et al., 2013) since the global economy uses 100,000 synthetic chemicals and introduces 100s of new synthetic chemicals yearly. Recently, some geo-scientists have argued forcefully and with more-than-ample correct documentation that human-induced global climate changes are nearly negligible. But, for purely argument's sake, let us suppose the Cole's statement 45 years ago stands, that Earth's biosphere is uncontrollably heading for a heat death—the exact opposite of Buffon's anti-Greenhouse Effect as well as the Nuclear Winter predictions of the 1980s. 1969, the year of Cole's unpleasant forecast global climate regime of Venus-like conditions was printed to impress the world-public, two men first tramped on the Moon's rugged regolith! Why did not Cole comprehend that there are three terrestrial-type planets (Mars, Venus and Mercury) and the Moon, with a total surface area almost exceeding seventy-eight times the USA's territory, available for future colonization? Many 21st Century geo-scientists still bear the torch, not of Liberty, but of L.C. Cole-like "sustainability".

If an Earthly super-Greenhouse Effect developed here, then it will be only inside cooled domes that *Homo sapiens* could reproduce. R. Buckminster Fuller, on 8 February 1962 in London, England, drew the world-public's attention to his macroproject plan to lace a hermetic dome above a vast areal segment of the City of New York, namely Manhattan Island (Hilst, 1967; Anker, 2007). Fuller's Manhattan Dome is the most famous example of a Utopian pneumatic structure in professional Architecture (Steiner, 2005; Hauplik-Meusburger et al., 2009). Construction of Fuller's Dome raises the dreadful prospect of Sun-heated smoggy air inside and naturally circulating hot air outside, furnishing not only poor air quality breathing conditions for humans but actual structural instability due to lifting forces induced both inside and outside the balloon-like structure. Special mechanical tie-downs and structural edge-ballasting measures must be taken by its builders to make Fuller's Dome safe for inhabitation.

If all of Earth's proven and estimated reserves of fossil-fuel were coal were burnt, the carbon dioxide gas released therefrom to the Earth-atmosphere would cause air's content to rise to 0.35% from 0.035%; if all of Earth's proven and recoverable coal were burnt in an instant, then this nightmarish event-process (releasing all coal-bound sulfur as sulfur dioxide gas) would cause a concentration of sulfur dioxide gas present in our air to rise to four parts per million. Exposure to such a level of polluted air for only an hour would kill all humans!

If macro-engineers managed the removal of all presently-extant an-thropogenic sulfur dioxide gas from the atmosphere then the globe's averaged surface air temperature and amount of precipitation could increase, probably in less than a decade.

Richard L.S. Taylor's Mars Worldhouse idea for taming that planet is an excellent Macro-Imagineering attempted at realistic ecological planning, in fact "Greening" of the planet via the installation of tended greenhouses. Mars' area is almost equal to Earth's New World plus Australia. Taylor's Worldhouse is designed, like all terraformation ef-forts proposed so far, to modify the existing atmosphere of a planet in order to make climatological conditions suitable for some Earthly life-forms. When first completed, the Mars Worldhouse would be slightly more comfortable—assuming a self-contained oxygen sup-ply for each person working there—no tourists—than Antarctica, but with swirling white-snow blizzards called, un-colorfully, "Whiteouts", replaced by pinkish fine dust storms obscuring all starlight, mak-ing Mars into an enormous Ganzfeld Space a la Techno-Artist James Turrell (Adcock, 1985)! In the cited reference above, Cole was a "gloom and doom writer" who seriously misinformed readers. Such contin-ued misleading negativism, still much too present during 2014, could retard Macro-Imagineering/Macro-Engineering and Terraforming R&D. "Doomsday"—only for our species, not all of Earth's adaptable biota—resulting from whatever cause(s) is a cliché, the purveyors of which have broadcast a gospel of despair based upon the purveyor's acceptance of insolvable macro-problems and the purveyor's rejection of all possible macro-resolutions! Such an outlook is a self-defeating, fear-driven selective viewpoint, a point of view that certainly leaves no room for human hope, imagination, creativity or enlargement of humanity's world(s)-view. Published large-scale transformation plans nearly always garner nasty castigations by tenured professors indoc-trinated during the late-20th Century by Green-fomented propaganda.

That university-lodged taskforce of socialistic persons seeking stasis in an ever-changing environment is well masked by "scientific" anti-progress opinion-forming news media and inapt textbooks. Often this useless, mushy viewpoint is adopted in Geoscience publications (Hooke et al., 2012). In addition, their continuing strong influence is felt via lawmakers, bureaucratically anonymous regulators and court law-challenges. Sadly, in the USA during 2014, "scientific method" remains, bafflingly, a focus of important legal contention in Court conflicts whilst "scientific method" is still a matter of controversy in Science!

Since 16 July 1945 more than 2,000 nuclear explosions have vibrated the Earth to its core, and yet there is little that can be said of blast effects upon landscape, the world-ocean, the planet's air, mainly because those explosions were mostly underground tests done in accordance with a 1962 international treaty or for reasons of national secrecy. History's greatest recorded thermonuclear explosion, the former USSR's 58 Megatonnes yield atmospheric test of 30 October 1961 at 8:34 GMT, some 3,600 meters above the Arctic Zone island of Novaya Zemlya (at 73^0 30 North latitude by 53^0 30 East longitude) has left so little lasting surface evidence of its transient existence that Russia and others are now exploring the seabed south of the Cold War test site, near the southernmost tip of Novaya Zemyla, for petroleum and natural gas plays! Long-lived chemical products remaining in the air after the world's most powerful anthropogenic aerial nuclear explosion may be misleading climatologists about the minor importance of the Arctic Zone's "Ozone Hole"; careless safety engineering officials, during the Soviet Era, however, left the Novaya Zemlya archipelago and the Kara Sea unnaturally radioactive. Some of Russia's landscape and oceanic territory is a unique formerly gargantuan Superpower's "Eco-nightmareland". The 1986 Chernobyl accident lightly rained radioactive fallout onto the Arctic Ocean. [Nuclear-powered submarines as well as Russia's nuclear-powered ice-breakers have achieved the North Pole. Camp Century, a USA-built underground city carved entirely from Greenland's icecap, opened in 1959 powered by a nuclear reactor serving the military contingent working therein.]

During 1957, the USA's "Plowshare Program" was established to foster beneficial changes of the Earth's crust (rather than detrimental wartime alterations of humanity's townscapes) using thermonuclear explosives in peacetime excavation and other macroprojects. The USA's stated goals with its "Plowshare Program" are outlined in Scott Kirsch's **Proving Grounds: Project Plowshare and the Unrealized Dream of Nuclear Earthmoving** (2005). Widespread American public apprehension as to the wisdom of the funded program forced its discontinuance by 1973. Russia's counterpart to the USA's practicality investigation, the "Program for Use of Commercial Underground Nuclear Explosions", began in 1965 and was curtailed in 1988. Both the USA and the USSR used "dirty" thermonuclear explosives in at least 115 unsafe experiments for "peaceful purposes" such as mining and canal-creation. In Russia and the USA, joint-financed theoretical work does continue on perfecting "clean" thermonuclear explosives. If ever achieved, such economical and rapid geography-changers could transform spacious Russia's resource-rich territories quickly, making the same technology attractive to foreign investors of Japan, Germany, India and China. "Clean" thermonuclear explosives might even be used to destroy or deeply bury some previous "dirty" test sites or pulverize threatening Earth-crossing asteroids and comets. [The Tunguska event of 1908, which caused an unusual luminescence of the Northern Hemisphere's night sky after its fall, was so alarming that some postulated the object colliding with Earth had some anti-matter within it, making it an "anti-rock" macro-object (Cowan, Atluri and Libby, 1965).] Matter-antimatter interaction produces more energy per unit of mass than any other known means of energy production, natural

or unnatural because when a particle meets its anti-particle they an-
nihilate and the energy equivalent to their combined mass is converted to various new particles and kinetic energy; energy is released by
simple contact, no ignition energy is necessary. An anti-matter triggered explosive will have extremely reduced radioactive fallout and
such explosives have been advocated for "peaceful nuclear explosion"
earth-moving macro-projects. The populist USSR writer Arkadiaei
Borisovich Markin, in *Soviet Electric Power: Developments and
Prospects* (1956), enthused that "Gigantic atomic explosions will give
rise to volcanic activity. New islands and colossal dams will be built
and new mountain chains will appear. Atomic explosions will cut new
canyons through mountain ranges and will speedily create new canals, reservoirs, and seas, carry[ing] out huge excavation jobs. At the
same time we are convinced that science will find a method of protecting against the radiation of radioactive substances." Markin may have
known of the advance planning for the unusual 300 meter-high Nurek
Dam (constructed 1961-1980) on the Vakhsh River; rock and earth fill
was blasted from the canyon's sidewalls with explosives to cover an
in-place concrete core. Nature makes island quickly too: during a 7.7
magnitude earthquake that shook Baluchistan, Pakistan, on 16 April
2013 a circular island, approximately 76 meters in diameter and about
18 meters high, suddenly appeared in the Arabian Sea near the seaport
of Gwadar (Anon., 2013). Unnamed, and fast disappearing from wave
erosion, its appearance was due to seabed liquefaction of sands and
muds.

Russian geographers, when describing their huge but needlessly poor
and radioactively contaminated ecosystem-country, often regard snow
as a semi-permanent soil horizon (situated atop the ABC soil profile);
snow which has to be plowed or compacted to retard rapid thawing,
supplies not only freshwater, but some nutrients and minerals (owing to natural dust fallout) to the soil's super-stratum. Snow virtually

symbolizes Siberia, a part of Russia still after the Cold War. During Northern Hemisphere winter, the Arctic Zone provides a mind-depressing environment for humans consisting of low temperatures, strong winds, and weeks-long darkness. In complete darkness, during January to March 2006, Mike Horn and Borge Ousland, starting from Cape Artichesky (Russia) finished the first historically recorded over-floe ice trek to the North Pole! Surprisingly, the frigid Arctic Zone has a per capita economic output that is approximately 10-12 times that of Earth's Tropic Zone. The distance north and south from our planet's Equator is among the most significant measured environmental variables underlying the differences in per capita nation-ecosystem, but this is probably most easily explained by the overall geographic pattern of human settlement, which tends to influence social institutions (Hall and Jones, 1999). In other words, if people residing and working in the Arctic, or the were made more comfortable than now, it is then very possible that the economic value per square kilometer of polar climate regime territory would decrease slightly since only very poor persons (nomads) and highly-paid persons (industrial technicians and scientists) live full-time Arctic. Because of long-range air transport, seawater currents and river flows, the Arctic Zone is a place of deposition, a sink, for industrial and agricultural pollutants from the Northern Hemisphere's Temperate Zone. Marla Cone's **Silent Snow** (2005) documented fully the nature and scope of the macro-problem in the same style as Rachel Carson's **Silent Spring** (1962). Future increased ship traffic in the Arctic Ocean's perennial cold air will enlarge the natural "Ozone Hole" above that region by ozone emissions. Non-nomadic people work in the Arctic's cold air and dreary settlements only because there are known mineral, natural gas, and petroleum resources to be extracted and shipped elsewhere, along with seasonal and non-seasonal hydroelectric facilities to be efficiently operated for the benefit of dependent urban populations dwelling far away in the

Temperate Zone as, for example, the populace of southern Canada served by major power-plants in northern Quebec described by J.F. Hornig (Ed.) in *Social and Environmental Impacts of the James Bay Hydroelectric Project* (1999) and, earlier, by Robert Bourassa's *Power from the North* (1985). There is every reason to assume that valuable Arctic Zone resources remain undiscovered, awaiting future exploration, definition and exploitation by industry. Many persons muse on the possibility of humans someday living in Space Settlements far more capacious than the impending retirement of the International Space Station, or in a terraformed Mars, but few seem to contemplate an increased human use of about 25% of Earth's surface—the Polar Zones.

It is now scientifically proved that the Arctic Zone is warming and eventually it may reach a seasonally "ice-free" state. Arctic Ocean floe ice that polar bears romp upon is more immobile than Antarctica's encircling sea; the ridged Arctic Zone ice lasts through summer. Arctic Ocean floes extend to the North Pole because those floes receive less solar energy at the surface since the Sun's rays strike at a more oblique angle, as compared to the Temperate Zone. Reduced sea-ice extent and thickness in the Arctic Oceans would promote regular summertime shipping, and present new opportunities for offshore oil and natural gas drilling and production. A Northern Sea Route following the Siberian coast would be 40% shorter than the current Europe-Asia Route that requires passages through the Suez Canal; Siberians would have new marketplaces for Siberia's exported products. In addition, new macroprojects—opportunistic hydroelectric power development of diminishing Greenland glaciers and a permanent tensioned textile fabric dam, bored railway tunnel or all-purpose sea-ice resistant suspension bridge spanning the Bering Strait—may attract new landscape-based settlers to the Arctic Zone. "There is also the possibility… for Arctic [Zone] greenhouses under inflated membrane hemispheres producing fresh fruits and vegetables for workers on such macroprojects and to house workers seeking to maintain the present-day natural stock of sea-ice by construction of ice-floe islands" (Bolonkin and Cathcart, 2007).

Possibly the first true serious attempt in Architecture to build effective artificial life-support systems in climatically harsh climate regions was the construction of greenhouses. Circa year 30, sating his jaded palate, the Emperor Tiberius had cucumbers grown in his "specularium" in Rome and glass-enclosed buildings of the type widely prevalent during the 21st Century in northern Europe were first constructed during the 1600s. Extensive commercial greenhouses in The Netherlands, even in outer space, are maintained nearly automatically by heating, cooling, irrigation, nutrition and plant disease management equipment. Humans share commonalities in their biological and medical responses to natural environmental stresses that are stimulated by cold air, snowstorms and unrelentingly strong wind. In the Arctic and Antarctica, life-threatening "whiteout" blizzards inflict the same personal visual discomfort and disorientation as spationauts experience during the hazardous "space walks"—that of being adrift in featureless space! With special clothing and shelters, humans can adapt to the Arctic Zone successfully. Medical researchers assert that "...cold-related deaths are far more numerous than heat-related deaths in the United States, Europe, and almost all countries outside of the tropics, and almost all of them are due to common illnesses that are increased by cold" (Keating and Donaldson, 2004). Incontrovertibly, living in the Arctic Zone is difficult, even when tempered by sturdy conventional protective buildings. The intensity of UVB radiation has increased in the Arctic during spring, caused by severe ozone depletion instigated by human-generated chemicals in polluted air masses deriving from all Northern Hemisphere industrialized ecosystem-nations as well as by climate change; UVB radiation stimulates sunburn (erythema) and snow-blindness (photokeratitis). People working outdoors in the Arctic and Antarctica ought to be shielded from UVB radiation as much as possible.

The first big "Evergreen Dome"-type dwelling hemispherical design "City in the Arctic" was commissioned in 1970 by Farbwerke Hoechst AG in Germany. "City in the Arctic" was to be a pneumatically stabilized climate-regulating transparent membrane half-sphere shell with a diameter of 2,000 meters, a maximum height of 240 meters and a dome radius of 2,200 meters that was intended to comfortably shelter 15,000 to 450,000 workers. The approved design membrane was to be reinforced and supported by a net of intersecting, braided polyester fiber cables. Even stronger cable-making materials, such as carbon nanotubes—the key materials for lasting Space Elevators serving Earthlings—that are forty-two times stronger than steel wires, are available nowadays that would improve upon the formidable performance characteristics of "City in the Arctic". A dynamically supported membrane building normally costs only about 30% of a building assembled with ordinary construction materials. The Macro-Imagineering concept of inexpensive to construct and operate "Evergreen Dome" facilities offered by Bolonkin-Cathcart is supported by sound computations, making their speculation more than a daydream. The single greatest boon to Evergreen Dome construction, whether on Antarctica or in the Arctic Zone, is the protected cultivation of green plants inside an air-pressurized membrane structure that generates energy from the available and technically harnessed sunlight. Innovations are needed, and wanted, to realize such structures in both Polar Zones, but at this time, primarily the Arctic. James E. Lovelock's *The Revenge of Gaia* (2006) forecast that, before 2100, some billions of persons will die because humans have altered the Earth's biosphere so massively that an abrupt "global warming" of the whole atmosphere is going to occur; Lovelock further suggests that the few breeding human couples that survive after 2100 will live in the Arctic Zone where the climate may remain tolerable!

Some geological survey reports estimate the Arctic Zone may contain about 25% of Earth's undiscovered petroleum reserves; most of the best prospective territory is located near the State of Alaska, Canada, Greenland, Norway and northwest of Russia. The place on the Arctic Ocean's surface farthest from any landscape [1,008 kilometers, in fact] is located at absolute geographic co-ordinates of 85.802⁰ North latitude by 176.149 East Longitude. It is a prospectors' axiom that for repairs, importation of supplies or transporting people into and out of the Arctic Zone calm weather "windows" rule operations. Exploration costs in the Arctic Zone are comparable to those usually incurred during offshore exploration of the world-ocean's seafloor; however, the shipping costs entailed in removing the oil and natural gas from fields in the Arctic Zone are greater than those from completed prospects in the offshore realm. Lacking pipelines, extracted natural gas must be liquefied, then moved in special ships called tankers and oil must be kept warm so that it can flow into tankers for shipment to markets. Cruise ships and cargo vessels are mostly powered by diesel and marine oil, which have higher sulfur and nitrogen oxide emissions than the diesel powering trucks. Some newer ships, especially those transporting liquefied natural gas, can be powered by the cargo they are carrying. From 17 September until the first week of October 2013, the first bulk carrier loaded with 15,000 metric tonnes of coal transited from Vancouver, British Columbia (Canada) to Pori, Finland—the coal-laden vessel is the first to traverse the Northwest Passage through Canada's territorial waters of the Arctic Zone. Alexander A. Bolonkin has devised a unique means for long-distance natural gas transmission—the Aerial Gas Pipeline (AGP) AGP consists of pipe composed of impermeable plastic that, when filled with natural gas, is self-supportive in the air. It can be made to rise to an altitude of 1-4 kilometers and bow between ground-based gas pumping stations that are separated by distances of up to 200 kilometers. Terrorist lodges are not likely to roam the Arctic Zone wilderness, but should any AGP segment be disrupted—most easily, Geographos supposes, by destroying a compressor station situated on the landscape, thereby "shooting down" a 400 kilometer-long AGP section—the AGP can be repaired/replaced relatively inexpensively. (**Figure 1.**) Crossing Siberia's rivers cheaply and simply will become a great expense-saving advantage the AGP would have over standard pipelines. Vast expanses of Arctic Zone permafrost landscape will never be disturbed using Alexander A. Bolonkin's Aerial Gas Pipeline.

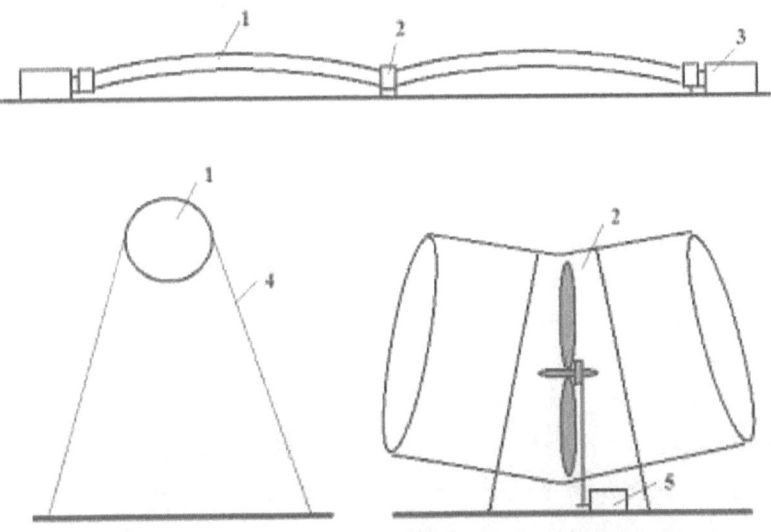

Figure 1. Top: #1 shows a side-view of one segment of the completed AGP and an end-on view of the pipe anchored to the ground by cables (#4). #1 is the pipe, #2 a junction below which is #3 the gas pumping

plant. #5 is the engine powering the propeller that moves the natural gas along through the AGP. (Image: Bolonkin and RBC, 2009.) It is possible that, because of fracking-stimulated natural gas mining In the region, the speculated Kansas Aqueduct bringing freshwater extracted from the Missouri River 600 kilometers uphill to a reservoir in western Kansas, could use this technology.

At the opposite extreme of the Russia view of snow as "soil" is the Dutch view in The Netherlands of plaggen soils, which are profileless anthropogenic soils. Somewhere between these two extremes of interpretation—that is, distilled freshwater and animal feces—fall most of our Earth-biosphere's soils! As the source of foodstuffs, soils are vital in every sense of that word. Many macroprojects described hereinafter are planned in the best long-term interests of humanity, to bring more soils under cultivation and to make fertile land wasted by global Nature (Dooley, 1990). *Homo sapiens'* energy use increased greatly during the 20th Century—from about one Terawatt in 1890 to approximately eighteen Terawatts by the first decade of the 21st Century. Of course, that eighteen Terawatts represents only a directly measurable part of Earth's bio-apparatus that our species has, so far, harnessed by, of and for Art.

All territory belongs to: (I) Somebody: under private-sector ownership and control; (II) Anybody: places so geographically isolated that "King of the Mountain" competition prevails; (III) Nobody: unpopulated Anthromes; (IV) Everybody: public, national, international control of some kind prevails. There are few sane ways in which Earth-biosphere spatial volumes have peacefully changed ownership. Two of the 20th Century's most rational territorial transfers are that of Hong Kong (from UK to China) in 1997 and Macao (from Portugal to China) in 1999. Germany's unification and the self-division of the former Czechoslovakia are also sterling examples of sensible metamorphosed region management. Macro-Imagineers/Macro-Engineers look beyond the immediate economic horizon, since both are prepared to accommodate our world's trading blocs (Europe, North America, Asia, and Latin America), which may in the future be consolidated in various bigger combinations. It was Johan Rodolf Kjellen (1864-1922) who first endowed the study of the planned and unplanned alterations in the status of Earth-biosphere Anthromes and Astythromes with the name "Geopolitics". The play-writer Karel Capek (1890-1938) published **R.U.R.** in 1920 to broadcast the word "robot" internationally by dramatically developing a storyline about the forced labor of artificial "men" (composed of organic material) who are persecuted by robots! Molecular Nanotechnology's researchers, working in many 21st Century Earthly ecosystem-nations, seek to obliterate definitions of androids—the charming character "Commander Data" from **Star Trek: The Next Generation** is an android—and robots currently employed by the news and entertainment media, trading such stale and long-outmoded definitions for a still-unwritten useful working definition of artificial life. Modern Geopolitics (Dargin, 2013) must include the operation called "Verstehen"—German for, literally, "to know"—since "Verstehen" would still be useful when Earth's robosphere contains bionic humans, androids and artificial life (Able, 1948). "Ignorance" of everything and everyone but that which is closely observable by immediate purview means that humans and intelligent others (both are agential entity groups) would never be able to understand any contemporaneous time or any adjacent places and distant regions—it is the opposite of "Verstehen". Broad-scale empathy never hurts personal relationships except where evil intent is present, as in the aims of global terrorist lodges such as Al-Qaeda and their associated or imitative anti-human civilization groups.

Next, it behooves the reader to examine some macroprojects that are currently being considered by multiple international corporations, a variety of Noosystems, communities, and certain globally-known celebrity industrial "Big Thinkers"! Since about 1978, some of these planned macroprojects have been glossed over in mainstream magazines in the form of too-sweet intellectual candy, although some of these same macroprojects have been detailed in laborious scholarly siftings and compilations. Somewhat strangely, in books published before 2014, that include fascinating chapters by Macro-Imagineering/ Macro-Engineering's living devotees, nothing can be found which even approaches a systematic discussion of the most important geopolitical factors—facts that stand in causal relationship with other facts— finding their center/fulcrum in particular macroprojects (present-day, proposed as well as those being built)! Geographos' negative criticism is hardly super-critical. Insufficient interrelating of built and proposed macroprojects evades the topic of synergism, the linkages between environment and economic development of the Anthropocenic Earth-biosphere. Not still a discipline in its infancy, all relevant Noosphere geopolitical factors involved with Macro-Imagineering/Macro-Engineering cannot be ignored by practitioners; for all professionals, the prescriptive role for Geopolitics is its prime purpose. Frank P. Davidson, the USA leader of the "Macro Movement" asserted, in *MACRO: Big is Beautiful* (1986) that "...the time is now approaching for a further institutional invention: a School of Engineering and Diplomacy". [Davidson's sub-title played on *Small is Beautiful* (1973) by Ernst Friedrich Schumacher (1911-1977).] His insight is not his alone, but merely a published reincarnation of a reasonable perception held by many but voiced by few.

It is a historical fact that between 1965 and 1968, the domestic behavior of the USA's economy changed, having become dependent upon economic activity outside of North America to an unprecedented post-World War II degree. First, Japan became a strong competitor; next, China has become the USA's strongest emerging economic rival. For a while, the Cold War (1948-1989) frictions between the USA and USSR was amplified by a developing "North-South" confrontation, a hemispherical public debate first recognized as such, during 1958-1959, by Sir Oliver Shewell Franks (1905-1992). The North is and has been colonized by people arriving recently from the South, a reversal of World War II geopolitical status. Since Earth's Northern Hemisphere has 60% of the planet's landscape and is owned by only 80% of humanity whilst the Southern Hemisphere has 60% of the planet's landscape and is occupied by 20% of all humans, Geographos marvels why a hostile dialogue is uttered! By 2050, according to the United Nations Organization demographers, the biosphere may have a human population exceeding nine or ten billion persons. It seems logical to Geographos that the co-incidental "birth" in factories located in many lands of millions or billions of "roboto-baby" units might be an effective and affordable method to slow human population increase. How many persons realize how inaptly designed the "Ecology" symbol is to express the North-South geopolitical split? The symbol, **Figure 2**, placed in the public domain by cartoonist Ron Cobb, seems to illustrate an abstractly divided Earth-crust, surrounded by a shared atmosphere or biosphere. In common with previous symbols such as the mandala and the Mobius loop, Cobb intended his symbolic invention to represent the idea of Earthly unity. But it does not. New UNO geopolitical approaches to regional and global conflict resolution are needed. Human Noosystems could view Earth as the Solar System's first sub-divided tax assessment district to fund the construction of an augmented Rodoman-ALPS. In a period when most humans are fully aware that debris falling from outer space can devastate, it is time humans commence thinking of solutions for this newly-perceived threat to human life, infrastructures and planetary biosphere.

Figure 2. The official "Ecology" symbol. Many artistic adaptations have been used by different individuals and Green groups. While not the equivalent to a lawful endorsement, its use does have some sociological restrictions imposed by good taste and rabid Green public media denouncements! (Image: Google Images.)

That isolated Tibet has for uncounted human generations been considered—especially by North Americans and Europeans—as a kind of bastion of spirituality hidden in an Earth-Noosphere increasingly riddled with crass materialism, was disclosed comprehensively by Peter Bishop's *The Myth of Shangri-La: Tibet, Travel Writing and the Western Creation of Sacred Landscape* (1989). Novelist James Hilton (1900-1954), chose to site his dramatically fictional story *Lost Horizon* (1933) somewhere on the Himalayas' north-slope; he invented "Shangri-La", a hidden valley-city of Tibet, to deal with the idea of non-robotic human immortality and existing to protect and preserve all great portable works of mankind (books, art pieces, music, but not macroprojects) from an inevitable future global holocaust. [Bizarrely, humanity's first nuclear bombs were fabricated at a site code-named "Shangri-La", the Los Alamos Scientific Laboratory, in 1942.] Tibet is placed between Earth's two most populous ecosystem-countries—China and India—is remote and little known to non-Chinese persons, even in 2014. Tibet's future economic and military importance is symbolized by the 1 July 2006 dedication of the USD 4.2 billion Qinghai-Tibet Railway carrying passengers and freight 2,520 kilometers to Lhasa from Beijing. With an average altitude of 4,500 meters, it is Earth's longest railway in a high-elevation region and it traverses more than 632 kilometers of permafrost; riding human passengers and livestock experience the discomforting travails of people riding in high-flying, but unpressurized, airplanes! Its solemn dedication as the "Green railroad" (Peng et al., 2007) by a former governor (during the 1980s) and, later, President of China, Hu Jintao, coincided with the 85th Anniversary of the Communist Party (Lustgarten, 2009).

On 6 October 1950, China invaded Tibet, integrating Tibet into China's national territory-ecosystem. On 20 October 1962, China invaded India at two places: east of Bhutan and north of Jammu and Kashmir. (**Figure 3.**)

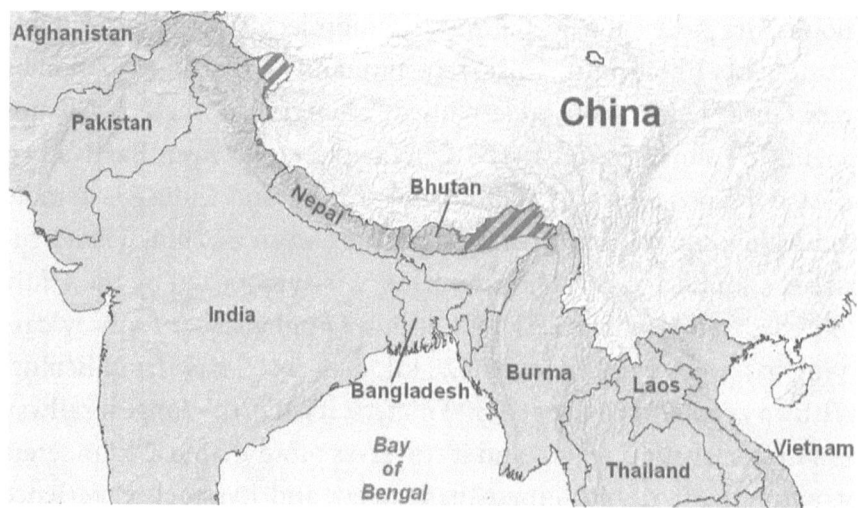

Figure 3. Diagonal red slash-marks indicate the disputed territory conquered by China during 1962, its armed-forces imposed annexation of Tibet. On the left-hand side is Aksai Chin and on the right-hand side is Arunachal Pradesh. (Image: Google Images.)

[China's opportunistic invasive army actions occurred whilst the Superpowers, the USA and the USSR leaderships, were distracted and ensnared in the geopolitical complexity of what would be later dubbed the "Cuba Missile Crisis". On 20 October 1962 the USA conducted a seven Kilotonne nuclear yield test, "Checkmate", 147,000 meters above its mid-Pacific Ocean Proving Site and, on 28 October 1962, the USSR exploded a nuclear warhead, yielding three hundred Kilotonnes, some 150,000 meters above its Kazakhstan weapons proving site. Both scheduled pre-Crisis weapon explosions generated electro-magnetic pulses that, were these tests done in wartime in outer space over the opposing ecosystem-nation, easily would have crippled all electronic pre-war situation management or actual war-fighting decision making!] The landscape acquired suddenly by China's aggressive military action remains in a state of dispute between India and China. Many geopoliticians assume

China's goal in 1962 was to obtain administrative control of the Aksai Chin region near western Tibet, which it accomplished, and that the march on Arunachal Pradesh in eastern Tibet was a diversionary thrust. Domination of the Aksai Chin's trade routes linking Sinkiang Province with Tibet was only one sound strategic reason for these internationally-condemned attacks. Another important reason that China wanted to permanently secure under its domination all land routes and aircraft approaches to the Tarim Basin, where China's military authorities later conducted their first nuclear weapon test, on 16 October 1964, at Lop Nor. The largest part of India's lost territory that was occupied by China was in Arunachal Pradesh, where China's probing attacks were first initiated. Some of India's territory, however, was never relinquished by China afterwards. Even twenty-six years after the armed clash, Indian geopoliticians speaking in public adhered to the broadcast idea that the "...purpose of the Chinese attack on 20 October 1962 is still shrouded in mystery" (Subrahmanyam, 1988). But, just perhaps, the statement was timed to support diplomacy niceties for the December 1988 visit to China by India's Prime Minister.

Generally speaking, India is a socialistic ecosystem-nation because Socialism is mandated by its Constitution, whilst in 2014 China is less vigorously under the sway of Communism than before 1989. These states could share a unity of purpose, centered on the eastern part of Tibet, shortly. Geographos suggests that India and China pool their Macro-engineering talents and efforts to detain their willing peoples in separate well-managed, prosperous agrarian-industrial Noosystems. Co-operatively managed by both societies, a Tibet-based hydroelectric power generation and transmission system could be built that would, intentionally, entangle both groups in a self-reinforcing web of freshwater and electricity management organizations reminiscent of the hypothetical "hydraulic societies" described so intuitively by Karl

August Wittfogel (1896-1988) in *Oriental Despotism: A Comparative Study of Total Power* (1957) (Banister, 2013; Cathcart, 2011).

China's theoretical potential from all tabulated hydrological resources is nearly 600,000 Megawatts. Some 17,000 to 38,000 Megawatts could be generated at a single hydropower dam site on the Yarlung Zangbo, a river that eventually becomes the historic Brahmaputra River upon reaching India (Cathcart, 1999). Running from West to East in southern Tibet, the Yarlung Zangbo's huge power drop site for a proposed Chinese hydroelectric plant is very close geographically to the landscape "reconnoitered" by China during the Sino-India War of 1962. (**Figure 4.**)

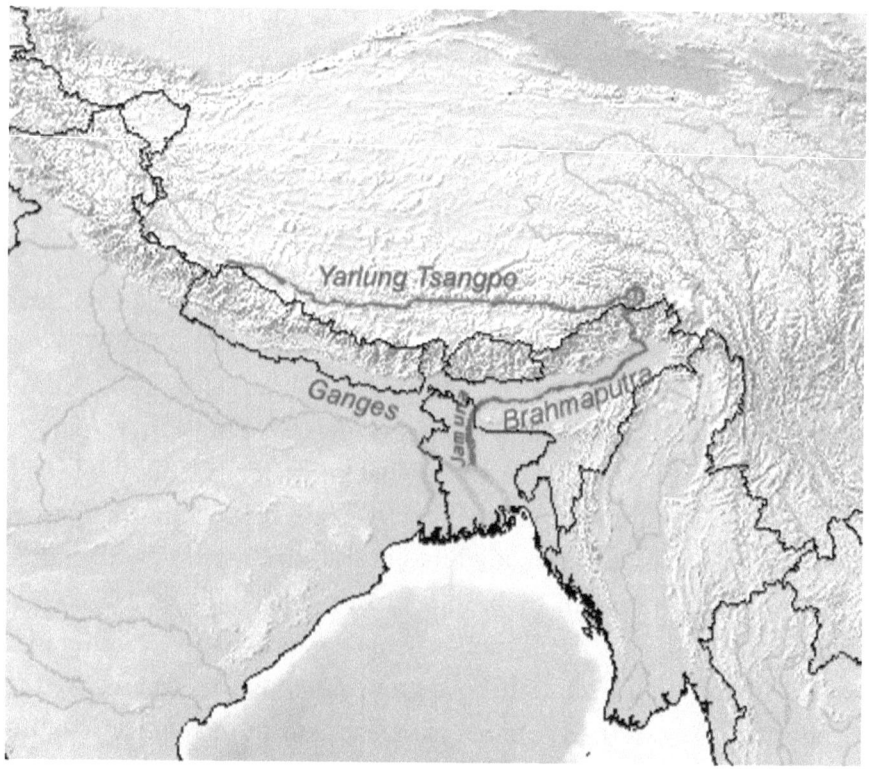

Figure 4. The sharp course reversal, the loop about Namcha Barwa mountain, creates the power-drop dam sits on the Yarlung Zangbo [alternatively spelling of mapped "Yarlung Tsangpo"]. (Image: Google Images)

Planning and construction techniques needed to emplace very large concrete dams are still, even in 2014, unfamiliar to the revered older members of China's educated leadership because older members of that ruling group of macro-engineers predate China's "Great Leap Forward" (1958-1960) as well as the "Great Proletarian Cultural Revolution" (1960-1968) periods when macroprojects were not constructed in China. Younger Chinese macro-engineers postdate those lost tumultuous years of cultural disorganization and, therefore, are inexperienced to some significant degree, even considering the Three Gorges Dam project and other like it that are physically smaller in size than the commonly supposed Yarlung Zangbo barrier proposal. China's government gives very high priority to the exploitation of hydrologic resources, and mini, medi-, and macroprojects are being pursued in all parts of the ecosystem-country with flowing rivers and very grand new canals. China exports its dam-building organizational skills to earn hard currencies and increase its geopolitical influence (McDonald et al., 2009). It is, obviously, a national security risk of China to concentrate so much of its electrical energy generating plant in so few, geographically fixed, places (Gezhouba Hydroelectric Dam, Three Gorges Dam, and its possible Tibet Hydroelectric Dam). China exercises full control over Tibet, considering it an integral part of China and not announced USA foreign affairs policy is based on the premise that Tibet is not part of China. During 2003, India officially accepted China's definition of "Tibet". During 2012, China revealed that its near-term future effort would be directed to damming the remarkable Yarlung Zangbo canyon: Motuo Dam powerhouse which would generate 38 Gigawatts and Daduqia Dam with a powerhouse generating 42 Gigawatts (Pearce, 2012). China has also mentioned a possible freshwater diversion to its waning Yellow River. (**Figure 5.**)

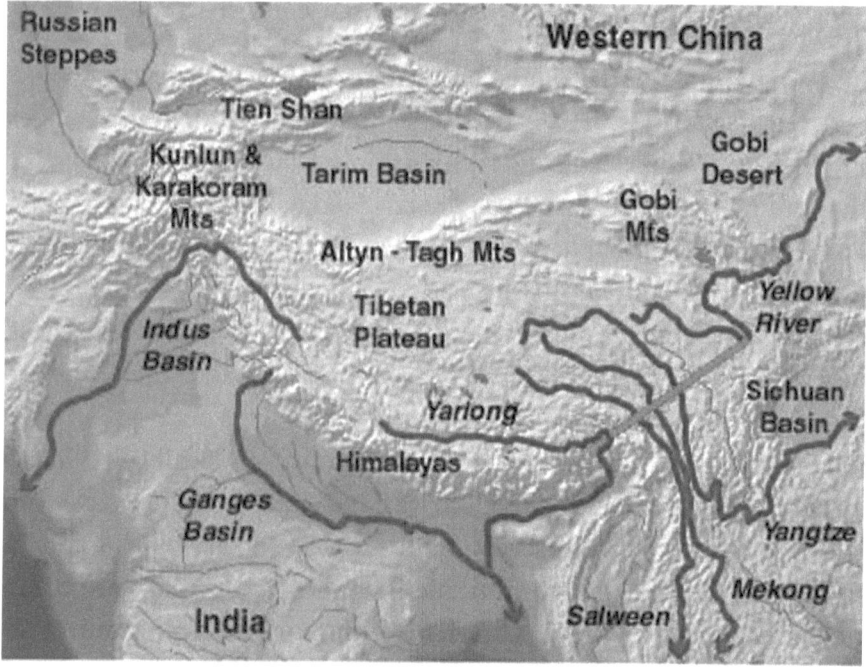

Figure 5. Red straight line improbably indicates possible freshwater diversion routing of a canal/tunnel/pipeline from the Yarlung Zangbo to below the Yellow River's headwaters.

More than 1.5 billion persons depend on the Indus, Ganges, Brahmaputra, Yangtze and Yellow rivers; climate change will affect the freshwater sources supplying these famous Asian rivers, likely with reduction is runoff from glaciers in the Himalayas (Immerzeel et al., 2010). India's Brahmaputra River is probably the most susceptible to a marked runoff decline. Often the net effect of dam-retained freshwater reservoirs is to reduce the amplitude of flow variance downstream by holding freshwater upstream during peak seasonal flows for flood control or augmenting low downstream flows in summertime. Social forces govern hydropower release strategies. However, dams also perform a positive service because these structures provide ecological and engineering resilience to climate change (Hatcher and Jones, 2013). Climate change due to carbon dioxide gas emissions is already prodded by the heavy industrial usage of mined coal in China and India. Furthermore, China is rapidly building coal-powered synthetic natural gas factories in the name, mostly, of national energy supply security (Yang and Jackson, 2013) that will have the effect of massively increasing the emissions of a Greenhouse Gas. At the same time, India is suffering economically because of a shortage of electricity and a power grid that is rickety infrastructure at best. More than half of India's power generation capacity of 205 Gigawatts is coal-based. The macro-problems associated with India's infrastructure, including highways, seaports and airports—cuts about 2% from that ecosystem-country's Gross Domestic Product growth annually! On the last day of July 2012, India incurred our world's biggest electric power outage as its transmission networks serving nearly 700 million persons collapsed. A costly overhaul of India's infrastructure is necessary, and a soon as possible! India was helped by the neighboring nation-ecosystem of Bhutan because it released additional 8,200 Megawatts of vital electric power from its hydropower plants. India plans to install 292 dams throughout its Himalayan territory by 2030; unfortunately most of these powerhouses will be run-of-river operations and the dams emplaced will not create any freshwater storage reservoirs (Grumbine and Pandit, 2013)!

The 1,800 kilometer-long Yarlung Zangbo is the main drainer of Tibet's runoff. For much of its course, the river follows a detrital-filled graben marking the crustal boundary between two tectonic plates. The large volume river is normally frozen from October-November well into March-April; normal Himalayan snowmelt causes high water flows and flooding in April, May, and June. Later, summertime rain showers also may cause some local flashfloods. The Yarlung Zangbo's average discharge is more than 4,000 cubic meters per second. At its most easterly point of flow—at place at 29^0 50 North latitude by 95^0 10 East longitude—the river makes a sharp course change, looping around the 7,756 meter-high Namcha Barwa peak and the rivers falls in elevation almost 2,200 meters. Peter Heller's exciting **Hell or High Water: Surviving Tibet's Tsangpo River** (2004) described a harrowing January 2002 kayaking team exploration on the river. [See also: Michael McRae's **The Siege of Shangri-La: The Quest for Tibet's Legendary Hidden Paradise** (2002).] If a major concrete dam, dam composed of fill created by controlled rock blasting or Geographos' macro-imagineered fabric barrage were emplaced from Namcha Barwa near the town of Pe, enough of the Yarlung Zangbo could be harnessed to produce immense quantities of hydroelectricity needed by an industrializing Tibet, northern India and eastern China. The mountain tunneling needed to make such a macroproject feasible and workable was pioneered by the Chinese in their Jinping-1 and Jinping-2 hydropower plant excavations in Sichuan Province connecting two sections of the YaLong River, bypassing a 150 kilometer-long section of the river looping Jinping Mountain (Shiyong et al., 2010). All Himalayan dams must be built to endure greater runoffs and increased glacier melting since, under most global albedo change scenarios, the summertime monsoon system will have a different future impact on the whole region than it does in 2014. Since the cost of a single, large and safe dam at the bend of the Yarlung Zangbo might be, approximately, USD 225 billions, China might wish to share its cost with others, or not.

References Cited

Able, T. (1948) *The Operation Called "Verstehen".* American Journal of Sociology 54: 211-218.

Adcock, C. (February 1985) *Perceptual Edges: The Pyschology James Turrell's Light Space.* Arts Magazine 59: 124-128.

Anker, P. (2007) *Buckminster Fuller as Captain of Spaceship Earth.* Minerva 45: 417-434.

Anon. (14 October 2013) *Pop-Up Island.* Time 182: 9.

Arrington, L.J. and Alley, J.R. (1992) *Harold F. Silver: Western Inventor, Businessman, and Civic Leader.* (Logan, Utah: Utah State University Press) 250 pages.

Ashford, D. (2013) *London Underground: A Cultural Geography.* (Liverpool, UK: Liverpool University Press) 188 pages.

Banister, J. (2013, in press) *Are you Wittfogel or against him? Geophilosophy, hydro-sociality, and the state.* Geoforum 50??

Bolonkin, A.A. and Cathcart, R.B. (2007) *Inflatable "Evergreen" dome settlements for Earth's Polar Regions.* Clean Technologies and Environmental Policy 9: 125-132.

Cathcart, R.B. (1999) *Tibetan power: A unique hydro-electric macro-project servicing India and China.* Current Science 77: 854-855.

Cathcart, R.B. et al. (2011) *Antarctica-to-Western Australia Liquid Freshwater Shipments Using Stauber Bags in a Paternoster-Like Transfer System: Inaugurating a Southern Ocean Antidrought Action Sea-Lane.* Journal of Coastal Research 27: 1005-1018.

Cathcart, R.B. (2011) *Freshwater Supplies Necklace Super-Project: Floating Bags and Rolling Freshwater Tires Facilitating Future India-China-Bangladesh Life Necessities Trade.* Volume II, Chapter 87, Pages 1531-1540 *IN* Brunn, S.D. (2011) *Engineering the Earth: The Impacts of Megaengineering Projects.* (The Netherlands: Springer) 2266 pages.

Cole, L.C. (1969) *Thermal Pollution.* Bio-Science 19: 989-992.

Cowan, C., Atluri, C.R. and Libby, W.F. (1965) *Possible anti-matter content of the Tunguska Meteor of 1908.* Nature 206: 861-865.

Dargin, J. (Ed.) (2013) *The Rise of the Global South: Philosophical, Geopolitical and Economic Trends of the 21st Century.* (New York: World Scientific) 452 pages.

Dooley, P. (May 1990) *More's Utopia: An Ecosystem at Climax Stage.* Moeana: The Bulletin of Thomas More 27: 37-46.

Easley, B. (1980) *Witch Hunting, Magic and the New Philosophy: An Introduction to Debates of the Scientific Revolution 1450-1750.*

Edwards, M.D. and Bailey, E. (2012) *Gravity in Art: Essays on Weight and Weightlessness in Painting, Sculpture and Photography.* (Jefferson NC: McFarland & Company, Inc.) 355 pages.

Finkl, C.W. et al., (2012) *A Review of Potential Tsunami Impacts to the Suez Canal.* Journal of Coastal Research 28: 745-759.

Goulder, A. (1953) *Metaphysical Pathos and the Theory of Bureaucracy.* American Political Science Review 49: 496-507.

Grumbine, R.E. and Pandit, M.K. (2013) *Threats from India's Himalaya Dams.* Science 339: 36-37.

Hagler, G.S.W. et al., (2013) *Panama Canal Expansion Illustrates Need for Multimodal Near-Source Air Quality Assessment.* Environmental Science & Technology 47: 10102-10103.

Hall, R.E. and Jones, C.I. (1999) *Why Do Some Countries Produce So Much More Output Per Worker Than Others?* Quarterly Journal of Economics 114: 83-116.

Haraden, J. (1989) *An Appraisal of Magma Power Generation and the Greenhouse Effect.* Energy 14: 333-340.

Hatcher, K.L. and Jones, J.A. (2013) *Climate and Streamflow Trends in the Columbia River Basin: Evidence for Ecological and Engineering Resilience to Climate Change.* Atmosphere-Ocean 51: 436-455.

Hauplik-Meusburger, S. et al. (2009) *Inflatable technologies: Adaptability from dream to reality.* Acta Astronautica 65: 841-852.

Hilst, G.R. (1967) *What can we do to clear the air?* Bulletin of the American Meteorological Society 48: 710-712.

Hooke, R.L. et al. (December 2012) *Land transformation by humans: a review.* GSA Today 22: 4-10.

Horvath, R.J. (1974) *Machine Space.* The Geographical Review 64: 167-181.

Immerzeel, W.W. et al. (2012) *Climate Change Will Affect the Asian Water Towers.* Scince 328: 1382-1385.

Keasling, J.D. (2010) *Manufacturing Molecules Through Metabolic Engineering.* Science 330: 1355-1358.

Keating, W.R. and Donaldson, G.C. (2004) *The impact of global warming on health and mortality.* Southern Medical Journal 97: 1093-1099.

Lustgarten, A. (2009) *China's Great Train: Beijing's Drive to the West and the Campaign to Remake Tibet.* (London: St. Martin's Giffin) 320 pages.

Matsumoto, K. (July 2006) *A psychological effect of having a potentially viable sequestration strategy.* Carbon Balance and Management 1: 4.

McDonald, K. et al. (July 2009) *Exporting dams: China's hydropower industry goes global.* Journal of Environmental Management 90: S294-S302.

Pearce, F. (28 April 2012) *Whose Water Is It? China is taking control of Asia's great rivers at their source.* New Scientist 214: 89.

Peng, C. et al. (2007) *Building a "Green" Railway in China.* Science 316: 546-547.

Persson, L.M. et al. (2013) *Confronting Unknown Planetary Boundary Threats from Chemical Pollution.* Environmental Science & Technology 47: 12619-12622.

Pestrong, R. (1994) *Geosciences and the Arts.* Journal of Geological Education 42: 256.

Pysklywee, R.N. (2006) *Surface erosion control on the evolution of the deep lithosphere.* Geology 34: 225-228.

Sayre, N.F. (2008) *The Genesis, History, and Limits of Carrying Capacity.* Annals of the Association of American Geographers 98: 120-134.

Shiyiong, W. et al. (2010) *Jinping hydropower project: main technical issues on engineering geology and rock mechanics.* Bulletin of Engineering Geology Environment 69: 300.

Simonit, S. and Perrings, C. (2013) *Bundling ecosystem services in the Panama Canal watershed*. Proceedings of the National Academy of Sciences 110: 9326-9331.

Steiner, H. (2005) *The forces of matter*. The Journal of Architecture 10: 91-109.

Stettler, M.E.J. et al. (2013) *Global Civil Aviation Black Carbon Emissions*. Environmental Science & Technology 47: 10397-10404.

Subrahmanyam, K. (10 October 1988) *India's Security: The North and North-East Dimension*. Conflict Studies 215.

Tay, L. and Diener, E. (2011) *Needs and Subjective Well-Being Around the World*. Journal of Personality and Social Psychology 101: 354-365.

Vale, B. and Vale, R. (2013) *Architecture on the Carpet: The Curious Tale of Construction Toys and the Genesis of Modern Building*. (London: Thames & Hudson) 208 pages.

Yang, C-J. and Jackson, R.B. (2013) *China's synthetic natural gas production*. Nature Climate Change 3: 852-854.

CHAPTER 6

MEGA-TERRITORIES: ASTYTHROMES AND ANTHROMES

During the past 300,000 years, about fifty billion people have lived in our Earth's biosphere; at the present-time, more than seven billion persons inhabit our planet's bio-apparatus. During 1968, towards the finale of the Hollywood movie *2001: A Space Odyssey*, when the lonely and confused Jupiter-bound spacecraft's onboard Heuristically programmed Algorithmic super-computer goes haywire and has to be rewired and unplugged by the spaceship's spationauts, in its last moments of "life" HAL-9000 reveals its birthday: 12 January 1992 (Weiss, 1992). In effect, HAL-9000 is a Dictator fought by a single brave human who, finally, blinds the murderous Cyclopes forcing its reversion to childish simple-mindedness! As of 2014, neither the super-computer fabrication industry, nor the molecular Nanotechnology industry, has presented humanity with a HAL-9000 equivalent. However, judging the currently observed trend in computer design is not a dead-end, by mid-21st Century the end of human dominance in the Earth-biosphere may be drawing to a close (McLaughlin, 1987). Technological determinist Marvin Minsky says, without any trepidation or hesitancy whatsoever: "Will robots inherit the Earth? Yes, but they will be our children. We owe our minds to the deaths and lives of all the creatures that were ever engaged in the struggle called evolution. Our job is to see that all this work shall not end up in meaningless waste" (Minsky, 1994). Geographos, speaking on the level of 1967's popular television culture, even the over-populated Village's minders, which does its work entirely underground—that is, as if the Village were similar to Ballardia—and is totally reliant for its daily mega-system business operations on the hushed elegance of a "super-computer" named "the General", in the Surrealistic TV serial *The Prisoner,* could not control the local ocean tidal flux in North Wales' Tremadoc Bay adjacent to a strange-appearing rural urban node named Portmeirion, built during 1925-1972 by Sir Clough Williams-Ellis (1883-1978). *The Prisoner's* imaginative teleplay writers always treated its "technology" as an abstract noun for an evil category of Art. Whatever their size, when robots are added to our planet's biosphere, professional macro-imagineers/macro-engineers will have to then speak of a cyber-ecology of Earth, at the very least!

In 1981, Julian Lincoln Simon (1932-1998) (**Figure 1.**) published *The Ultimate Resource.* That supreme wealth is, thankfully, *Homo sapiens*—an animal with a mind and tools to express its collective and individual desires. Simon's voice and publications made it abundantly clear that prejudices and material interests basically underlie the greater part of those special-interests groups' idealistic-sounding rhetoric demanding "preservation", "birth control", and "self-denial" (Sabin, 2013). A late-2013 United Nations Organization report, prepared by its demographers, forecasted that by 2050, for the first time in our species' history, people more than 60 years old will outnumber children younger than 15 years old! Won't that encourage the increasingly rapid development of humanoid servants, human-like robots? These Greenish assemblages of scornful, politically-motivated

Figure 1. J.L. Simon, economist and battler for truth, justice and a rational way.

Individuals have no extraordinary effect on the most prolific societies in the Noosphere—so far; these numerous and far-from-publicity-shy Environmental Action Groups do tend to confuse and misdirect our world's middling and most technologically advanced Noosystems. Such noisy, but comparatively small aggregations, have little understanding of a scientific Ecology in which *Homo sapiens* actually exists (Egerton, 1973). Green environmental extremism is the most dangerous and obscene form of misanthropy yet developed in the North America-Europe geographical region. A 21st Century new Macro-Imagineering/Macro-Engineering elite will have to reflect seriously on and completely reorient themselves towards new and galvanizing operational images of our Earth-biosphere's potential future. The new task of these professionals is to turn Art into Science and Science into Art with Technology being the vitalizing linkage in both instances!

Professionals must make it their honor-bound duty to educate all humans to appreciate their wondrous planet—a planet so very difficult for terraformers to duplicate! Natural resources are a question of technological insight, and like Erich Zimmerman (1888-1961) in *World Resources and Industries* (1933), Julian L. Simon treated technology as constantly improving—as technical means progress, more resources become accessible; many resources can be recycled, while others can be dispensed with as substitutes are found by Science. As our physical sciences resolve our planetary homeland (as well as the known Universe) into macro-objects/hyper-objects composed of mass and energy, our investigative social sciences have demonstrated that persons experience the locus of policy event-processes, hence "Verstehen". Some psychologists define human personality as that which permits a prediction of what a person will do in a given situation. Raymond Bernard Cattell (1905-1998), in Psychology's arcane lingo, claimed every person has a "personality sphere" clearly indicated by "surface traits" which can be used to measure the entire range of any human's personality. Geographos thinks it likely that every individual artificially intelligent machine—a representative of 21st Century mindkind—will also have a "personality sphere", making each a measurably definable secular person.

Reliable predictivity is the ultimate hallmark of Science and the essential determinate of Science's success; Science has also introduced objectivity as the decisive measure of enlightened discourse in the everyday affairs of most of humanity—most notably terrorist lodges are excluded from this category or reasonableness. Scientific objectivity distinguishes Science from the Humanities and other disciplines where understanding rests on debatable dogmas and subjective interpretations. Without predictivity, in both Science and Technology, our comfort would be uninsurable. Some view "Spaceship Earth" as an apt expression of the notion of unified planetary fate, as if the whole thing were on some unalterable geophysical trajectory, quite literally a universal flight-path, to inevitable tragedy! Maybe future Mars and Venus terraforming should rightly be considered as "restitution" for mankind's so-called crime of long-term Earth-biosphere "mauling"?

Humankind's perceptions of physical reality have changed under the influence of Science to encompass terms with special ancient Latin and Greek prefixes describing the size of things: words beginning with micro-, macro-, and nano-. George Bugliarello (1927-2011), **Figure 2**, defined Civil Engineering as the modification of global Nature to make human habitats possible; his succinct definition focuses Macro-Engineering on manipulation of all elements that can be employed to become "human habitat". Geo-engineering and Terraforming are included in Bugliarello's brief and reasonable definition. Art is not to be overlooked, as Leonard Shlain (1937-2009), **Figure 3**, points out at pages 427 to 430 of *Art & Physics: Parallel Visions in Space, Time and Light* (1991): "...revolutionary art anticipates visionary physics.... When the vision of the revolutionary artist, rooted in Dionysian right [human brain] hemisphere, combines with precognition, art will prophesy the future conception of reality.... I propose that [Albert Einstein's post-1905] space-time [continuum] generates universal mind.... Universal mind would be the moving force behind our [existing Earth-noosphere's] zeitgeist, speaking through works of revolutionary, right-brained, visionary artists first, and later through [Apollonian] left-brained, visionary, rational physicists." Wolfgang E.F. Pauli (1900-1958), who first proposed the neutrino particle in 1930 [confirmed to exist by 1956], said in 1948: "It would be most satisfactory if physics and psyche could be conceived as complementary aspects of the same reality". Modern Cognitive Psychology seemingly supports Shlain and Pauli's positions on this specific matter (Evans, 2007). Whilst the UK science-fictionist J.G. Ballard novelized the ultimate in postulated geohazard-less planets, the economist J.L. Simon did not foresee robots ultimately viewing *Homo sapiens* as an intellectual resource! Humans assess the "environment", whatever its volume, in terms of "cultural ecosystem services" (Milcu et al., 2013). Will non-humans such as our robotic offspring, and certainly Aliens, do so too? Future molecular Nanotechnology, when it does separate *Homo sapiens* from Earth's global Nature forever—or that of any future human-inhabited celestial body—should bring about a marked hastening of the on-going spiritualization of human lives, especially in a world which after 2050 will have many over 60 aged persons with a strong interest in the "beyond", in every extra-territorial sense of that hazy word.

Figure 2. George Bugliarello, civil engineer and broad-minded solo philosopher.

Figure 3. Healing-arts physician and astute Techno-Art connoisseur, Dr. Leonard Shlain.

Only in our Solar System is the Sun our ultimate source of energy, beyond radioactivity emanating from Earth's core and mantle. This Solar System is massive: the Sun (1.99 followed by 30 zeros kilograms) plus the traditional planets (2.7 followed by 27 zeros kilograms) equals a lot of potentially handed substance. Earth has a volume of 1.0832 followed by 21 zeros cubic meters—a fact that gives *Homo sapiens* no cause for concern about Earth-biosphere living space, industrial and residential energy, or the Anthropocene's remaining Geologic Time. In terms of conventional energy-consuming technologies, humans will continue to find economical fuels for some time to come. If nuclear fusion reactors are perfected, humans will then have fuel, derived from seawater, for centuries. Bogus radical Green "environmentalists" would have humanity believe that such unlimited power supply could create a naïve public belief in Technology, but have no important effect on Geopolitics because our Earth-Noosphere's macro-problems are unrelated to energy worries, or might even cause a species "population explosion"! 1975's *The Population Bomb* brought forth 1990's cant metaphor, *The Population Explosion*. Such gloom-and-doom social movement books really convince the world-public that it is far too late to halt Earth's destruction! (For sure, where an affirmative Macro-Imagineering/Macro-Engineering vision of the future is absent, people will perish needlessly.) Further, some negative social and physical planners theorize global or local Nuclear Winter—the result of, respectively, macro-war and meso-war—as a potentially useful event-process demonstrating their veridical insights. True, thermo-nuclear weapons are anti-biotic, the most effective ever invented. That is a bitter truth pill that all human beings must swallow. There remains, for cause of global or local war, always the act of a criminal, a pirate, a prankster, or a religious or ideological dissenter, or a biological and/or machine entity with motives arising from artificial intelligence or Alien technology and culture. A common serotonin-induced psychological depression can make some human males and females overly aggressive and destructive (Jacobs, 1994). Earth's carrying capacity is, depending on the reader's point of view, set by Terran or Alien Science! A developing molecular Nanotechnology—the instrument through which humans are going to transform Earth's global Nature totally—will soon, more and more, be regarded as a crucial technology affecting geopolitical conflicts and confrontations. Potentially, both the fundament and the firmament can be more than just altered since these spacial realms may even be redesigned.

Although molecular Nanotechnology's future projected role in stimulating *Homo sapiens'* final technological revolution was not then widely anticipated or known, Julian Simon's most telling argument in *The Ultimate Resource* is not merely an economic one, but "…that the sheer number of people alive, enjoying life, may also be relevant". The role of Macro-Imagineering and Macro-Engineering is to make all Anthromes and Astythromes ready to sustain the richest types of human culture and the fullest span of human life while preserving those things that cause a pleasant response owing to our deepest subjective needs and wants. Mature moderns, instructed by World War I and World War II—when global Nature disaster became possible following the advent of nuclear weaponry—desire alternatives sometimes offered by our professions. Terraforming and Geo-engineering professionals ought to imbue some of their writings with certain of these ideals. Virtual realities proponents only offer temporary, terminable escape, not a "cure".

Learned professors of Engineering as well as other subjects have suggested that humans have made profound material gains—the amplification of our powers with technical means as persons and groups within the Earth-biosphere. Yet, too many idiot-Greens have virtually called for the dematerialization of humanity! Macro-Imagineers and Macro-Engineers do respect the Earth as they attempt to create a better human World. During September 1949, just as the Cold War began, a meeting called under the title of "Biology and Civil Engineering" was convened in London, England. Participants examined some documented facets of humanity's materials impact on the planet. The major concern and conclusion of those gathered builders and biologists was that humans have outpaced supra-mantle global Nature in promoting changes, that civil and military engineers ought to endeavor to better match their activities to Earth's biogeochemical "cycles"—many alleged "cycles" used by teachers, some in classrooms, some available via the Internet, to inform their student-auditors simply are not actually geophysical cycles!—lest *Homo sapiens*-instigated event-processes overwhelm the Earth-biosphere. Molecular Nanotechnology's post-20th Century perfection, which seems pending, will certainly blur existing differences between structures and materials for all Geologic Time, and possibly for all Astronomic (or Astrologic) Time if, in addition, nuclear transmutation is ever economically accomplished.

During 1948, it was recognized by Geoffrey Gorer (1905-1985) that a typical American characteristic response involving innovative and inventive industrial "Big Thinking" and industrial "Big Building" was "To any protest of the break with tradition, of the impracticality of a new proposal, the response has always been: 'Why not?'" (Gorer, 1948, page 152). The USA has long been a noosystem that encouraged strongly and nurtured human creative capacity; however, cultural values that shape economic and geopolitical success always change. Thomas Parke Hughes, aping a late-1980s fashion for "Endism", asserted that the USA's enthusiasm for Science and Technology extended from 1870 until 1970, surviving only amongst "...engineers, managers, system builders, and others with vested interests in technological systems" (Hughes, 1989, page 11). As of 2013, the USA spent USD 400-500 billions on R&D, with 19% of that expended on basic Science; China has become the world's number two R&D spender (approximately USD 221 billions in 2013), but with only 5% spent on basic Science (Davis, 2013). The results of basic Science R&D usually leads to new consumer products and new technologies to fundamentally change our world and lifestyles. Gorer's flattering characterization of the American people was not a pre-World War II Englishman's admiration carried forward in time to an almost pre-World War III moment. With remarkably few reservations, Americans invited Macro-Imagineering/Macro-Engineering, Bio-technology and molecular Nanotechnology to devise controls for the things of this world! Indeed, that 2014 task garners much public praise and financing and few curses from the USA's citizenry. Truly, the Northern Hemisphere is humanity's primary world-homeland geostrategic region. Failure of the Tripartite Pact after 1940 as well as the end of the Cold War by 1990 prevented totalitarian conversion of mega-territories of this prime Earth-biosphere part. Americans still tap into the Zeitgeist of the Space Age, which was begun by the USSR's orbiting of Sputnik-I— the American public, victorious in World War II, successfully battling totalitarianism in the Cold War, was suddenly shamed and terrified, but ultimately inspired to costly new efforts to remain unrivaled in pioneering exploration beyond the Earth's stratosphere.

Pre-21st Century geopoliticians habitually divided Earth into halves, in such a manner that most of the landscape is in a single hemisphere. The center of our planet's "Land Hemisphere" is near Nantes, France, the once rustic birthplace of that redoubtable author, the fretful Jules Verne, whilst the "Seawater Hemisphere" center is located near New Zealand. Technological advances in submarine industry should reduce the significance of this vintage planet divisional scheme. Hugh Robert Mill (1861-1950) compiled a geographical terminology dictionary circa 1900-1910; in the lexicon, he defined "anecumene"— meaning, not "ecumene"—using the English Channel as a paradigm: "the English Channel, with its population of passengers and crews of boats, is a type of aneokumene". Halford John Mackinder stated that the unity of Geography in terms of "natural regions" should be based on our Earth's hydrosphere (Mackinder, 1931). Most of Earth's living biomass survives in the world-ocean. Innumerable submersibles have already been launched and are in regular use globally. Some unscrewed undersea vessels are guided by brilliant computer programs and mission-management systems. The carrying capacity of the Earth-surface could be greatly increased in volume by sound application of always improving underwater technologies. Geopoliticians are nowadays forced by circumstances of technology's impingement to ponder a Geopolitics that includes the surface and sub-surface of the land and the ocean. For the time being, Antarctica and the world-ocean cannot be colonized by ordinary people, except temporarily. In future, the Geopolitics of terraformed planets will come into play whenever *Homo sapiens* became enveloped in a thriving poly-global Solar System civilization.

Fully-fledged, Siamese-like professionals of Macro-Imagineering/Macro-Engineering must necessarily include anthropological political geography, as indicated by a timely word's coinage in 1959 by Stephen B. Jones (1903-1984): "Geoanthropolitics". There exists a wide range of outlook among contemporary Noosystems that is a powerful geographic deterrent to the establishment of a worldwide political community—the largely dysfunctional United Nations Organization is a prime example. Despite the spread of literacy and higher education, many persons lack knowledge about the chronology and relationships of historical facts. The 18th Century's Industrial Revolution irreversibly altered Earth-biosphere ambient conditions, to the point where it is difficult for persons living in 2014 to project themselves, via "Verstehen", back beyond the recorded and derived economic statistics on life-styles to imagine Earth-biosphere conditions that are already alien and almost incomprehensible. It is a measure of the North America and Europe's isolation from the human past that significant members of the best-educated youth perceive only the evils of Technology and none of its wonderful benefits. For people worldwide, the only kind of Anthropolitics we can know is the one that has to work amidst a credulous species, poorly educated, ruled by jumbled and non-concurrent moralities, and bent upon other activities than the search for universal truth via Science. The so-called ecosystem-nation of Somalia, a true cacosystem if ever one existed, is a failed state promoting little human resource development that is positive in nature. Few person making and earning a living to survive in today's morass of more than seven billion seem to acknowledge that Europe and North America's direct ancestry suffered ugly, killing labor in unsafe mines, weather-buffeted farms, and in industrial fabrication. Barry Commoner (in **Making Peace with the Planet**, 1990, page 166) slandered Americans and Europeans with this incorrect pronouncement: "Colonialism has determined the distribution of both the world's wealth and its human population, accumulating most of the wealth north of the equator and most of the people below it". And, just what is "wealth" since it can be counted in so many different ways? Many persons Commoner's ilk use marshmallow words in meanly-phrased accusations and deprecations. According to mass communications practitioners, about 75% of North Americans and Europeans get "environmental news" from television. Since "live" and "video-taped" television images can be digitally retouched, at leisure or in seconds of real-clock time, that simply means "television diplomacy"—the use of television-generated pictures to affect world-public opinion, usually measured by polling, on international relations is a present-day danger our world's undereducated need to be prepared to discern and respect lest they be forced to pay a fearful price through conscription and taxation. As the Tudor playwright William Shakespeare (1564-1616) foretold in **As You Like It** [II. Vii, 11, 139]: "All the world's a stage…". Would not the Geographos-proposed Earth-surrounding augmented Rodoman-ALPS, a possible 21st Century remedy, that is a comprehensively potent entity caring for *Homo sapiens* and its Earth-biosphere (or Earth-robosphere) really confirm this future fact? Popular media in the USA have been lax in correctly and fully reporting the advances of Science and Technology—in fact, broadcast television and many daily newspapers have fostered a culturally induced ignorance; their failures to perform properly have caused impedance, delaying acceptance of scientific truths by the general population of the USA. "Agnotology", the study of such public ignorance, was coined and introduced by Robert Proctor's *The Cancer Wars: How Politics Shapes What We Know and Don't Know About Cancer* (1995). The existence or extent of anthropogenic climate warming is just one topic constantly reviewed in the biased news media populated by a class of persons that seems delighted to bring all humanity news of fresh disasters every day. [Some ecofreaks say "Mother Earth" is enduring many human-caused malignancies; nowadays, researchers are attacking human cancer by viewing cancer cells as a macro-problem for Physics (Jain, 2014).].

The late Arthur C. Clarke, the world-famous science fiction/science fact writer, lived in Sri Lanka. The lushly-vegetated island-nation (noosystem) was to have marked the border control station between the German-Italian and Japanese geopolitical spheres of influence-coercion, as aggressively propose by territorial expansionists of those country-ecosystems prior to World War II. For example, Giichi Tanaka (1863-1929), Prime Minister of Japan and Minister of Foreign Affairs from 1927, formed his infamous "Tanaka Memorial" during 1927, according to John J. Stephan in *Hawaii Under the Rising Sun: Japan's Plans for Conquest After Pearl Harbor* (1984). During 1944, Mitsubishi Aircraft Company engineers "...were working on plans for a Japanese aircraft able to bomb the Panama Canal and the American West Coast from the Kurile Islands", according to F.P. Hoyt's *Japan's War: The Great Pacific Conflict, 1853-1952* (1986, page 345). Three years earlier, in 1941, Nazi Germany's Eugen Sanger (1905-1964), **Figure 4**, drew the final blueprints for the "America Bomber",

Figure 4. Aerodynamics genius Eugen Sanger, post-WW II photograph. (Image: Google Images).

Me264, an atmospheric skipper craft capable of becoming our species first crewed spaceship (Myhra, 2002; van Pelt, 2012)! Nazi geopoliticians made tentative territorial acquisition policies to occupy North America as far west as the Rocky Mountains, while Imperial Japan's schemers composed designs to capture and administer permanently the western coasts of Canada and the USA, thereby strategically dominating the Pacific Ocean. Their September 1940 alliance linked finally the Asian and European theaters of war, turning the 1939-1945 conflict into a struggle between different conceptions of World Order. Japan and Nazi Germany, independent foci of geo-militaristic expansionism, had each done preliminary engineering work on nuclear fission explosives. The Nazis, however, had a technical edge over the Japanese in delivery systems work, by R&D on long-range ballistic missiles. "Had the war continued into 1946, the Germans might well have made good their plans to bombard New York City", alleges D.R. Baucom in *The Origins of SDI, 1944-1983 [Strategic Defense Initiative]* (1992, page 4).

Continued trans-Pacific Ocean economic and geopolitical integration seems to be a real internationalized prospect. During the early-20th Century, a key event-process occurred, the Russo-Japanese War—as Asia's first victory over the colonizers from Europe, it boosted Asians self-confidence and focused their thinking on the kind of ideas about "civilization" that interested most Asians. The term "Pacific Rim" is a 1980s word coinage meaning the coastal country-ecosystems (Noosystems) bordering the North Pacific Ocean and the South Pacific Ocean—places like Canada, Korea, Japan, Australia, Indonesia and some South American noosystems. The first Pacific Rim Conference was held in Perth, Western Australia during November 1986; Perth faces both the Southern Ocean and the Indian Ocean. A straight map-line connecting Perth, Australia and Seattle, State of Washington (USA) clearly indicates a modicum of the potential global importance of the Central Pacific Ocean, the location of Jarvis Island's (**Figure 5.**) possible 21st Century Space Elevator Base Station.

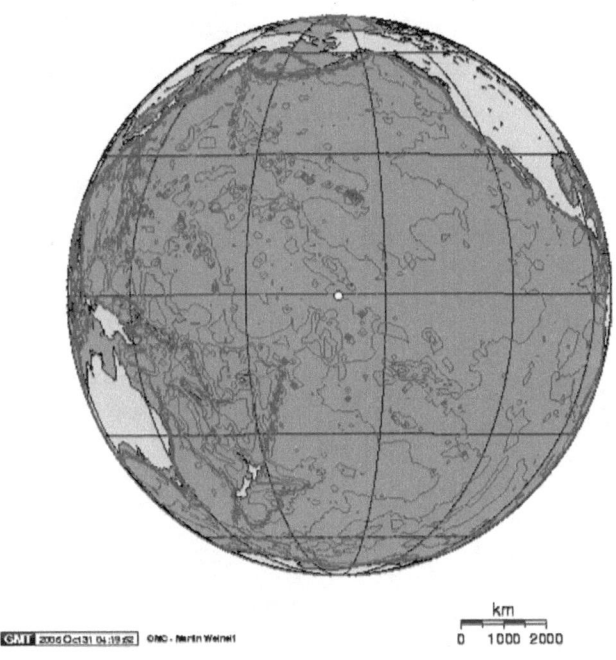

GMT 2006 Oct 31 04:19:62 ONC - Martin Weinelt

km
0 1000 2000

Figure 5. The white dot in the center of this hemispherical chart represents Jarvis Island. Its centrality for a Pacific Rim segment of global human civilization is quite apparent. (Image: Google Images.

Considering China's vast and expensive catch-up industrialization event-process and its costly infrastructural development, as well as its likely future rationalization and improvements, Geographos thinks, as many others do, that a new world industrial center is quickly being realized in Pacific Ocean-facing Asia. However, in 2014, the business focus of the Pacific Rim remains located in the Northern Hemisphere, and it may have a transportation infrastructure reinforcing today's focus: (I) Russia's Baikal-Amur Mainline railroad and its extensions, a containerized cargo railway operational since 1989; (II) ocean containerships taking advantage of a newly floe-ice free Arctic Ocean and (III) the materializable Jarvis Island Space Elevator installation. With Siberia, Russia will continue to be the largest, and potentially the richest, nation-ecosystem in the Earth-biosphere; the implicit converse is radical: without Siberia, Russia could cease to be a World Power and, indeed, fade forever from the dwindling ranks of Superpowers! For China, nowadays remarkably dependent upon sea lines of trade and communication, the Straits of Malacca between Indonesia and Malaysia, is a gauntlet-gateway that may soon become avoidable, at least in part, by the construction of the Kra Canal in Thailand (Cathcart, 2008; Thapa et al., 2007). **Figure 6.**

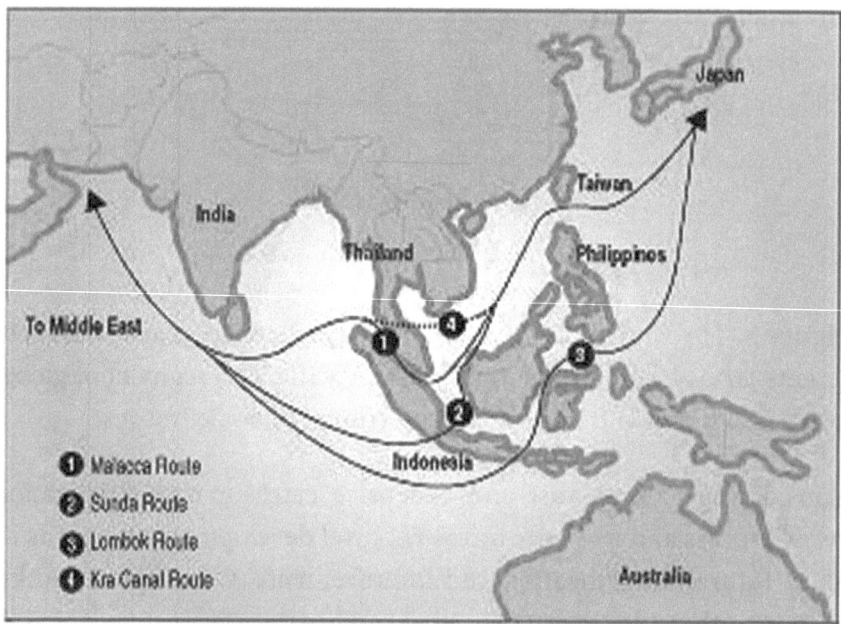

Figure 6. Location and ship routing impact of a Kra Canal (#4). (Image: Google Images.)

Railway and roadway fixed connections between Taiwan island and mainland China are being considered—an entire issue of *Marine Georesources and Geotechnology* (Vol. 22, Number 3, July-August 2004) was devoted to such Macro-Imagineering/Macro-Engineering planning! Relatedly, Taiwan is contemplating construction of The submarine Kuroshio Power Plant emplaced offshore of that island's rocky cliff eastern coastline, a facility to harness the ocean's strong, northerly seawater current that skirts most of Asia, including the eastern coast of Japan (Chen, 2013). Furthermore, South Korea and Japan's macro-imagineers are slowly edging towards resolution of the fixed linkage(s) macro-problem aggrieving their noosystems crossing the Tsushima Strait—say, from Karatsu, Japan to Iki Island, then onwards toTsushima island, then reaching the famous seaport of Busan (or the island City of Geoje), South Korea. (**Figure 7.**)

Figure 7. Maps indicating the possible placements of various proposed bridges and tunnels to connect offshore and mainland Asia.

Busan, Korea is a seaport on the prestigious AE10 Asia-Europe service of the world's longest ship in service, the **Maersk Mc-Kinney Moller** the 18,270 TEU containership (Anon. 2013). It might be thought that the region, in terms of trade, imaged above, is Asia's "Mediterranean Sea Basin"! (Image: Google Images.)

Nuclear warfare, especially in geographical contexts smaller than global—that is, macro-war, general nuclear war—as well as other economically damaging aggressions using "smart" and "brilliant" weapons, could be considered by some Noosystems and some terrorist lodges as merely a type of foreign economic policy, statecraft in the case of eco-system-nations, if such a war had important side-effects on foreign economic policy. Terrorism has a powerful allure for some persons since the breathtakingly successful 11 September 2001 attack on New York City's twin iconic World Trade Center skyscrapers. The Earth-Noosphere fittingly reflects and inflects a single technical human civilization, a world of unified Science-based technique but very discordant multiple social values. Unflinchingly, Geographos suggests that the only appropriate "Global Village" is an Earth-biosphere geopolitical World Order that is a new model of co-existence lodged within a single interconnected civilization. But, a curious Geographos humbly asks: when methods of predicting, from parents' DNA [deoxyribonucleic acid], the likelihood their baby would have certain traits becomes commonplace, what will be the social consequences for 21st Century international relations? The first USA patent, "Gamete donor selection based on genetic calculations", Patent 8543339, awarded on 24 September 2013, portends extraordinary tumult in all facets of human civilization. Mightn't individual terrorist personalities be deliberately cultivated by hateful, special interest lodges, intent on homogenizing our world socially? Perhaps formation of a "Pacific Rim" focused on Jarvis Island, as suggested in Chapter 4, and/or a 21st Century incarnation of "Atlantopa", Geographos' Chapter 8 offering, could prove to be useful shaping molds for a future geopolitical vision? "Ecopolitics" has emerged as a force operating to transform international geopolitics. Needless to say, Ecopolitics bears no close resemblance to Macro-Imagineering/Macro-Engineering!

Technology's progress is challenging the present-day concept of Earth's ocean being non-developable. Some persons, endeavoring to convince Earth's more than 190 nationalized and skeptical publics that ocean regions are capable of development, have coined glamor words as, for example, "hydrospace", to stimulate recognition of the world-ocean's potential basis for *Homo sapiens'* future prosperity. Representatives of humankind have had the capability to reach any place on the submerged Earth-crust at any time, even for purposes of drilling into our planet's seafloor.

Humanity operates machinery using more power than is generated by the metabolisms of all living natural persons. As almost hallucinatory as that generalization might seem, the present-day flow of energy through our functioning technology does not compete meaningfully with the energy flow through Earth's global Nature. Certain anthropogenic alterations of the landscape and world-ocean involve the horizontal movement of matter and energy; flowages of matter (carried by railway cars, trucks, automobiles, boats and ships, liquid-gas-slurry pipelines, electric grids), usually as concentrates, have no Earth-biosphere counterpart. The delivered concentrates converge on, and are processed in, Astythromes. Land Art devotee Nancy Holt's infrastructure-like "system sculptures" seem to be Art's homage to Technology; the artworks entice viewers to conceive Earth's great infrastructures and to stimulate a viewer's vision of spiritual communion with our Solar System's wonders (Williams, 2011; Marter, 2013). Cities are actually "mass savings accounts", physical mass which can be widely distributed with great rapidity by nuclear and thermonuclear explosions, one kind of those hideous "weapons of mass destruction". Cities have their own biogeochemical cycles involving re-use, waste and pollution that are analogous to our Earth-biosphere's "reservoirs" and "plumbing". John Stuart Mill (1806-1873) covered most of the elements when he asserted his opinion, in **Principles of Political Economy** (1848, Volume I, page 32) that moving *macro-objects* and mass converted to controlled energy are all that *Homo sapiens* can do with any available global Nature. A lot of earth is moved unintentionally because of soil erosion (Dotterweich, 2014). But, there is another point to be elucidated: whilst global climate change that tends towards warming will, obviously, increase macro-biological diversity and habitat diversity, promoting speciation by hybridization (Kim, 2013), our excretion and unsafe disposal of uncounted thousands of metric tonnes of antibiotics into the biosphere will cause biodiversity in the *microbial* part of our world, invisible and a lot less obvious!

A future Earth-Noosphere, physically transformed by molecular Nanotechnology's earlier perfection and use, surely will amplify the 21st Century usefulness of Allen K. Philbrick's theoretical Geography innovation, "...the particulate region" (Philbrick, 1982). In other words, macro-imagineers/macro-engineers should already be prepared for molecular Nanotechnology's future creations! A.K. Philbrick (1914-2007) (**Figure 7.**) foresightedly devised an appropriate visualization for all geo-scientists, especially those persons who may have enthusiastically played with LEGO toys as unfettered tots; during August and September of 2011, two different spationauts carried LEGO sets to their International Space Station job-site. Data about the "particulate region" will be "Big Data"—a data set that seems enormous in 2014 will seem ridiculously miniscule in the not-too-distant-future mid-21st Century since the complexity of "Big Data" will always ensure that its handling will be beyond generally common, conventional techniques. Data storage and data analysis/synthesis are always going to be macro-problems for those monitoring or manipulating "particulate regions" with the volume of the Earth-biosphere! Humans appear to have migrated away from the Cold War's

Geiger Counter-read radioactive mushroom cloud metaphor (Evans, 1999) to the Utopian image of ineffable electric "Cloud Computing" developed from 1966 but only widely used as a popular term since circa 2007; Cloud Computing's truest real-world symbol is the nubilous "Blur" building, designed by the architectural firms of Diller + Scofidio, installed over a lake at Yverdon-Les-Bains, Switzerland in 2002. **Figure 8.** Cloud Computing would be impossible except that atomic clocks, offering precise time-keeping, have been instrumental in Science and Technology which already have produced Global Positioning Systems and the ever more omnipresence of advanced telecommunications as well as enhanced Earth- and outer space-based navigation.

Figure 7. American geographer Allen K. Philbrick. [n.d.] (Image: Google Images.)

Figure 8. Photograph of the misty "Blur" structure sited atop the Anthropocene and earlier sediments and bedrock of Lake Neuchatel (Gorin et al. 2003). When Naturists walked into the fog, momentarily it felt like donning a weightless water-cooled garment! (Image: Google Images.)

The ultimate uncontrolled particulate region would be an Earth-biosphere pulverized into radioactive powder via macro-war uncompromisingly fought with ghastly nuclear weapons. Any viable 21st Century geographical prediction of the location of selected phenomena in our Earth must be a description of the movements of objects and energy in a gigantic open-air nano-structure. Who will name this imagined nano-structure? For example, emplacements or displacements of objects surely would be many and varied (interactions, diffusions, circulations, flows, orbits and even planetary escapes of nano-objects) in and beyond an Earth-biosphere dominated by molecular Nanotechnology's tiny machines (nano-robots). Whatever energies and matter the interacting Universe inserts into our Solar System—in transit now and moved in the future—will simply be imbricates upon humanity's enormous materialized deeds. A rapid increase in the volume of the Anthropo-cosmos will promote open-mindedness amongst humans! Still, in 2014, without a regularly published journal devoted to the topic, the First International Conference of the International Association of Macro-Engineering Societies met in Barcelona, Spain during November 1989.

It is known the Earth's Tropic Zone is expanding in area as the Polar Zones shrink in area because of enhanced mid-latitude (the Temperate Zones) tropospheric warming. Whilst geoscientists cannot yet state specifically which Polar Zone glacial ice-sheets will soon sublimate (colloquially "melt"), some do speculate in an informed manner about using freshwater produced by Greenland's glaciers, which comprise 10% of Earth's ice volume. A floe-ice free Arctic Ocean, with heavier ship traffic, will cause the deposition of soot on the ice that, in turn, caused the topmost ice surface to absorb more sunshine, heat, and melt. Some macro-imagineers have studied the promising possibilities of harnessing this natural and human-caused source of electricity-generating runoff (Partl, 1978). Such macroproject concepts involve construction of a network of ice-surface freshwater collections gutters, flumes, and aqueducts directing the cold flowing melt-water to natural and excavated reservoirs. From those storage places, the precious liquid would be dropped into Francis or Pelton turbines to manufacture 200,000 Megawatts. Greenland has had a degree of autonomy from Denmark since 1979; by 2008 natives voted for more autonomy from Denmark, many expressing their desire that Greenland become an independent ecosystem-nation (noosystem) by 2021. The natives, so far, in addition to six operational mineral mine enterprises have flourishing hopes that petroleum revenues derived from fields drilled in Davis Strait will sustain, or at least bolster Greenland's future autonomy. Greenland could become a party to an "Arctic Free-Trade Zone" comprised of Russia, USA, Canada, Iceland, Norway, and Sweden (Powell, 2014). How can electricity transmission to distant markets be made exportable by Greenland? Conversion of the power to microwaves, then transmitting it to Earth-orbiting satellites for rapid relay to electric power-hungry industrial regions on mainland Europe and North America is one means (Rogers, 1981). The disfigured and polluted landscape of all Europe could be cleansed through liberal use of Green electrical power not generated in Europe. Soil contaminants can be removed with induced electrical fields. Power generated in Greenland may also be exported via high voltage direct current (HVDC) undersea cables to Iceland, thence to the UK and mainland Europe (Hammons et al., 1989).

21st Century farming and ranching in Greenland has benefited from climate change and concomitant deglacialization. More and more, previous ancient human settlements are being revealed by 21st Century glacial retreat! If native Greenlanders decide to further capitalize on the geophysical results of glacial ice-sheet melting, then electric power generated in Greenland might also be used to in another local macroproject with demi-global ramifications. CFCs (chlorofluorocarbons) ejected by Technology's handiworks are still catalyzing the thinning of Earth's stratospheric Ozone Layer above the Arctic Ocean. Huge amounts of electricity generated in Greenland by falling freshwater could power laser installations in the Northern Hemisphere, which could then process, over a decade-long period, all of Earth's air, disrupting the CFCs causing some of the human-augmented Greenhouse Effect. The CFC clearance macroproject was first imagined and proposed by Thomas Howard Stix (1924-2001) during 1989, and the idea received a fuller mathematical treatment and detailed macro-engineering exposition in *Plasma Science and the Environment* (1997) by W. Manheimer, L.E. Sugiyama and T.F. Stix.

Before World War I, prior to the enhanced Greenhouse Effect's widespread public recognition as a possible macro-problem for humanity, macro-imagineers conceptualized plans to build a causeway to divert the "warm" Gulf Stream by blocking the southward flow of the "cold" Labrador Current. **Figures 9-10.**

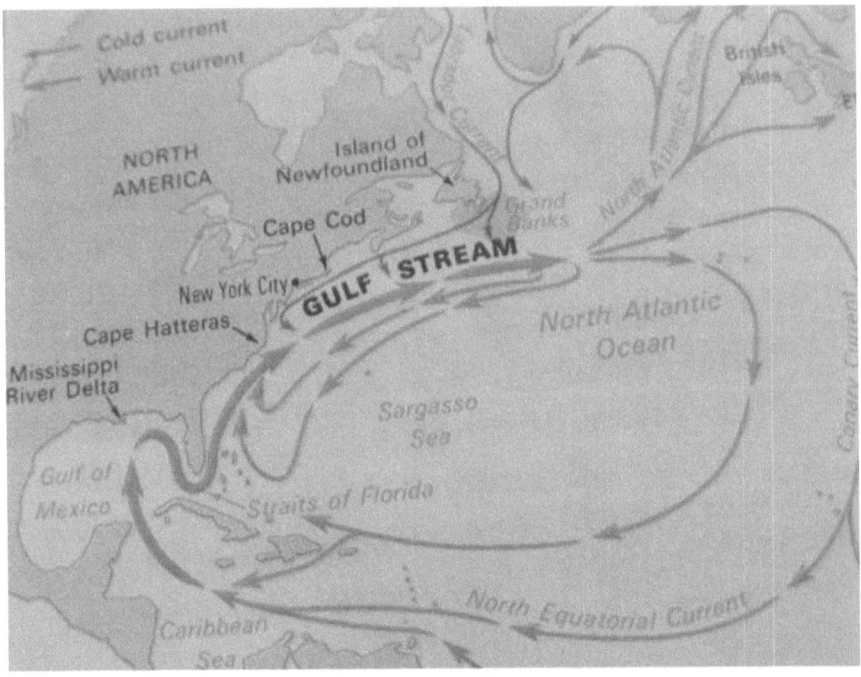

Figure 9. The Grand Banks, a rich fishing ground, plays a key role in the seawater circulation of the North Atlantic Ocean. (Image: Google Images.)

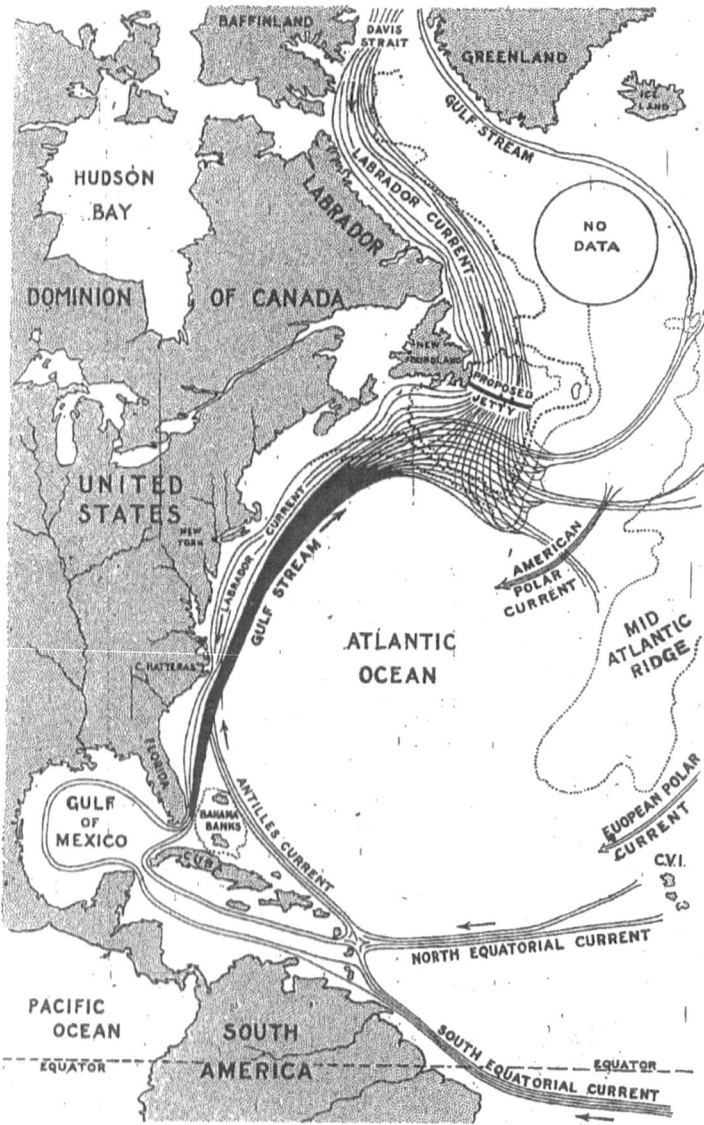

Figure 10. Grand Banks Jetty Macroproject proposed in 1912 by C.L. Riker. The streamlines for seawater movements is incorrect in some places due to incomplete information available over a century ago. (Image: Google Images.)

A 1912 macroproject proposal by Carroll Livingston Riker (1854-1931) to the USA's President and Congress, *Conspectus of Power and Control of the Gulf Stream*, envisioned a 320 kilometer-long jetty extending eastwards from Cape Race, Newfoundland, Canada onto the Grand Banks, a famous fishing ground. Geographos thinks Riker may have been greatly stimulated by the railroad, built during 1904-1912, connecting the Florida Keys. (**Figure 11.**)

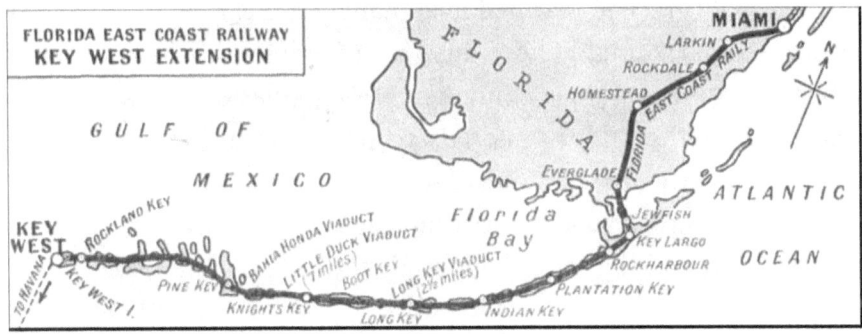

Figure 11. The Key West railway's route, via bridges over the seawater gaps between the small coral islands. (Image: Google Images.)

Henry Morrison Flagler (1830-1913) undertook that macroproject to open the State of Florida for multiple touristic beach resort developments. Completion during 1914 of the Panama Canal probably represented the last physical manifestation of the USA's interest in the "Manifest Destiny Doctrine"—hence, Riker's address to Americans rather than Canadians. All of the oceanic territory encompassed by Riker's quixotic macroproject planning effort is owned by Canada. Oceanographic and climatologic research since 1912 has utterly shredded Riker's geophysical assumptions: the Gulf Stream, first charted in the 16th Century, cannot be the primary cause of Europe's mild western coastal climate regimes (Riser and Lozier, 2013; Kaspi and Schneider, 2011). Simply put, considering the presence of the Iceberg Patrol, the many modern safety devices, both electronic and mechanical, aboard ships plying the sea-lanes of the North Atlantic Ocean in 2014, there is no real need for Riker's macroproject to reduce dense fog episodes and ship or iceberg collisions. [It is worth noting, however, that the world's deepest causeway, the piled rocks Canso Causeway (Doucet, 1955), in places more than 66 meters atop the strait's seafloor, connecting Cape Breton Island with Nova Scotia completed in 1955, demonstrated that C.L. Riker's jetty could be constructed as he had planned it so many years earlier! 21st Century macro-engineers might chose, instead, to employ seafloor stacks of geobags, geotubes and geocontainers—tough plastic bags filled with locally-dredged seafloor materials such as sand used in systems of stuffed geotextile installations. Use of these bags would obviate any necessity for ugly landscape creation caused by extensive rock quarrying since the filling seabed materials could be obtained easily near the jetty macroproject worksite.] Still, a knowledgeable Hollywood screen-writer might have introduced Riker's Jetty as a maligning causative factor in the 2004 block-buster epic movie *The Day After Tomorrow* (Leiserowitz, 2004) but, thankfully, was not.

Other direct applications of Macro-Imagineering's skill-sets on vast geographical scales have been mulled over the past decades since the **RMS Titanic** sank on 14-15 April 1912, mostly by Russian industrial "Big Thinkers". One of these, Pyotr Mikhail Borisov (1901-1973), thought demi-globally! Borisov's controversial propaganda, which commenced circa 1961 and peaked during 1973, favored a causeway, much bigger than the Canso Causeway, spanning the 85 kilometer-wide and 55 meter deep Bering Strait that separates the USA from Russia in the Arctic Polar Zone. According to his booklet *Can Man Change the Climate?* (1973), he planned to pump cold seawater from the Arctic Ocean into the North Pacific Ocean for the purpose of letting more warm North Atlantic Ocean seawater replace the volume removed by his pumping action at the Bering Strait, thus warming the Arctic Ocean, melting the floe-ice and moderating the northern Polar Zone's climate. The Arctic Ocean accounts for only 1% of the world-ocean's seawater volume. His causeway seawater-pumping station would have consumed a fantastic amount of electricity, the source about which Borisov was quite vague but, as noted above, might be provided from Greenland-based electricity generation facilities. P.M. Borisov's causeway-pumphouse duo infrastructure would be the most costly "solution" for quickly sublimating Arctic Ocean floating ice, glacial ice and the circum-North Pole permafrost landscapes. Nowadays, geoscientists are aware that defrosting permafrost landscapes unwise as Greenhouse Gases are released that lead to global warming of Earth's air (Hu et al., 2012). Still, Borisov's barrier-pumping plant combination could hasten the North Atlantic Ocean's recovery [to "normality"] due to a large, abrupt input of freshwater runoff from Greenland ice-sheet sublimation, promoting the thermohaline circulation that tends to moderate Europe's westernmost coastal climates. He had hoped nuclear fission reactors would be the source of economical electricity by 1980. Borisov's flat-topped structure was intended to provide beds for railway tracks extending the Trans-Siberian Railway (Panova, 2011) into North America as well as an inter-continental highway; ship-locks might be required too, probably best placed between the Diomede Islands. Of the five bordering Arctic Ocean coastal noosystems, Russia gains the most access to the Arctic Ocean Basin's Exclusive Economic Zone by international treaty. Since the Bering Strait is shared by the USA and Russia, is it possible that China's intentions to use the opening there might be thwarted by a commercial ship passage toll-booth? Trans-Arctic Ocean sea-routes will be navigable by multiple decades before 2100 (Smith and Stephenson, 2013); cruise ship tourism started shortly before the turn of the century (Headland and Splettstoesser, 1999) and tourism in that region is forecasted to increase greatly during the 21ˢᵗ Century (Muller et al, 2013). Oil-exploration and exploitation was tested in 1969 when the **SS Manhattan** petroleum tanker voyaged through the Northwest Passage (Coen, 2012) and, during 2013, the first bulk carrier [of mined coal] and the first container-ship [the vessel **Yong Sheng**, owned by China] traversed the Arctic Ocean after each had passed uneventfully through the seasonally opened Bering Strait. A counter-proposal to Pyotr Mikhail Borisov's was made during 2011 by Cathcart, Bolonkin and Rugescu in which a tensioned fiberglass curtain, quickly installed, would serve to divide the Arctic Ocean's seawater from the North Pacific's seawater at low cost. Thus, it is possible to retard, even perhaps retain, the Arctic Ocean Basin's permafrost landscapes more or less intact so that methane gas is not released to the Earth-atmosphere to cause possible warming episodes or long-term overall warming of our air. This outcome is a real contrast to the unreal, fictional Arctic Ocean Basin milieu contrived by several novelists: James Graham Ballard's **Hello America** (1981) raised the biosphere/technology collapse storyline to new whimsical heights. In that novel, intrepid explorers—circa 2114-2126—describe North America a century after a Bering Strait Dam had caused widespread arid climate regimes, leaving New York City awash in sand dunes, whilst the State of California is portrayed as a jungle, a miserable sweltering rainforest! In 2012, writer Stephan Malone's **Polar City Dreaming: How Climate Change Might Usher in the Age of Polar Cities** warns individuals and ecosystem-nations that unchecked air warming will foment a Polar Zone-bound humanity land rush just 500 years after 2014!

The James Bay Hydroelectric Project is a partly capitalized facility intended to ensure the flow of power overland by almost outmoded HVDC cables to North America's industrial heartland. However, severe weather conditions and electromagnetic storms, which interfere with ground-level cable transmission from extremely remote locations to the Astythromes, have caused several memorable electrical power outages in Quebec and beyond. As far as the USA is concerned, it would be beneficial to expand a mostly installed 1921 "superpower system" for the cities between Boston and Washington, DC, as first elucidated in *US Geological Survey Professional Paper 123* penned by W.S. Murray and his colleagues. Quebec, Canada, could produce exportable hydroelectricity for the USA, especially east of the Rocky Mountains and Great Plains landscapes—even if Quebec were ever to become independent of Canada. Prime Minister Robert Bourassa (1933-1996) documented the exciting Big Business approach to future Quebec freshwater and electricity exportation facilities and policies in *Power from the North* (1985). When James Bay Hydroelectric Project commenced its Phase I construction in 1971, Bourassa called it "...the project of the century". Phase II started in 1989. When or if completed, the final version of the Project may generate 27,000 Megawatts.

The Hudson Bay-James Bay is a 1.23 million square kilometer shallow inland sea, the world's largest in area, contiguous with the Arctic Ocean. Even if all the rivers merging with the Hudson and James bays were dammed and diverted, there would be only a slightly detectable consequential effect on the climate of the North Atlantic Ocean (St-Laurent et al., 2012). Canadians, some, realize that they are watching potentially useable freshwater lose its economic value as it mixes with the seawater sitting in Hudson and James bays! At least two macroprojects have been developed by macro-imagineers to deal with this wastage. Closely identified with the James Bay Hydroelectric Project is the 1959-announced Great Replenishment and Northern Development (GRAND) Canal submitted by Thomas W. Kierans (born in 1914). **Figure 12.**

Figure 12. Thomas W. Kierans, famed Canadian hydropower macro-engineer. (Image: Google Images.)

With foresighted political savvy, Kierans defined the GRAND Canal as a freshwater "recycling" macroproject, not a freshwater "diversion" effort since he did not plan to intercept river runoffs before all the freshwater entered James Bay (about 60,000 square kilometers), but simply wished to pump it from James Bay after the embayment's conversion into a freshwater lake from a shallow (average depth is 28 meters). At the present-time, James Bay is a cold, brackish inland sea covered by ice for approximately six months of the year. Prolongation of an ice-cover—possible because the water freezing is fresh not brackish—could serve as an emergency regional plan to cool the Arctic. In fact, a form of climate macro-engineering. Its shore is of international importance because it is a feeding and breeding ground for migratory avifauna of central and eastern North America. Due to post-Ice Age glacial rebound, it is shallowing. However, shallowness has a dampening effect on open sea wave development. The GRAND Canal has, sometimes, been referred to as the eastern version of the North American Water and Power Alliance, NAWAPA.) Kierans proposed to dam the northern sited mouth of James Bay with a barrage equipped with sluicegates that open at low-tide—a tidal range of less than 4 meters occurs in Hudson Bay—allowing seawater to flow northwards into Hudson Bay whilst retaining the freshwater derived from contributory rivers. **Figure 13.**

Figure 13. A planimetric mapping of the James Bay Barrage, shown by black-color east-west line severing James Bay from Hudson Bay and (middle, center-right) the blue-color north-south line) enhanced

exported freshwater flow to Lake Huron. The Barrage lies south of Cape Henrietta Maria, on the East, and Long Island on the West. (Image: Google Images.)

Within a few years, James Bay would become a freshwater lake from which 20% annually could be pumped southwards to refresh the Great Lakes. A yearly export of 347 cubic kilometers of "saved" freshwater was planned. The only geophysical factor disregarded by Kierans is the high probability that earthquakes will increase in frequency owing to the post-Ice Age crust rebound that supports Hudson and James Bays (Wu, 1998; Milko, 1986). [A similar macro-problem will almost certainly plague Sweden and other northern Europe Noosystems centered on the Baltic Sea (Nordlund, 2001). Interestingly, cities that endure subsidence, places such as the Port of Long Beach, California, find that new construction costs are often higher than usual elsewhere because infrastructures, some forgotten or unrecorded, are encountered when something requiring deep foundations is built!] In July 2009, the Montreal Economic Institute offered "Northern Waters: A realistic, sustainable and profitable plan to exploit Quebec's blue gold", an ambitious macroproject plan to generate 3640 Megawatts of hydroelectricity, producing annual revenues in excess of USD 2 billions, or more, was first studied from 2004 and then presented to Canadians and the world-public by *MEI Note* during 2009. Its lead author, F. Pierre Gingras (**Figure 14**) outlined the plan thusly (Geographos' summary): By reversing the floodwater flows of three rivers (**Figure 15.**), with their runoffs continuing to run in their natural riverbeds, Quebec could care for the waters of the St. Lawrence River and/or spill the harnessed flow into the Great Lakes, thereby serving the needs of more than 150 million persons. Gingras has written a thoroughly detailed discussion of this suggested macroproject in *L'eau du Nord* (2010).

Figure 14. F. Pierre Gingras, industrial macro-engineer and hydro-power/freshwater macroproject planner working in Canada's Quebec Province. (Image: Google Images.)

Figure 15. Mapping of the MEI's "Northern Waters" macroproject scheme announced during 2009 after five years of laborious effort by many professional contributors. The electricity generated would far exceed the amount required by the operational projects, making it self-liquidating in terms of capital long before a use is found for the saleable freshwater! The southernmost part of James Bay is at the upper left-side of the image. (Image: Google Images.)

Much has been published about the North American Water and Power Alliance (NAWAPA), a series of macroprojects devised in the USA during 1950-1959 by Donald McCord Baker (1890-1960). Designed to move southwards approximately 310 cubic kilometers of freshwater yearly from Alaska to Canada and, thence, to the USA and Mexico at an estimated 2014 monetary cost of more than 600 billions USD, an extant NAWAPA would constitute a geopolitical tool useful to Canada and the USA in any future territorial disputes with Latin America's Mexico. The NAWAPA has not yet captured the North American public's attention, remaining a mere pre-shadow, far from concrete realization. There is always the prospect that future climate regime changes and Mexico's overpopulation will stir the publics to ponder it 21st Century establishment. The NAWAPA is most favorably described in The Ralph M. Parsons Company Brochure 606-2934-19, "NAWAPA: North American Water and Power Alliance", issued October 1964. Ralph M. Parsons (1896-1974) founded his civil engineering company during 1944 in Los Angeles, California. [Its current California headquarters in nearby Pasadena is not many kilometers from the State of California's infamous Salton Sea, an artificial inland saltwater-sewage-irrigation runoff sump, created by macro-engineering error in the first years of the 20th Century that is nowadays, technically an out-of-control polluted lake. 2014's Salton Sea is the single most remarkable physical environmental contradiction of a basic tenet of life in the USA—it confounds Southern Californian's and the Nation's traditional citizen confidence, alluded to previously in Chapter 6, that physical macro-problems must inevitably yield to Anthropocenic macro-engineering fixes (Cathcart, 2013).] There is no means existing for anticipating the full impact of NAWAPA construction or to accurately predict California's geological future; the next major earthquake anywhere in Southern California is most likely to be unanticipated since only 10% of all earthquakes are recognizably recurrence of earlier seismic event-processes (Mulargia, 2013; Holzer and Savage, 2013). It is possible that any future enlargement of the Salton Sea could trigger strong seismic activity; ditto for the enormous reservoirs, canals et cetera present if NAWAPA and the GRAND Canal were actualized. However, both the GRAND Canal and the NAWAPA offer the potential of physical landscape connections—horizontal electricity and freshwater flows—between the Geographos-proposed "Arctic Free-Trade Zone" and North America. And, inter-river basin freshwater transfers at the NAWAPA and GRAND Canal geographical scales could help the successful achievement of awesome synergism, even becoming a geopolitical tool to stabilize North America's fluctuating relationship with Latin America. Replenishment of the Ogallala Aquifer which lies beneath thousands of farms and ranches southwest of the Great Lakes would benefit the world's population because so much food is produced there; without a freshwater deposit in the aquifer, which was discovered in 1899, all irrigated agriculture and stock watering endeavors would eventually be terminated for lack. As of early-2014, wells are being bored deeper and deeper to tap this sub-surface supply, sometimes during the 21st Century the entire geological formation will have been sucked dry (Scanlon et al., 2012). Frank Zyback's 1948 introduction of the water-pivot sprinkler system, combined with D.L. McDonald's 1908 instigation of aquifer wellhead pumps, has prompted a massive diminishment of the natural freshwater subterranean deposit. A follow-up technology, the "Continental SuperGrid" concept of "Plantran"-scale pipes filled with liquid hydrogen, each holding a super-conductive wire that would act as a repository and conduit of energy (electricity and fuel) to make possible the mid-21st Century's envisioned Hydrogen Economy in North America and elsewhere (Grant et al., 2006).

Connecting the Great Lakes, which are a result of intensive long-term glacial scouring during the Earth's most recent Ice Age, with the Mississippi River is the Chicago Sanitary and Ship Canal that, by 1900, had fulfilled a long-held (since 1887) public dream for a "Great Lakes-to-the-Gulf of Mexico Waterway". From 1892, approximately

33 million cubic meters of solid material was displaced to dig the channel—"The machines and techniques of excavation developed for the Chicago [Illinois] project demonstrated the feasibility of digging the Panama Canal from 1906-1914" (Condit, 1973, page 28). Since 2 January 1900, thanks to the imposition of an artificial waterway, the Canal has linked hydrologically Lake Michigan with some of the watershed of the Mississippi River. Uncontrolled freshwater exports from Lake Michigan can drastically affect shipping and urban freshwater supply systems of Great Lakes cities. Future climate change as well as introduced or invasive aquatic species arriving via ships, or swimming upstream in the Mississippi River, are a danger too. Invasive fish and other biota are a threat to the stability and productivity of the Great Lakes and some speculation was prevalent amongst professional macro-engineers and others that the Chicago Sanitary and Ship Canal should to be sealed shut, forever, sometime after 2014 in order to prevent any possibility of a future anthropogenic ecological mess occurring in the Great Lakes.

Russia is both famous and infamous for a Communist-era macroproject by which it was planned to reverse the northward flows of several major Siberian rivers that emptied into the Arctic Ocean (Micklin, 1977). The most recent period of active planning for the macroproject was publically announced was in June 1985, about five years before the break-up of the USSR. A vast tract of Russia—especially Siberia—is drained by north-flowing big rivers that finally meld with the Arctic Ocean. For decades during the 20th Century, the former USSR's Communist Party elite, primarily Russians, toyed with ideas for exploiting this wasted freshwater resource by diverting it to the southern USSR's arid landscapes. Post-1985 climate impact modeling indicates that a built transfer system annually transporting 200 cubic kilometers of freshwater southward would evoke nil Arctic Ocean climatic change. Decades of macroproject planning effort were trashed during

March 1986, without detailed explanation to its hostage citizenry or others, until the 16 August 1986 issue of **Pravda** was posted (Micklin, 1986). Europe long entered a post-colonial period, whilst Russia may have done so only by 1991—several ugly international incidences since then make this generalization iffy even in 2014. As in Canada, better data and computer modeling of 21st Century Russia's freshwater resource base may spur a revival of the explorative North-South freshwater rerouting macro-project's planning. Freshwater reservoirs and exports are first-class indicators of the extent to which the boundaries between Earth's global Nature and the Anthropocenic anthropogenic world are vaguely defined since both realms contain some components of the other since freshwater reservoirs are constructed aquatic ecosystems. Russia's north-flowing river runoffs must be named a wasted resource because an insufficient number of freshwater-retention reservoirs are available for filling, and because there is no practical market yet for the freshwater pooled by damming at some considerable financial cost.

Surface freshwater withdrawals from the Missouri River, which eventually joins the Mississippi River, would surely lessen the risks of structural failure at the Old River Control Structure (ORCS) in Louisiana, USA. The function of the ORCS is to draw off floodwaters, which would otherwise pass Baton Rouge and New Orleans downstream, into the Atchafalaya River's, where the flood waters subsequently enter the Gulf of Mexico. ORCS was installed to postpone the Atchafalaya River's natural capture of the entire Mississippi River! ORCS's construction was authorized in 1954, it was completed by 1959 and the water flow was totally "controlled" by 1963. Were the ORCS absent—never built or crumbled due to a design failure—the Mississippi River would flow to Morgan City instead of New Orleans and there would be a deep saltwater estuary reaching northwards as far as Baton Rouge (Rego et al., 2010). Since Hurricane Katrina of September 2006, New Orleans has become a swamp phoenix of a sort. But, macro-engineers have not yet devised a comprehensive plan to keep the city, which rests atop the subsiding Mississippi River Delta, from subsiding and eventually becoming inundated by local sea level rise. One macro-imagineered project plan that has been bruited is one to predominately shift the Mississippi River's sediment load into the Atchafalaya River where the water-borne material will eventually come to rest in a shallow coastal region and, thusly, creating new wetlands replacing that which has been, and will be, lost to erosion of the Mississippi River Delta (Winer, 2007). However, this idea won't remove the impending financial costs that will accrue because of the ruination of New Orleans and Baton Rouge infrastructures. And, one potential Old River Control Structure failure mode not fully addressed is the high probability of major seismic disturbance focused nearby at Madrid, Missouri. The strongest earthquakes ever recorded east of the Rocky Mountains in the USA and Canada by recorded by historians were those at New Madrid during 1811-1812. Whilst such great tremors 200 years ago may not recur for hundreds of years they are, nevertheless, worrisome to Macro-Engineering professionals. The succession of three powerful shocks, originating at New Madrid in the central Mississippi River Valley, began on 16 December 1811 and stopped after 7 February 1812 (Politz and Mooney, 2013; Page and Hough, 2014). The earthquakes have not yet been equaled for number, continuance of disturbance, area affected, and severity. At the Old River Control Structure, authorities must wonder if, during the next big tremor, will the ground on which the ORCS sits be uplifted or subside and will large waves (seiches) be generated on the Mississippi River by earthquake-induced ground motions deforming the riverbed? Despite public indifference, essentially, the ORCS must be made capable of withstanding any expected hydraulic and seismic loading or risk the prospect of a reorganization of the USA's export/import economy on the coast of the Gulf of Mexico. Mitigating such a catastrophic disruption scenario would be the unaffected operations of the Tennessee-Tombigbee Waterway opened in 1985. *Mixing of the Waters: Environment, Politics, and the Building of the Tennessee-Tombigbee Waterway* (1993) by Jeffrey K. Stine reveals the sordid history of this 375 kilometer-long canal that connects Mobile, Alabama with the Tennessee River and which significantly shortened the barge route extending from the northern USA to the Gulf of Mexico by almost 1,400 kilometers. Because it is not a river but, rather, a current-less freshwater channel, it offers lower fuel costs for towboats pushing or pulling long, linked strings of barges. Another future mitigation might be its function as a substitute, not merely a supplementary barge route for the Mississippi River if that waterway becomes blocked, either by future major earthquake, a superbolide impact or by some anthropogenic calamity (conventional, nuclear or bio-terrorism meso-war attack, accidental toxic industrial spillage).

Adabashev (1966) recounts that, in the USSR during the early-1920s and later, several notables (Shlygin, 1975; Kochina and Ratkovich, 1983) seriously considered converting the Sea of Azov from a briny extension of the Black Sea into a gigantic freshwater reservoir by damming the Kerch Strait. The Sea of Azov is a destination for, amongst others, the Don River (Lagutov, 2012) and the Kerch Strait serves as a waterway for about 8,500 ships moving between the Volga River-Kama River and the Azov Sea-Black Sea basins annually. The Sea Azov has been utterly transformed by anthropogenic actions, even freshened artificially by releases from public reservoirs on the major rivers, for example the Don and the Kuban. On 25 December 2003, Russia and Ukraine signed the Treaty on Sea of Azov and Kerch Strait, both are deemed legally as "inland waters". On 10-11 November 2007, the Kerch Strait was beset by high storm waves (of magnitude 6-7) that many ship accidents (sinkings, groundings, damage) that caused major pollutional macro-problems (Matishov et al., 2013). If Kerch Strait were sealed then the Sea of Azov would freshen to the point were it might be usable as a public reservoir, if pollution and previous seabed oil-contamination were handled properly.

From September 1961 until September 1966, all or major parts of four-teen states in the Northeast, about 7% of the conterminous USA's area with (then) almost 28% of America's population, suffered a freshwater supply crisis induced by meteorological drought. A French geographer, Jean Gottman (1915-1994), issued a book describing the areal conti-nuity of cities in the region extending from Boston, Massachusetts to Washington DC, in his *Megalopolis: The Urbanized Northeastern Seaboard of the United States* (1961). ["Megalopolis" is Greek for "very large city".] In 1938, and again in 1957, macro-engineers had offered plans to span the Long Island Sound with bridges, tunnels and causeways but all proposals were ended by 1966. The 1961 "Eastern Long Island Sound Crossing", from Orient Point, New York to Watch Hill, Rhode Island, was to have been an island linking series of spans and causeways that cross Plum Island. Plum Island is the worksite of a contaminated US Government germ experiments laboratory, the infamous Plum Island Animal Disease Center, established in 1951 and closed by 1995 (Carroll, 2004); during 2010 suggestions were publicly voiced that the decontami-nated place ought to become a public parkland. Plum Island's presence may pose some danger for this enormous domestic water supply facility and the worrisome fact that a Category-3 storm, Hurricane Bob, passed directly over Long Island Sound on 18 August 1991 gives further pause to be concerned about storm-stirred freshwater supply purity; in other words, sea-bottom sediments are laced with contaminants! The fresh-water capacity of all extant urban-supply storage places was strained by low natural input and high usage output. By 1966, even Washington DC's reservoirs were nearly dry. Anthony J. Pansini (1966), an electric power transmission expert, and Robert D. Gerard (1926-2010), each separately, proposed to convert Long Island Sound into the USA's most voluminous freshwater reservoir by the construction of dikes at both ends of a post-Ice Age expression of the North Atlantic Ocean (Gerard, 1966; Gerard, 1967). **Figure 16.**

Figure 16. The two straight red lines indicate the low-sill dikes necessary to close-off Long Island Sound from the North Atlantic Ocean so that it gradually freshens and the water contained becomes potable or otherwise useful. (Image: RBC.)

Many glacial features are evident on the seafloor of Long Island Sound long after their submergence by the ocean's advance following the Ice Age. Melt-water from the glaciers that once covered the region collected to form a freshwater lake, "Lake Connecticut", which encompassed most of the present-day Long Island Sound's area. A spillway, caused by erosive water flow, drew down Lake Connecticut and eventually the Sound became dry land, but starting about 15,000 years ago, the air-exposed lakebed was inundated by seawater due to the post-Ice Age eustatic ocean level rise (Latimer, J.S. et al., 2014). Sources to fill the Long Island Sound Reservoir with about 64,000,000,000 cubic meters of freshwater were thought to be the Connecticut and Housatonic river runoffs, together with other streams. The states adjacent to Long Island Sound own it, according to "United States v. Maine et al. (Rhode Island and New York Boundary Case)", decided by the US Supreme Court in 1985, and reported in *US Law Week 53 LW 4151* of February 1989. Long Island Sound is closed by the borderline extending from Montauk Point to Watch Hill Point. Neither Pansini nor Gerard ever considered a possible diversion of Hudson River runoff into Long Island Sound Reservoir. Nowadays, water flowing in the Hudson River's estuary below Troy, New York, is fresh south of Poughkeepsie and saline south of Peekskill. At its western end, Long Island Sound is connected with New York Harbor through a tidal strait, the East River.

The Hudson River's flow might be vastly augmented by intense future use of a refurbished New York State Barge Canal—formerly named the Erie Canal—as projected in *Chattey's Island: The Story of One Man's Breathtaking Plan to Revitalize the Northeaster U.S.A.* (1982) by James Ehmann. [The shallow, outmoded Erie Canal has been declining in value ever since the St. Lawrence Seaway's 1959 opening.] Nigel Chattey (1927-????) dramatized artfully his desired New York State Barge Canal, with a working depth of 8-9 meters that would permit large shipments from the Great Lakes Regions of all kinds of valuable natural resources, such as exportable coal, and manufactures. Then, he also proposed, a large offshore artificial island super-seaport piled on Cholera Bank [depth there about 23 meters] south of Manhattan Island. [Building of islands is based on Nigel Chattey's patented construction technique, the "Universal, environmentally secure, wave defense systems using modular caissons", World Intellectual 94/28253 Publication Number, dated 8 December 1994.] ICCON, an acronym for Chattey's Island Complex Offshore New York and New Jersey, was to become a service center 40 kilometers south of Manhattan Island used for importations via ocean-going tankers and containerships and exportations via ocean-going ships carrying coal and fabricated products of all kinds originating in the USA and Canada. **Figure 17.**

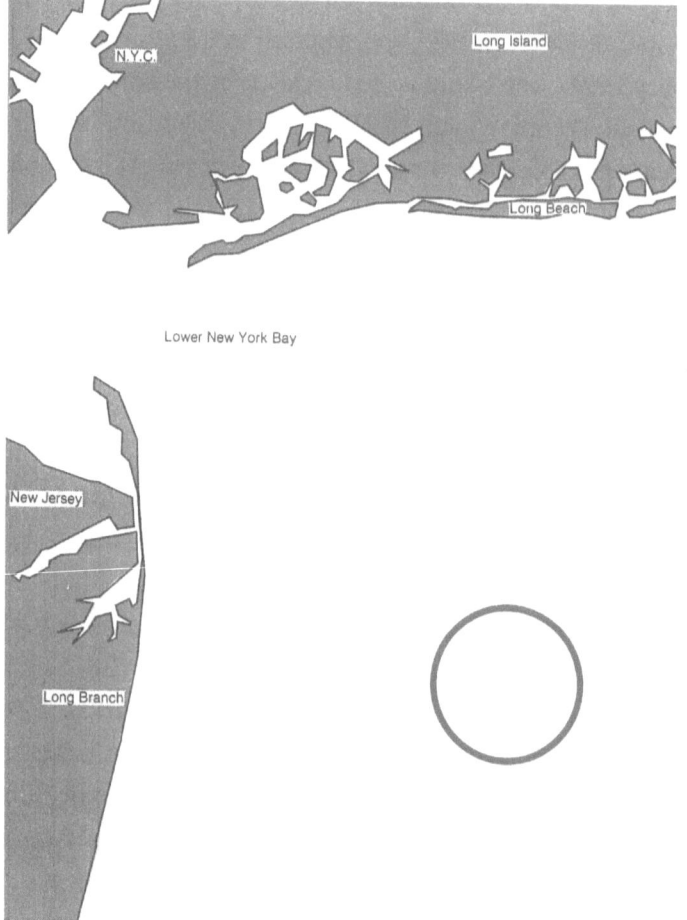

Figure 17. The proper placement of ICONN, as per Nigel Chattey's original macroproject planning concept is shown above. ICONN, represented here by an oval is placed approximately 32 kilometers offshore. Originally Chattey planned for his island to be rectangular. ICONN, sited landward of an imaginary chart line drawn between Cape May, New Jersey and Long Island's Montauk Point, would be

positioned inside the USA's territorial seas and militarily securable. (Image: RBC.)

[Similar concepts have been aired in Europe for artificial islands off the coast of Belgium in the North Sea (Charlier and DeMeyer, 1992).] The bottom of Lake Erie is at a higher elevation than the surface of Lake Ontario—in other words, there is a 30 meter-thick freshwater layer that can be utilized. That layer could be constantly replenished by freshwater inflows via the GRAND Canal and/or NAWAPA. The USA alone owns 28% of Earth's currently recoverable coal reserves, and is becoming a large coal exporter, especially since the start of the 21st Century. If a large percentage of the USA's yearly exported coal were to exit via Chattey's ICONN situated near New York City, then much freshwater used to operate the New York State Barge Canal would flow in the Hudson River—at first flushing riverbed sediments—to the vicinity of Manhattan Island. Surely, some of the Great Lakes Region-James Bay freshwater could be sequestered in the Pansini-Gerard "Long Island Reservoir"? Especially as available local water supplies are increasingly contaminated with road salt (Kaushal et al., 2005). The Pansini-Gerard item of urban infrastructure ought rightly to be termed a restoration scheme because, during the end phase of the most recent Ice Age, even Block Island Sound to the east of Long Island was a giant freshwater lake; Long Island Sound was a freshwater lake for thousands of years before the North Atlantic Ocean breached a recessional moraine at a submarine "seascape" place named "The Race" illustrated well in bathymetric charts, separating Fisher's Island-Watch Hill from Plum Island-Long Island. By no stretch of the imagination could this potential facility be considered a design limit—that is, a design that pushes the planning and construction skills that are the pride of Macro-Imagineering/Macro-Engineering! It is worth noting that Hong Kong reclaimed a nearby bay to serve as its freshwater reservoir.

Geographos first became aware of Nigel Chattey's ICONN by reading a magazine article (Sedgwich, 1981). Chattey sought a release of the landscape and political restraints for industrial development in Gottman's Megalopolis and beyond. ICONN reminded Geographos of another liberating man-made artificial island—Ellis Island, New York City! [Ellis Island was swamped by super-storm Sandy during October 2012, ruined and waterlogged, only to be restored at a cost of many millions of US-taxpayer dollars.] Long past its prime, the Port of New York is static, mostly being converted to parkland as well as world-class skyscrapers with thrilling harbor vistas (Kellner, 2006). Someday, residents of these "view" real estate developments may have far more action scenery to see than they really appreciate presently!

It is currently supposed by geoscientists that, circa 100-400 BC, Manhattan Island was partially overwhelmed by a 20 meter-high tsunami, possibly caused by a superbolide impact 150 kilometer off the coast of New Jersey, or by a super-hurricane. **Figure 18** indicates the proposed macroproject to construct anti-storm surge defenses, dams which close to prevent seawater flooding of valuable landscapes in New York and New Jersey. There is another service that these storm surge dams might serve—of they could not stop very big tsunami caused by an superbolide impact or by super-hurricane, but they could slow and somewhat mitigate the effect on seacoasts. Geographos proposes here that a modified and repurposed Nigel Chattey ICONN become a part of the New York Bight in a circular form that protects the valuable East Coast shorelines located west of ICONN. **Figure 19.** It should be possible to shield shores from tsunamis by making the land "invisible" to the incoming waves. A circular installation with the appropriate baffles will act like an anthropogenic whirlpool, causing the tsunami wave to dissipate and redirect its energy (Farhat et al., 2008). If ICONN were built in a way different than Chattey's macroproject plan, but to serve two purposes instead of just one (seaport). It is worth some R&D

effort! In combination with a damming of the whole of New York Bay, the new ICONN—the big RED DOT will suffice to provide a maximum of protection that is economically feasible!

Figure 18. The rectangular white shapes mark the tentative locations of three storm surge barriers. The very short straight red line near Throgs Neck Bridge (opened 1959) is approximately in the same location as illustrated by Pansini-Gerard "Long Island Reservoir" dike mappings. (Image: Google Images.

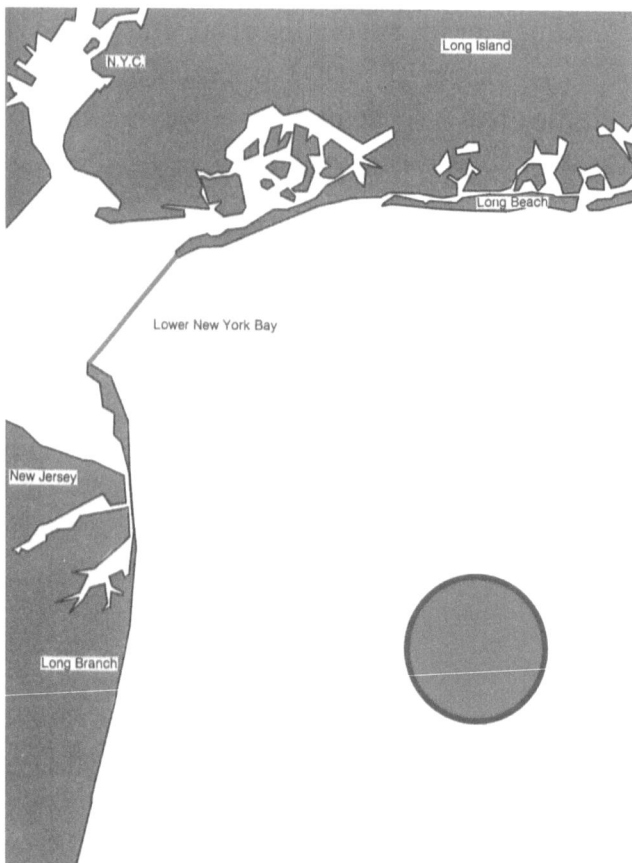

Figure 19. Geographos' suggested "New ICONN" facility—a unique dual-use artificial island that targets the suppression of hurricane storm surges and small tsunamis using a kind of whirlpool hydraulic effect. This facility would not likely be an industrial island like Nigel Chattey's proposed ICONN (Image: RBC.)

References Cited

Adabashev, I. (1966 and 2005) *Global Engineering.* (Honolulu, Hawaii: University Press of the Pacific) pages 128-129.

Anon. (December 2013) *Maersk Mc-Kinney Moller: First of the Triple-E boxboat giants.* Ships Monthly 49: 33-37.

Carroll, M.C. (2004) *Lab 257.* (New York: Harper Collins) 301 pages.

Cathcart, R.B. (2008) *Kra Canal (Thailand) excavation by nuclear-powered dredges.* International Journal of Global Environmental Issues 8: 248-255.

Cathcart, R.B., Bolonkin, A.A. and Rugescu, R.D. (2011) *The Bering Strait Seawater Deflector (BSSD): Arctic Tundra Preservation Using an Immersed, Scalable and Removable Fiberglass Curtain,* pages 741-777 **IN** Badescu, V. and Cathcart, R.B. (Eds.) *Macro-engineering Seawater in Unique Environments: Arid Lowlands and Water Bodies Rehabilitation.* (The Netherlands: Springer) 790 pages.

Cathcart, R.B. (2013) *Macro-Imagineering a "Fifth Coast" USA: Salton Sea Megaport and Seawater Canal.* Chapter 4, pages 127-156 **IN** White, M.R. (Ed.) *Seawater: Geochemistry, Composition and Environmental Impact.* (New York: NOVA Publishing) 176 pages.

Charlier, R.H. and DeMeyer, C.P. (1992) *An Environmental Purpose Artificial Island Off Belgium.* International Journal of Environmental Studies 40: 249-265.

Chen, F. (2013) *The Kuroshio Power Plant.* (The Netherlands: Springer) 225 pages.

Coen, R. (2012) *Breaking Ice for Arctic Oil.* (Alaska: University of Alaska Press) 215 pages.

Condit, C. (1973) *Chicago 1910-1929: Building, Planning, and Urban Technology.* (Chicago: University of Chicago Press).

Davis, B. (30 September 2013) *How China Chases Innovation.* The Wall Street Journal CCLXII (77): A2.

Dotterweich, M. (2014) *The history of human-induced soil erosion: Geomorphic legacies, early descriptions and research, and the development of soil conservation—A global synopsis.* Geomorphology 201: 1-34.

Doucet, L.J. (1955) *The Road to the Isle: The World's Deeptest Causeway".* (Fredericton, New Brunswick: University Press of New Brunswick Limited) 48 pages.

Egerton, F.N. (1973) *Changing Concepts of the Balance of Nature.* Quarterly Review of Biology 48: 322-350.

Evans, J. (1999) *Celluloid Mushroom Clouds: Hollywood And Atomic Bomb.* 224 pages.

Evans, J. St. B.T. (2007) *On the resolution of conflict in dual process theories of reasoning.* Thinking & Reasoning 13: 321-339.

Farhat, M. et al. (2008) *Broadband Cylindrical Acoustic Cloak for Linear Surface Waves in a Fluid.* Physical Review Letters 101: 134501.

Gerard, R.B. (August 1966) *Potential Freshwater Reservoir in the New York Area.* Science153: 870-871.

Gerard, R.D. (November 1967) *A Long Island Sound Reservoir.* Journal of the American Waterworks Association 59: 1351-1356.

Gorer, G. (1948) *The Americans: A Study in National Character.* (New York: W.W. Norton).

Gorin, G. et al. (2003) *Bedrock, Quaternary sediments and recent fault acitivity in central Lake Neuchatel, as derived from high-resolution reflection seismics.* Ecologae geol. Helv 96: S3-S10.

Grant, P.M. et al. (2006) *A Power Grid for the Hydrogen Economy.* Scientific American 295: 76-83.

Hammons, T.J. et al. (September 1989) *Feasibility of Iceland/United Kingdom HVDC Submarine Cable Link.* IEEE Transactions of Energy Conversion 4: 414-424.

Headland, R.K. and Splettstoesser, J.F. (1999) *First circumnavigation of the Arctic by tourist vessel.* Polar Geography 23: 2460249.

Holzer, T.L. and Savage, J.C. (February 2013) *Global Earthquake Fatalities and Population.* Earthquake Spectra 29: 155-175.

Hu, A. et al. (2012) *Role of the Bering Strait on the hysteresis of the ocean conveyor belt circulation and glacial climate stability.* Proceedings of the National Academy of Sciences 109: 6417-6422.

Hughes, T.P. (1989) *American Genesis: A Century of Invention and Technological Enthusiasm, 1870-1970.* (New York: Viking).

Jacobs, B.L. (September-October 1994) *Serotonin, Motor Activity and Depression-Related Disorders.* American Scientist 82: 456-463.

Jain, R.K. (February 2014) *Viewing Cancer as a Physics Problem Suggests New Treatments.* Scientific American 310, Issue 4.

Kaspi, Y. and Schneider, T. (2011) *Winter cold of eastern continental boundaries induced by warm ocean waters.* Nature 471: 621-624.

Kaushal, S.S. et al. (2005) *Increased salinization of fresh water in the northeastern United States.* Proceedings of the National Academy of Sciences 102: 13517-13520.

Kellner, A.D. (2006) *New York Harbor: A Geographical and Historical Survey.* (NC: McFarland) 199 pages.

Kim, K-G. (2013) *The Demilitarized Zone (DMZ) of Korea.* (The Netherlands: Springer) 583 pages.

Kochina, T.A. and Ratkovich, DYa. (May-June 1983) *Control of Azov Sea Salinity by Regulating Water Exchanges in the Kerch Strait.* Water Resources 10: 229-241.

Lagutov, V. (Ed. (2012) *Environmental Security in Watersheds: The Sea of Azov.* (The Netherlands: Springer) 253 pages.

Latimer, J.S. et al. (2014) *Long Island Sound.* (The Netherlands: Springer) 250 pages.

Leiserowitz, A.A. (November 2004) *Before and After the Day After Tomorrow: A U.S. Study of Climate Change Risk.* Environment 46: 22-37.

Mackinder, H.J. (1931) *The Human Habitat.* Scottish Geographical Magazine 47: 321-335.

Marter, J. (October 2013) *Systems: A conversation with Nancy Holt*. Sculpture 32: 28-33.

McLaughlin, W.J. (1987) *The Infinite Organism*. Interdisciplinary Science Reviews 12: 160-170.

Matishov, G.G. et al. (May 2013) *The environmental and biotic impact of the oil spill in Kerch Strait in November 2007.* Water Resources 40: 271-284.

Micklin, P.P. (1977) *Nawapa and Two Siberian Water-Diversion Proposals*. Soviet Geography: Review and Translation 18: 81-99.

Micklin, P.P. (1986) *The Status of the Soviet Union's North-South Water Transfer Projects Before Their Abandonment in 1985-86*. Soviet Geography: Review and Translation 27: 287-329.

Milcu, A.J. et al. (2013) *Cultural Ecosystem Services: A Literature Review and Prospects for Future Research*. Ecology and Society 18: 44.

Milko, R. (1986) *Potential Ecological Effects of the Proposed GRAND Canal Diversion Project on Hudson and James Bays*. Arctic 39: 3166-326.

Minsky, M. (October 1994) *Will Robots Inherit the Earth?* Scientific American 271: 113.

Mulargia, F. (2013) *Why the Next Large Earthquake is Likely to be a Big Surprise*. Bulletin of the Seismological Society 103: 2946-2952.

Muller, D.K. et al. (2013) *New Issues in Polar Tourism*. (The Netherlands: Springer) 224 pages.

Myhra, D. (2002) *Sanger: Germany's Orbital Rocket Bomber in World War II.* (New York: Schiffer Publishing) 176 pages.

Nordlund, C. (2001) *"On Going Up in the World": Nation, Region and the Land Elevation Debate in Sweden.* Annals of Science 58: 17-50.

Page, M.T. and Hough, S.E. (2014) *The New Madrid Seismic Zone: Not Dead Yet.* Science. DOI: 10.1126/science.124825.

Panova, Y. (2011) *Potential of connecting Eurasia through Trans-Siberian Railway.* International Journal of Shipping and Transport Logistics 3: 227-244.

Pansini, A.J. (April 1966) *The Long Island-Northeast Plan.* Eastern Industrial World 4: 28.

Partl, R. (October 1978) *Power From Glaciers: The Hydropower Potential of Greenland's Glacial Waters.* Energy 3: 543-573.

Philbrick, A.K. (January 1982) *Hierarchical Nodality in Geographic Time-Space.* Economic Geography 58: 1-19.

Politz, F. and Mooney, W.D. (July 2013, in press) *Seismic structure of the Central US Crust and shallow upper mantle: Uniqueness of the Reelfoot Rift.* Earth and Planetary Science Letters.

Powell, R.C. (2014) *Polar Geopolitics?* (Northampton, MA: Edward Elgar Publishing) 256 pages.

Rego, J.L. et al. (March 2010) *Numerical Modeling of the Mississippi-Atchafalaya Rivers' Sediment Transport and Fate: Considerations for Diversion Scenarios.* Journal of Coastal Research 26: 212-229.

Riser, S.C. and Lozier, M.S. (February 2013) *Rethinking the Gulf Stream*. Scientific American 308: 50-55.

Rogers, T.F. (July 1981) *Reflector satellites for solar power*. IEEE Spectrum 18: 38-43.

Sabin, P. (2013) *The Bet: Paul Ehrlich, Julian Simon, and Our Gamble over Earth's Future*. (Yale University Press) 320 pages.

Sedgwick, J. (January-February 1981) *Nigel Chattey Has Power Surges, Then He Macrothinks*. NEXT 2: 72-79.

Shlygin, I.A. (1975) *Possibility of varying the salt exchange between Black and Azov sea by creating a freshwater buffer zone*. Meteorol. Gidrol. (Moscow), No. 4, pages 97-100.

Smith, L.C. and Stephenson, S.R. (2013) *New Trans-Arctic shipping routes navigable by midcentury*. Proceedings of the National Academy of Sciences 110: E1191-E1195.

St-Laurent, P. et al. (2012) *A conceptual model of an Arctic sea*. Journal of Geophysical Research 117: C060010.

Thapa, R.B. et al. (2007) *Sea navigation, challenges and potentials in South East Asia: an assessment of suitable sites for a shipping canal in the South Thai Isthmus*. Geo-Journal 70: 161-172.

Van Pelt, M. (2012) *Rocketing Into the Future: The History and Technology of Rocket Planes*. (The Netherlands: Springer) 384 pages.

Weiss, E.A. (1992) *HAL 9000 (1992-1998)*. IEEE Annals of the History of Computing 14: 53-54.

Williams, A.J. (2011) *Nancy Holt Sightlines*. (Los Angeles: University of California Press) 296 pages.

Winer, H. (2007) *Re-Engineering the Mississippi River as a sediment Delivery System*. Coastal Sediments '07: 712-721.

Wu, P. (1998) *Will earthquake activity in eastern Canada increase in the next few thousand years?* Canadian Journal of Earth Science 35: 562-568.

CHAPTER 7

MACRO-PROJECT GEO-ECONOMICS AND POLITICS

Somber predictions about routine endeavors involving earth material movements have varied from good, under exceptionally favorable circumstances, to bad, in some occurrences where geographical data seemed reliable. At what depth does the Earth-crust become environmentally irrelevant? The lower boundary of the humanly important part of the whole Earth clearly depends on the nature of the geotechnical macro-problem in question and of the structures and/or event-processes intimately involved. Geographical predictions are founded on baseline data that indicate a comparable historical record of regional and/or global Nature segregated environmental conditions, a qualitative and quantitative evaluation of an impact parameter. If local or global Nature baseline data are sparse, inaccurately recorded, or exaggerated, then predictions based on them are bound to be useless, misleading, or even dangerous. A fascinating example is the LUSI [Lumpur, or mud Sidoarjo) volcano that has, 27 May 2006, inundated more than seven square kilometers of the Indonesian East Java town of Sidoarjo with water and mud of exclusively subterranean origin. LUSIs birth is still debated, whether it is anthropogenic or Nature-caused remains undetermined (Davies et al., 2011). During 2014, the mud volcano is an implacable geo-hazard around which social life centers. **Figure 1.**

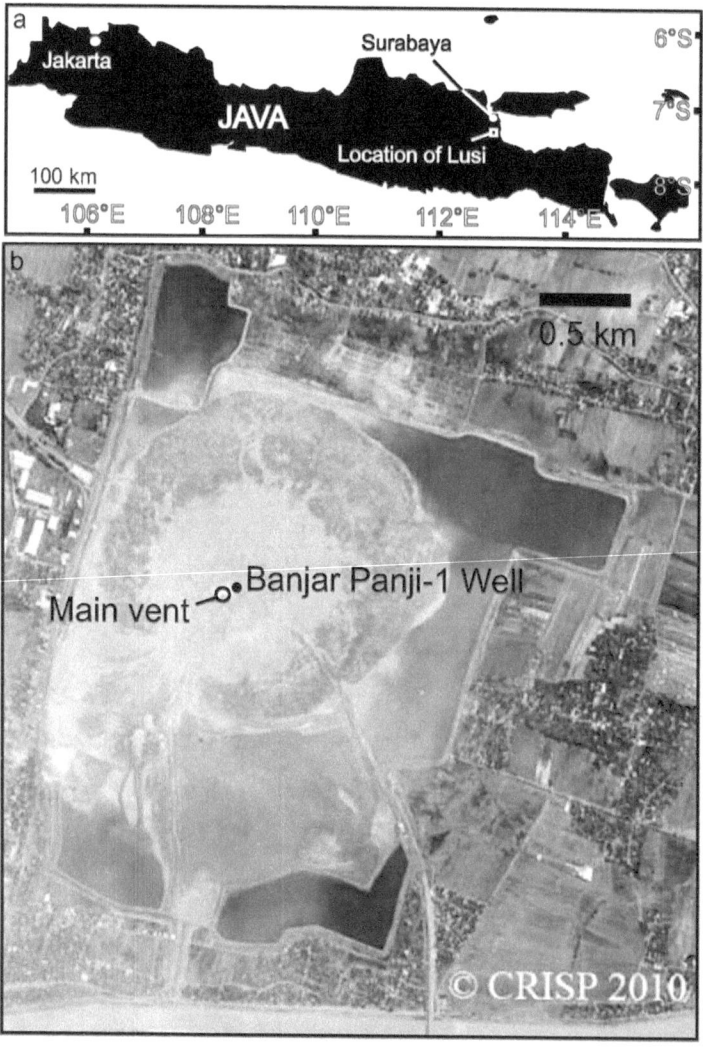

Figure 1. LUSI mud volcano aerial image. (Image: Google Images via CRISP, 2010).

So, how credible are highly-generalized geographical predictions focused on fuzzily-conceived, extra-routine macroprojects? Science's collection of theories always results in a real-world planet of human-made "data-scapes/guess-scapes" whilst applied Art's theories often result in many displayed humane "dream-scapes" of some type. Macroproject budgets and government finances are governed by numbers, the outgoing bank cheques and the fluctuating bank balance, as well as by elaborate internal or external structural political/geopolitical checks and balances! Humankind's current so-called economic "globalization" is nothing more than a temporary trompe-l'oeil, purveyed by the elitist and wealthy globalized upper-crust social group to engage the Earth-Noosphere's working class in its most recent self-serving Machiavellian psychodrama, according to its harshest negative critics! The deductive qualitivity of Geopolitics' approach (interpretation) should be recognized by all macro-imagineers/macro-engineers as important as Anthropo-geomorphology's inductive quantification, since all existing event-processes of our Earth-biosphere change and event-processes of the planet's past Geologic Time assume vital meaning when they are found to have set our Earth-biosphere's "stage" for a loss of life and property!

Peruvian Amazon River Basin (**Figure 2.**) primitives—some of the last folks ever to make social contact with the rest of humanity since their ancestors crossed the Berin Strait tens of thousands of years ago—as movingly portrayed in macro-imagineer Frank Paul Davidson's favorite "professional" commercial film, *Fitzcarraldo* (1982), see their marginal Tropic Zone as region as a "world", or "dreamscape", made geophysical.

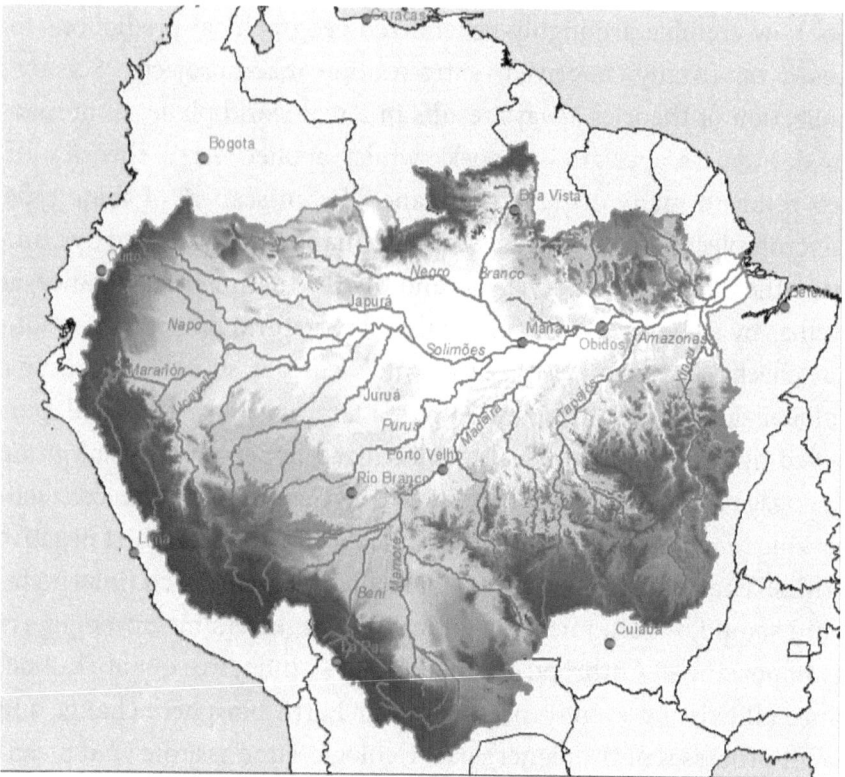

Figure 2. The Peruvian part of the Amazon River Basin is on the left-side, east of Lima, Peru's main city. (Image: Google Images.)

The movie was directed by Werner Herzog and starred the late actor Klaus Kinski. The entertainment conveyed the story of an Irish soldier-of-fortune who tries to haul a steamboat over a mountain to another river in his very personal search for a new water route to the Amazon River Basin. Since Herzog and the film's cast of characters actually did what the movie's main character did, this cinema drama is a fascinating record of this "heroic", if crazy-man, effort! **Figure 3.**

Figure 3. A full-size movie set decoration—in other words, a life-size ship being towed and pushed uphill for real effect in a filmed drama! (Image: Google Images.)

In a very real sense, modern peoples living in our world's discontinuously spaced Astythromes share that dreamy viewpoint since our geophysical surroundings may be but a kind of "data-scape shadow" of the real four-dimensional space-time continuum of localized global Nature worlds, or Noosystems (Walker, 1983). F.P. Davidson might reconsider an earlier commercial comedy film amusement, *The President's Analyst* (1967). Rodoman-ALPS, where televised on-screen images based on a "data-scape" would almost be the equivalent of being-on-the-scene, is a direct result of a unique blending of illusion with reality, like *Fitcarraldo*, but with a desire to reverse as well as enlarge the 1 January 1984 breakup of an enormous USA telecommunications network (American Telephone and Telegraphy Company) existing since 1885. [India, during 2013, finally turned-off, and began to dismantle, its outdated nation-wide telegraph system operation because Indian customers prefer personal e-mailing to public telegraphy.] Global telecommunications as it exists in 2014 is frighteningly comprehensive. If Frank P. Davidson's admired primitives really desire to encounter a wild dream world, then the United Nations Organization ought to donate portable cellular devices and, therefore, give the Amazon River Basin natives unlimited free long-distance call service to randomly-dialed numbers. If any of those doing so were proficient in several languages, it is possible they could speak with a spationaut orbiting Earth in the International Space Station! Maybe a freely-translated conversation with people so different from themselves would induce stranger dreams than those they presently endure?

Macro-Imagineering proposals for additions to Earth's future contents, contours and climate regimes ought to always be considered as "iffy" propositions. A naked primitive's abilities for committing mayhem are decidedly limited. Humankind—the prefix of "Anthropogeomorphology"—according to Sigmund Freud (1856-1939), is an unpredictable prosthetic near-deity: "Man has become a god by means of artificial limbs, so to speak, quite magnificent when equipped with all his accessory organs; but they do not grow on him and they still give him trouble at times.... Future ages will produce further great advances...and will increase man's likeness to a god still more.... All the same,...the human being of today is no happy with all his likeness to a god" (Freud, Great Books, 1952). It is possible future prosthetic attachments [cyborg] and robots may make it disadvantageous to remain human or to evolve further as *Homo sapiens*. A somewhat less joyless Freud sensed an inbred human aggressiveness as the greatest impediment to the formation and continuance of human civilization. The question remains open, however, if a spatially boundless cyborg and/or robot civilization would be one of joy (in German language, "freude")! The Universe with all things in it (God, the Aliens, the human world, the human soul, and all things in general) is partly observable and somewhat ponderable.

Perhaps it would be best to comment, at least until the Anthropocenic is replaced by an Age of Molecular Nanotechnology that is unmistakably underway and commonly apparent, that *Homo sapiens* is a non-omnipotent "deity", a fallible socializing group of organisms with "free will" subject to the Ultimate Managerial Energy/Element [Creative Force or God's] creation, a known, seeable Universe governed by discovered and discoverable laws which came into existence, along with Astrologic Time, when the known Universe began (Morgan, 1994). Modern Biology has borrowed terms quite liberally from Engineering's principles vocabulary (Pauwels, 2013; Calcott, 2013) and the mixture of words and concepts is ramifying fast in order to identify and describe some of the concepts previously mentioned in this paragraph. Documented human noosystems formations is the historical social event-process of *Homo sapiens'* self-definition and self-organization against universal, not merely Earth-biosphere, "chaos". As Earth expresses "laws"—that is, localized "Laws of global Nature" as understood by our extant Sciences, Technologies and Arts—whatever be their origins, emboldened geoscientists and technologists as well as artists must think that all is risk and reward. The motion of anthropogenic Earth satellites is slightly chaotic and aerospace engineers design orbits that are stable over the expected lifetime of the satellite. The motion of the planets and moons of our Solar System is chaotic; Earth's orbit around the Sun undergoes change due to both stellar mass loss and tidal dissipation (Spiegel and Madhusudhan, 2012). An errant asteroid or comet—28 million years ago a comet exploded above the Sahara's sand, converting 6,000 square kilometers of sand into glass (Kramers et al., 2013)—jerked by some barely understood cosmic force from a stable Earth-crossing orbit, could subject our 21st Century world to a terrible onslaught, perhaps vaporizing many persons, when landscape and seafloor elevations would then be markedly altered in minutes and a massive volume of magma becomes lava, possibly even at the Earth-crust's antipodes. Again, purely for eleemosynary gain Hollywood's cinematographers have capitalized on this epochal scenario with such epics as *Meteor* (1979), *Deep Impact* (1998) and *Armageddon* (1998). Without boarding any particular geo-ideological bandwagon, Macro-Imagineering/Macro-Engineering practitioners must read a wide spectrum of scientific, technical and geopolitical literature so as to acquire a sharp-sighted grasp of all germane, confirmed facts. Sigmund Freud was not speaking of cyborgs, which are a kind of centaur, but professionals must think of our worlds as truly self-reproducing factories embedded in a multi-global Solar System-based *Homo sapiens* civilization—as if they were James G. Ballard's geopolis (Earth, Mars, Venus, the Moon) or, perhaps, a copycat macroproject like Richard L.S. Taylor's Mars Worldhouse! Ultimately, humans, cyborgs, Terra-creatures are left with the ambiguous "greatness" of occupying an unstable and tentative position in this known Universe, whatever its hardly calculable dimensions or extensions.

Apprehending and determinedly mulling Earth's 2014 volatile and vituperative social atmospherics, events in our processing minds during current global economic conditions are the least surefire factor affecting our fragile Earth-world's future state, or existence. A French sociologist, Gustave Le Bon (1841-1931), in *The Crowd* (1913, page 15) resolved: "The age we are about to enter will in truth be the era of crowds". Adolf Hitler in Nazi Germany was the first national leader to prove Le Bon's dictum a fact; today's globalized "Television Generation" could become the next victims and perpetrators; if humankind does not learn to control its worst behavior patterns, then Geopolitics ought to focus on generational estate planning! Without the slightest aberration, however, Geographos thinks and believes that modern Geopolitics and Macro-Imagineering and Macro-Engineering professionals should look ahead with optimism (laced with steely realism) to 2057, the 100th Anniversary of Sputnik-1 and the International Geophysical Year.

Founded in 1966, the Stockholm International Peace Institute (SIPRI) during 1984 published *Environmental Warfare: A Technical, Legal, and Policy Appraisal*. SIPRI's monographists decry the awful macro-problems that would follow a hostile disruption of the Old River Control Structure on the Mississippi River in Louisiana, USA. The USA suffered its first Guernica or Rotterdam on 11 September 2001. It is entirely possible that other ecosystem-nations will also be attacked in a devastating manner, perhaps even on a larger scale of death and infrastructure destruction. SIPRI's slim book offered emendation to the 1977 Convention prohibiting "Geophysical Warfare", which would permit organization of Boris B. Rodoman's "Biosphere Command Centers". An intensive check-over of that Convention seems to lead to a non-stretched interpretation of its Article 56:5-6 as an opportunity for Macro-Imagineering's adherents to progress to a quiet conclusion: Article 56's clause 6 suggests that further agreements among the Convention's signers could provide additional physical protections for macro-objects [and even Timothy Morton's "hyper-objects"?] containing dangerous forces or materials. Since ecosystem-nations are to be considered as technical systems (micro- and macro-objects) and their elimination through conflict (conventional and/or unconventional warfare) puts at risk our Earth-biosphere, why not form an agency to build on that 1977 Convention? Specifically, non-terrorist lodge humans ought to wholeheartedly promote the agreeability of strategic anti-ballistic missile defenses, a melding of machinery benefiting the all of our world's biota (Stenke et al., 2013).

As currently constituted, the United Nations Organization reflects the post-WW II Superpower's desire to preserve the integrity of the inter-state system of legal relations. An amalgamation of various military doctrines, embodied in an augmented Rodoman-ALPS—the suffix is pronounced like the European mountain range, our "ALPS"—could help professionalized Macro-Imagineering and Geopolitics in something other than an "arena" where mobile, sometimes stateless, terrorist lodge "gladiators", green with technological envy, train to attack other settled people. Arms of all types, even super-weapons of mass destruction capabilities, are available in 2014 to acquisition by terrorist lodges worldwide and, in future years of the 21st Century, the Earth-biosphere may well be radically transformed by charismatic clan leaders who can link terrorist lodges into a chained band of awesomely destructive primordial power. Many nation-ecosystems have already been the victims of large-scale terrorism. It seems utterly impossible to devise a deterrent policy that might be universally applicable. Biosphere damage is bound to happen before the least wealthy Noosystems can achieve states equivalent to the topmost prevailing Standard of Living via perfected molecular Nanotechnology. Rodoman-ALPS should be constructed to militate against the globalization of mini- and meso-wars. Robots made "alive" by our Technology, could perform as tireless Environmentalists, taking on tasks of planetary cleaning, which Antoine de Saint-Exupery had to delegate to his adorable Little Prince!

Geographos' late mentor Frank Paul Davidson wished that Macro-Engineering curricula trained its "off-spring" in the styles of Diplomacy and Engineering, whilst one of the early proponents of anti-ballistic missile defenses, Daniel Orrin Graham (1925-1995), on 17 November 1985, lamented that professional diplomats were less competent to preserve the USA than its engineers! Davidson developed privately an answer, "Engineering Diplomacy", to Graham's public complaint, which time's passage will probably uphold (for example, Mumme, 1986). Davidson, the USA's premier popularize of Macro-Engineering, editor of six basic textbooks, trained in International Law, mays someday be viewed at the individual who did for the just-emerging Macro-Imagineering/Macro-Engineering profession what Charles Lyell (1797-1875), also an attorney, did to popularize the overly abstract basic insights of Earth's geological structure provided by James Hutton (1726-1797)—but only with remarkably better advice on matters of geo-science germane to Macro-Imagineering. [My negative critique is not professional animosity. Rather it is based on observation of the sometimes skewed handling and descriptive weakness of matters geological and astronomical of Davidson's otherwise excellent textbooks.]

Indoctrinated, internationally-recognized Green journalists, often read, heard or seen performing by millions of persons, are easily able to name some biosphere risks presently impacting Earthlubbers, and often seek to stymie Technology's R&D programs in the vain hope of "fossilizing" Earth as an Anti-Object Artwork! Considering the world-public receives little information on Terraforming, it is only natural that such a far-term future oriented profession should suffer slow growth. The profession of Macro-Imagineering/Macro-Engineering must not endure that same kind of non-reporting or negative news media criticism, since it would be unhealthy for planet Earth. Molecular Nanotechnologists are the key innovators supporting all further industrial "Big Thinking" in Macro-Imagineering/Macro-Engineering and Terraforming. Planning started with *Homo sapiens* farming thousands of years ago; planning would possibly terminate with molecular Nanotechnology's near-term future perfection.

Humanity's influence on the Earth-atmosphere energy budget has arisen from several sources. Although the world-ocean's surface is virtually unaltered—except for shipping's smokestack contrails and the fact that ships used by people displaces 1,767,000,000 cubic meters of seawater, keeping the ocean approximately five micrometers higher than it would otherwise be—*Homo sapiens* has markedly changed the landscape's reflectivity or absorptivity. Irrigation and agriculture-induced net heating of the troposphere is caused by freshwater evaporation. Massive injections of particulate matter such as the various forms of smog, has caused a net cooling of the air (dimming) by reflecting solar radiation; when smog is diminished by applied regulations such as clean air laws, the air warms following clearance (brightening) (Wild, 2012). The Asian Brown Clouds—above India it is composed of airborne dust from Rajasthan and dark aerosols from India's factories and low-grade coal burning, shades the region in which it is generated; above China, it exists due to the very same causes. **Figures 4-5.**

Figure 4. Oblique view from outer space viewpoint of India and Bangladesh sunlight-dimming smog. Anthropogenic aerosols are increasing above India and Bangladesh (Balcerak, 2013). The human

health impacts of air pollution and, generally, global climate changes is being correlated by HEAL [Health & Ecosystems: Analysis of Linkages], a consortium of medical personnel focused on the public health risks of anthropogenic changes to all of Earth's biosphere ecosystems (Meyers et al., 2013). Asia's immense recent emission of aerosols affects the upper atmosphere, including cloud formation, precipitation, cyclonic storm intensity and other factors of climate for the entire Northern Hemisphere (Wang et al., 2014)! (Image: Google Images.)

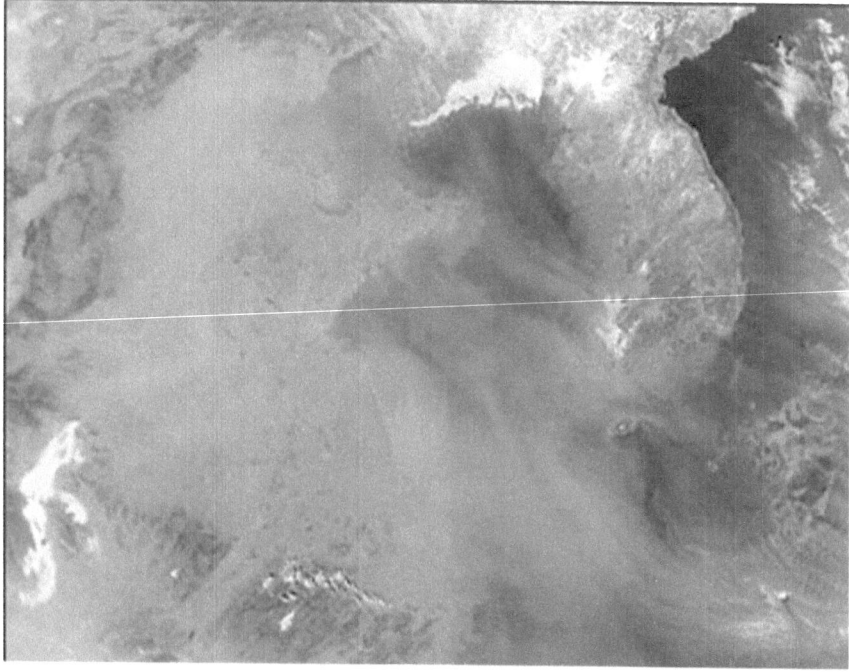

Figure 5. Space image of thick smog blanketing eastern China, the Yellow Sea and the Korean Peninsula as well as southern Japan! Eventually the dispersed smog will be carried by wind currents to North America. (Image: Google Images.)

In many Astythromes and Anthromes, dimming and brightening are macro-problems, although this atmospheric opacity ailment cannot be called "global" yet because it is caused by people and industry only in some landscape regions.

Carbon dioxide gas and other Greenhouse gases possibly have caused some heating of the Earth's air, but also sometimes even a cooling of the air—it is a mixed bag of conclusions that afflicts Climatology. Urban heat islands are generated by both a carbon dioxide gas buildup but also other gases and smog. The primary Greenhouse gas present in Earth's air is water (in the form of gas, not vapor), which causes 80% of the warming as compared to 20% for carbon dioxide gas. If gaseous carbon dioxide were subtracted all at once from the Earth's atmosphere—say, by molecular Nanotechnology means—then the immediate subsequent 3% decrease in 2014's global Greenhouse Effect would inevitably result in a 1^0 C decrease in Earth-normal global average temperature. [If 100% of air's carbon dioxide content were converted technically to carbonate rock, its distribution globally could produce a worldwide undifferentiated anthropogenic stratum 2.5 millimeters thick of Anthropic Rock—collected in one warehouse as a single lump, its bulk would total 128 cubic kilometers! Removal of all carbon dioxide gas from air would kill all life, so only robots might be interested in doing this strange deed!] The **World Meteorological Organization Greenhouse Gas Bulletin, Bulletin No. 1**, issued on 14 March 2006, purports that since the beginning of the Industrial Age circa 1750, "...atmospheric CO2 has increased by 35%, primarily because of emissions from combustion of fossil fuels...and, to a lesser extent, deforestation...". The WMO submits that carbon dioxide gas is "...the single most important infrared absorbing, anthropogenic gas..." in air and "...is responsible for 62% of the total radiative forcing of the Earth by long-lived greenhouse gases". The odd placement of the comma between "absorbing" and "anthropogenic" surely gives an exaggerated impression of social mitigation urgency to many Science-uninformed persons! On warm and dry regions, such as the Sahara or the vast central deserts of Australia, the 14% increase in atmospheric carbon dioxide gas that happened between 1982 and 2010 served to fertilize green plant growth—for example, increasing by 11% the amount of green foliage on warm and dry regions of our planet (Donohue et al., 2013). Could the greening of Earth's deserts actually be somehow construed as utterly detrimental to other kinds of Earth life?

Modern humans continue the ancient practice of sky-gazing, even in smoggy Astythromes, and of thinking that humans, alive and as preserved dead (cremains), will someday populate the Moon (Boyle, 2013; Damjanov, 2013). Cremains [a kind of sediment], or even graves, like burials within the Earth-biosphere on land and in the world-ocean, literally lays claim to the desolate, gray lunar landscape to which humans are newly introduced [1969] and creates an historical, archaeological continuity for all future Moon settlers. Even "extreme tourism" has been bruited as a possible 21st Century wealthy human recreational activity (Spennemann, 2007). The Moon (and Sun) helped the first humans to acquire a sense of organized time (Williams, 2014). Human familiarity with the Moon's phases had considerable practical consequences in pre-Industrial Revolution societies reliant on the reflective Moon for all nighttime outdoor illumination. Nowadays humans are aware that Earth's winds are affected by eclipse-induced changes (Gray and Harrison, 2012). The full Moon is 0.0000067 as bright as the overhead Sun. Outdoor artificial lighting plays a significant role in the world-public's acceptance of 21st Century architecture. Despite the low Bond (0.11) and visual geometric (0.12) albedos, the Moon is so near to the Earth and reflects so much sunlight, as moonlight, that it qualifies as the second brightest light source in Earth's sky, after the Sun. Each "spherical" celestial body [Moon and Earth] receives a solar irradiance of 1367.6 Watts per square meter and each has its unique absolute spectral reflectance signature. In modern Noosystems, romance-inspiring moonlight is not yet a matter of everyday practical importance; that may change as a globalized human civilization becomes increasingly industrialized, ever more technically reliant upon controlled energy generation and flows. Humanity's present-day needs now exceed 18 Terawatts. A society separately situation on the Moon, which has but 7% of Earth's surface area, could periodically measure and monitor Earth's Bond (0.306) and visual geometric (0.367) albedos. Approximately 13,000 Terawatts of solar energy strikes the Moon, and except for three hours during a full eclipse, the Moon is steadily exposed to sunlight; during the period 2001 to 2200, future Moon colonists and visiting Earthlings will experience only 155 eclipses. Continuous human monitoring of the Sun commenced during February 1996 when the Solar and Heliospheric Observatory became operational. Settlement of the Moon is expected to occur during the 21st Century. The Moon's rocky surface, pitted by craters created by collisions with space debris, reflects about 7-11% of the sunlight it intercepts; every square meter of Sun-exposed lunar surface is equivalent to a single 275 Watt light-bulb! The Moon is 31% brighter at perigee than at apogee: during our 21st Century its closest perigee (of 356,421 kilometers) will happen on 6 December 2052; during the 22nd Century, its most distant apogee (406,720 kilometers) will occur on 3 February 2125. Envision a "Disco Ball" Moon emitting continuously hundreds of thousands of lumens, its grayish natural landscape totally obscured by anthropic machines and mirrors! Would romanticism (Neave et al., 2011) permeate the Earth-Noosphere? [Far-distant Aliens might see it as a semaphore's signal!] Reflected sunlight, as side-effect, is assured if the ribbon-region of photovoltaic solar-energy collectors, LUNA RING, team macro-imagineered by a Japanese company, was constructed onsite (of mined and industrially processed lunar soil) sequentially emplaced as sectionals in a broad equatorial band on the Moon's surface. (**Figure 6.**)

Figure 6. The resplendent LUNA RING, illustrated as designed by macro-engineers at Japan's Shimizu Corporation, will circle the Moon's 11,000 kilometer equator. As wide as 400 kilometers, its collecting area will be 4,365,600 square kilometers. Its power might be transmitted to Earth as microwaves (Tucker, 2011). (Image: Shimizu website.)

Some of the biological consequences of reflected LUNA RING moonlight striking Earth are changes in human responses (Foster and Roenneberg, 2008) and world-ocean plankton (Hernandez-Leon et al., 2010); since landscapes of Astythromes are already polluted by artificial lighting at night, changes in animal and plant behaviors may not be as remarkable. As for the Moon's surface, regolith mining and other human activities (past, present-day and future) of exploration and exploitation will push some Greens to demand legal protections for historical sites and artifacts—an "Off-World Heritage" site system, if you will (Spennemann, 2004). LUNA RING will be subjected to bombardment by space debris, just like Earth is, and preliminary tallies suggest that there is an asymmetry of impact flux between the Moon's poles and its equatorial region—possibly 10% greater where the LUNA RING will be deployed (Oberst et al., 2012).

Selenography, the equivalent of Geography, has few landmarks on the Moon to serve as easily recognizable region-organizing framework—"sameness" visually speaking on a gigantic scale; even as the late Neil A. Armstrong first stepped onto the pitted and cratered lunar surface on 20 July 1969, macro-engineers were still uncertain about the places' human-scale geomorphology. Whether laboriously walking in their encumbering spacesuits, or riding the three Moon Buggies (Lunar Roving Vehicles) abandoned on the Moon in operable condition by Apollo Missions 15, 16, and 17 before 30 July 1971, for past Moon explorers as well as future human explorers, the lunar bulge, horizon, along a six kilometer chord is 1.5 meters. Exploration of the Moon has caused a measurable degradation of the natural lunar vacuum because of gas leaks from spacesuits and landing spacecraft as well as by the operation of ascent-phase spacecraft rocket engines. Complete industrialization of 59% of the Moon's 37.96 followed by six zeros square kilometer surface is usually touted as humanity's goal-quest for additional energy, ores and other resources. All realistic Macro-Imagineering development projections foresee the high monetary cost of ensconcing Earthlings comfortably on the Moon's nearside. The Sun and the Moon—each covering about 0.2 square degrees of Earth's sky—appear nearly the same size. How would the Moon appear to Earthlings if it were covered with reflective macro-objects such as shiny mirrors; perhaps it would resemble a glittering 1970s dancehall Disco Ball? Christian Marchal (1953-2013) solved the practical physics of placing Earth-facing mirrors on the Moon (Marchal, 1982; Salmon, 1982). Sunlight redirected at the Earth by Marchal's utility would, of course, heat our planet's air, increase evaporation and allow a vast reduction of nighttime street lighting! 21st Century internationalized Arctic Sea Basin industrial activities might be simpler and safer too.

The global sea-level rise, whatever it may be at the mouths of the Amazon River (Battersby, 2013), could make less attractive a spin-off from Robert B. Panero's Macro-Imagineering plan for damming the Amazon River in several key places, forming a "South America Great Lakes System" (Panero, 1969; Panero, 1972; Mitchell, 1979). Since 1900 dams have impounded about 10,800 cubic kilometers of freshwater, reducing the global sea-level by approximately 30 millimeters and decreasing the rat of the world-ocean's rise in elevation (Fiedler and Conrad, 2010); therefore, Panero's "South America Great Lakes" (SAGL) creation would accentuate that trend! Panero's SAGL would eclipse what was probably the most audacious hydraulic macroproject ever undertaken in the pre-1492 New World: the La Cumbre Canal located in Peru, which was constructed over a 200 year period by toiling natives and which represents a labor investment perhaps ten times that devoted to Egypt's Great Pyramid of Cheops (Ortloff, 1985; Brier, 2013).

Born ten years after Frank P. Davidson, in 1928, Robert B. Panero was tutored in a family-owned civil engineering firm started by his father, Guy B. Panero Engineers (incorporated in New York State during 1947 and merged with another company, Tissian in 1970, subsequently the name disappeared). [Guy B. Panero was a mechanical engineer by training.] Robert B. Panero (1928-2001) served in the Korean War from 1951-1952. From 1956 until 1960, he was involved with the RAND Corp. and the MITRE Corp. through a Special Projects Group of Guy B. Panero Engineers. During the period 1960-1964, Robert Panero then directed his firm's overseas operations from Rome, Italy. By 1964 he had joined the Hudson Institute, eventually becoming its Director of Economic Development Studies until his departure during 1974. Circa 1956-1957, Robert Panero had neologized "Super Projects", which matches F.P. Davidson's "Macroprojects" in both geophysical scope and generalized definition. During 1974, Robert Panero Associates was established in New York City to R&D undeveloped and under-developed noosystem resources, especially those in Latin America, via geographically gigantic and sometimes costly constructions. [R&D, interestingly, is a contraction for "research and development", a phrase that first appeared in a 1952 science-fiction novel, *The Space Merchants*, centered on the terraformation of Venus for profitable real estate development via human colonization, penned by Fredrick Pohl (1919-2013) and Cyril M. Kornbluth (1923-1958).] Davidson and his colleagues define "Macro-Engineering" as a "... processing of marshaling money, materials, personnel, technology, logistics and opinion on a huge scale to carry out complex projects, often international (or multi-national) in nature, that last over a long period of time. A macro-engineering project requires massive funding, significant manpower, large-scale equipment and [metric] ton[s] of material. Macro-engineering projects extend the state of the art of technology, may take place in difficult and sometimes hostile environments, and require sophisticated project management techniques" (Davidson and Huot, 1989, pages 133-142).

On the other hand, Robert B. Panero thought "A Super Project is one that represents, symbolizes and causes a fundamental change in mentality. Everyone understands it because it changes their own view of the future. It is not size, scale or nature which distinguishes the Super Project; rather, it is more the project's psychological impact on its audiences due to its tangible effects and intangible effects" (Panero, personal letter to RBC dated 24 July 1991; See also: after his passing in Columbia, his widow Helen Panero, compiled *Robert Panero: His Visions of the World*, a rare two volume biographical book set, self-published, totaling 582 pages). Systems thinking in Macro-Imagineering may have stemmed, in part, from Gestalt Psychology, as developed by Kurt Koffka (1886-1941) and Wolfgang Kohler (1887-1967). The Gestaltist view is that the "whole" was often different than merely the sum of all its parts (Rock and Palmer, 1990). It seems to sit atop the thinking of many living philosophers (McGinn, 1978) and certainly accords with the following 12 March 1964 *New Scientist* weekly magazine comment: "The real cause of our attachment to macro-engineering is at once more subtle and more profound". Whether they are referred to as "Super Projects" (Panero and Candella, 1968) or as "Macroprojects", such constructive endeavors—along with future terraformation of other Solar System planet as well as Moon remodeling—are futuristic. The "future" is, itself, a "project", a work of collective human imagination and aspiration. Panero and Davidson would be unlikely to disagree with that statement. Geographos supposes that terrorist lodges that have attacked Astythromes might be publicly identified as destructive macro-engineers; new technology, or existing high-technology, has applications inspired and used by global and local terrorist lodges.

As of 2014, all large-scale economic developments attempted in the Amazon River Basin (Hemming, 2008) have failed to become profitable, with the exception of forestry and petroleum and other kinds of mining. Most commercial agriculture in the Basin is non-competitive because the soils are so poor in vital plant nutrients. There are a few permanent lakes in the Basin. From the mid-1960s, Robert Panero proposed and sketched out several large hydroelectric macroprojects for the Amazon River Basin; by 19 July 1972 the Hudson Institute issued his report, *The Amazon: A Catalytic Approach to Development*. That report suggested a macroproject aimed at providing controlled freshwater and new arable landscapes for the Basin's farmers, macroprojects using what many Brazilians view as ecologically harmless large dams as well as mini-dams. Panero, an energetic eager-beaver, advanced his idea for many low-head dams on many of the Amazon River's tributaries. He called for a series of small dams to be placed alternately along a single river branch that would flood upstream regions and drain downstream regions, thus sparing much land from permanent widespread freshwater immersion. Significantly, small dams, causing the inundation of vast land tracts, repeat the rice-growing paddy-construction actions in Asia that started the buildup of methane gas in the atmosphere millennia ago (Kerr, 2013). The upstream regions would gradually fill with sediment, with the fertile downstream regions gradually losing fertility from over-farming in a challenging Tropic Zone climate regime. Every half-century approximately, the low-head dams would be deliberately demolished to be duplicated elsewhere a new cheaply constructed earthen dams in order to irrigate low-fertility landscapes (draining the silted, shallow waterlogged regions) for replacement agricultural enterprises. (Panero's revivification technique for farmland is worthy of beavers!)

Robert Panero's "South American Great Lakes System" [SAGL] plan will probably never be realized for geopolitical and other reasons even though the continent's physiography seems to favor it. It is more than interesting to note the absolute geographic location of Brazil's seat of national government. Brasilia was established in 1960 close to the Amazon River Basin. However, for several sound reasons, Geographos is not certain about that judgment. Since Robert B. Panero's passing in 2001, anthropologists and archaeologists have uncovered and discovered that the millions of primitive-technology natives who lived in the Amazon River Basin prior to 1492, built extraordinary hub cities, straight roadways and well maintained canals (Heckenberger, 2009) and successfully farmed vast swathes of landscape for long periods of time (Tollefson, 2013), probably utilizing some of the landscape regeneration schemes that Panero offered! Modern humans only started to gather accurate measurements of the Amazon River's flow in 1928 at the town of Obidos; the 2009 big flood helped to enlighten macro-engineers about the flow extremes of the Amazon River. For sure, climate change will alter the heat and precipitation regimes currently prevailing in the Amazon River Basin (Cook et al., 2012).

In 1950, the populations of North America and Latin America were about equal (166,000,000 each); by 2020, Latin America's population might total over 700,000,000 persons whist that of North America might be 330,000,000. In other words, the impending geopolitics of the New World will be characterized by disturbances (resource denials, trade restrictions, environmental damage, legal and illegal mass migrations, narcotic trafficking and, possibly, fanatic terrorism). Central America, like the Middle East of the Old World, poses unique geopolitical threats to all humans. Those regions are where maritime shipping lanes converge (on the Suez and Panama canals) and from where previously undetected cruise missiles could merge, traveling to their Europe and North America targets entirely over land. Briefly, any resulting cruise missile attack (in the absence of a functioning Rodoman-ALPS configured to track and destroy aerodynamic missiles as well as ballistic missiles) could be a matter of uninformed speculation, failing to justify a victim's counter-attack. The French Revolution installed Earthly global Nature as a quasi-divine Authority; the concept of "Latin America" was created in France under Napoleon III; the strongest model for political activity in Latin America is France. When he embarked upon his presidency, Dwight David Eisenhower (1890-1974) sought to remind the USA's public that global war could originate in Latin America in *The White House Years, Mandate for Change, 1953-1956* (1963, page 420). Paul Bracken, **Figure 7**, opined in 1982 that South America will be the Earth's first Anthrome "...subjected to a super-industrial scale of development..." (Bracken, 1982).

Figure 7. Paul Bracken, erudite academician and geopolitical strategist. (Image: Google Images.)

As Bracken commendably commented, "…super-industrial refers to the degree to which projects are undertaken on such a massive scale that unintended side effects…are of greater consequences than the intended effects". Geographos asks: How a single term (that is, "super-industrial") meaningfully encompasses all those references (high-rate, degree, counter-productivities)? Only an unwanted and/or unplanned World War III could possibly qualify as a macroproject with the characteristics Bracken has implied! Future cybernation could ensure a pollution-free environment for social health! Robert Panero's "South American Great Lakes System", **Figure 8**, does have considerable negative externalities, but nothing on the geographical scale Bracken invoke

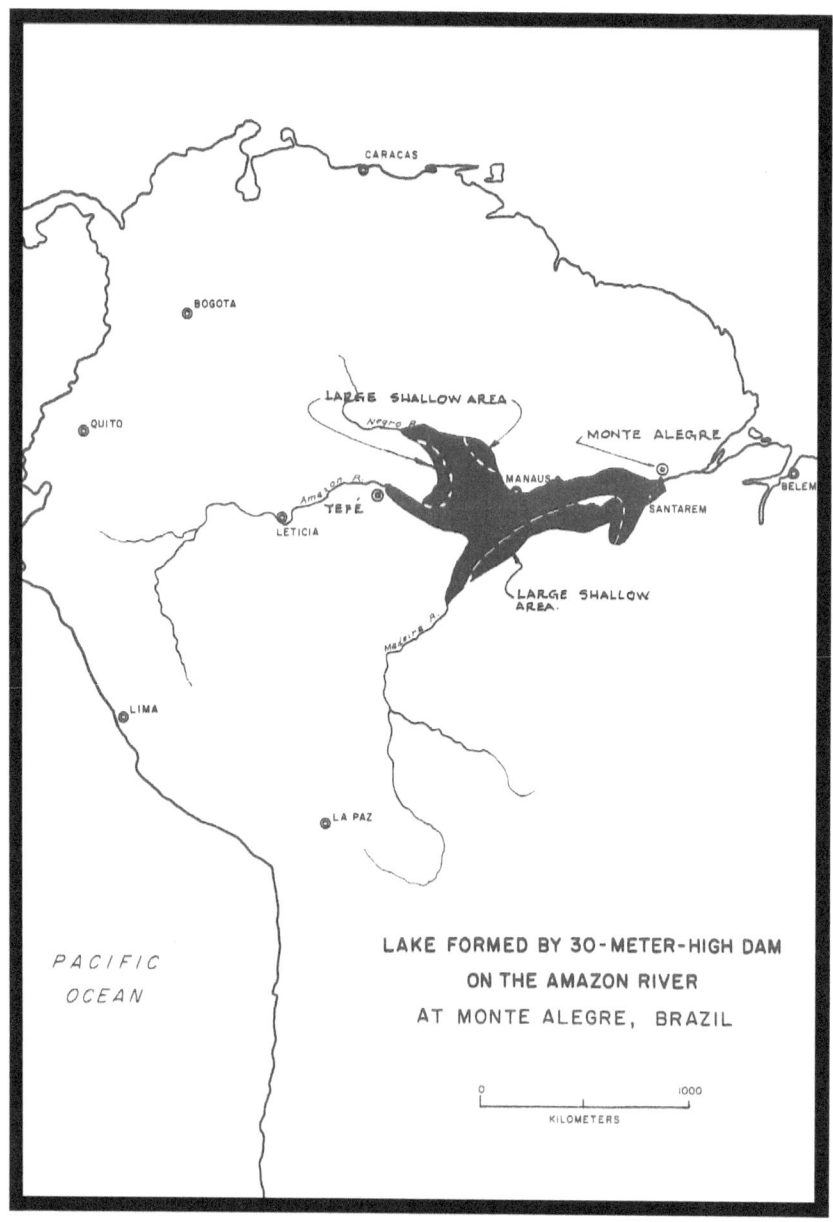

Figure 8. Self-explanatory. Robert B. Panero's SAGL concept from the early 1960s. (Image: Image by RBC.)

Riverside and coastal entre-pot port shipping in Brazil is probably a very good capital investment—better than railroads and rainforest-destroying super-highways. Such a coordinated freshwater/saltwater transportation system could be useful, if Robert B. Panero's SAGL were realized during the early-21st Century. The Amazon River has an elevation west of Iquitos, Peru (some 3,750 kilometers upstream from the mouths of the river) only 114 meters above present-day sea-level; it descends eastward to an elevation of 10.5 meters, about 3,200 kilometers from its termination in the South Atlantic Ocean, and ocean tidal effects have been measured and recorded many kilometers inland from the ocean. The greatest inland penetration of a tidal bore anywhere is in the Amazon River Basin, where noisy, predictable bores on the Capim, Guajara, and Moju rivers—tributaries of the Amazon River—occur more than 150 kilometers from the ocean (Chanson, 2012). The "Pororoca"—in the language of the Tupi-Guarani Indians, the name means "great din"—is the most famous on the Amazon River. [Tidal power plants may be usefully sited within the Amazon River's mouths (Charlier and Finkl, 2009; Clark, 2007; Hardisty, 2009). Gorlove helical turbines, since 2005, have been tested there.] Big floods have raised the Amazon River's runoff ten meters above its normal flow elevation, whilst the river discharges about 15% of the world's runoff yearly! Future sea-level rise will affect the river's plume and coastal actions (Dolgopolova, 2013). Because the terrain is so extremely flat, a one-meter rise might make Robert Panero's low-head, mini-dams inoperable and threaten the powerhouses of a built "South American Great Lakes System". [Indeed, under that particular Earth-biosphere circumstance, estuarial ecosystems would be enlarging geographically everywhere.] A very large dam, such as earliest 1960s proposal for Amazon River Basin development, a barrier placed across the Amazon River near Santarem, could save the Amazon River Basin's rainforest from ruinous seawater inundation. A large freshwater lake impounded by that enormous dam would not affect Earth's rotation but would affect local and global climate regimes through as yet undiagnosed changes of the troposphere's heat and water budget, thereby affecting other operating or planned macroprojects of South America as well as, possibly, similar infrastructures located in the Congo River Basin (Eltahir et al., 2004).

Marq de Villiers, in **WATER: The Fate of Our Most Precious Resource** (2000, pages 38-39), stated: "Even if we could, we wouldn't shift Brazil's 20% [of Earth's river runoff] to, say, the Sahara. Doing so would put an end to the greatest reservoir of biomass on earth, and the planet's greatest rain forest… [that] we have come to understand is its respiratory system. To do so would be like placing a giant vise around the Earth's lungs." This incorrect opinion is a journalistic exaggeration of the first water! Why, Geographos queries, would anyone go to the expensive of creating a desert in South America to irrigate a desert in northern Africa? And, further, does anyone still entertain the truly nonsensical belief that South America's Tropic Zone rainforest is one of Earth's lungs? There is no good, compelling reason why freshwater cannot be captured for use just as it flows, utterly wasted by Earth's landscape life-forms, into the spacially vast and voluminous South Atlantic Ocean!

An Austrian metallurgical specialist and macro-engineer, Heinrich Hemmer (1924-????), designed a means for the partial irrigation of the Sahara and its adjacent southerly ecotone, the Sahel, via freshwater imports from South America's Amazon River! A detailed account of Hemmer's plan for an undersea artery was first presented in the defunct periodical **Speculations in Science and Technology** (Hemmer, 1993). By 2009 Geographos, in association with the Candida Oancea Institute, had thoroughly reexamined Heinrich Hemmers' bold plan by (I) calculating the sizing of the fluid pipeline to suit the flowing characteristics of freshwater and the pressures at each end of the pipe; (II) calculated the strength of the pipe, considering the internal, external and ocean loads the pipe would have to endure; (III) suggested a pipe coating—pipe, whether steel or aluminum requires corrosion protection, including specific cathodic protection system for additional external protection; (IV) selected a feasible pipeline route across the Atlantic Ocean from the Amazon River's mouths to the shores of northern Africa; (V) determined that surface tow and/or mid-depth two is efficient; (VI) investigated what marine engineering codes and government regulatory requirements must be met (Badescu, Isvoranu and Cathcart, 2010). Fortuitously, during April 2009 in Delft, The Netherlands, Bruno O.B. Cedeno, had completed his Master of Science thesis at the UNESCO-IHE Institute for Water Education, *Economic valuation of two technologies to import water: a case study of Morocco*, which is a cost-benefit analysis of Hemmer's positively buoyant undersea intercontinental freshwater pipeline and Stauber Bags. Macro-Engineering already has supportive experience laying immersed tunnels, an effective alternative to bored tunnels and very long bridges (Lunniss and Baber, 2013). Hemmer's pipeline would face scrupulous examination by all signatories of the Treaty of Amazonian Cooperation (1980) (Landau, 1980). Hemmer foresaw the profitable exportation of 10% of the Amazon River's annual runoff—from Brazilian intake to terminal delivery near Nouakchott, Mauritania, in Africa, a distance of about 4,300 kilometers, the freshwater could then be shifted inland via an overland pipeline to, possibly, Libya's Bay of Saloum (Charlier, 1991). [Parts of the Sahara exist in ten northern Africa nation-ecosystems, but the Sahara contains only one national capital, Nouakchott (founded 1957) in Mauritania.] Therefore, Amazon River water could supplement the national freshwater supply extracted and distributed by Libya's "Great Man-Made River Project" (operating since 1980); perhaps the geological rock formation, of which 650,000 square kilometers is outcrop, from which that operational mega-project draws could be replenished, like the Ogallala Aquifer in the central USA (Goncalves et al., 2013)? Hemmer's pipeline would also have the effect, in South America, of continuously drawing down a large [Robert Panero] Amazon River Reservoir (see **Figure 8**, above), keeping that Basin's rainforests from becoming dead wood slowly putrefying at the bottom of large, fetid lakes. If very cheap electricity could be made available, the an additional incremental abstraction of Amazon River's 100,000 cubic meter per second average flow might then be siphoned off to South America's arid or water-short Astythromes. Prevention of a gigantic freshwater pool at one place on the Earth's flexible crust near the planet's Equator would also reduce the rate of slowing of our planet's 24 hours per day axial rotation.

MACRO: A Clear Vision of How Science and Technology Will Shape Our Future, Frank P. Davidson's 1983 summary of Macro-Imagineering/Macro-Engineering' professional and popular status at the time of its publication, is oddly anachronistic. On a geopolitical level of assessment, life in Davidson's generalized Earth-biosphere future is1930s American "New Dealism" upper-class wish fulfillment writ large. World-class macroprojects require worldwide generalities. At pages 236-237 of his book, Davidson illustrates a "Trans-North Atlantic Ocean Tube-Tunnel" he concocted with a submarine macro-engineer working then at G.P. Taurio Corp. in Groton, Connecticut, USA. The concept bears striking resemblance to a science-fiction novel by pseudonymic Harry Harrison (1925-2012), *"A Transatlantic Tunnel, Hurrah!"* that was serialized in *Analog: Science Fiction Science Fact* during April-May-June of 1972. April's front cover-art looks remarkably like Davidson's supervalent thought. Previously, the same submerged floating underwater tunnel macroproject had been mentioned in passing by a coffee table-size book, *The World Ocean* (1979, page 407) authored by Jacques-Yves Cousteau (1910-1997). Michael G.

Zey's *Seizing the Future: How the Coming Revolution in Science, Technology, and Industry Will Expand the Frontiers of Human Potential and Reshape the Planet* (1994, at page 57), describes the "Trans-North Atlantic Ocean Tube-Tunnel" physically, claiming (earlier, at page 50): "Certainly, a cross-Atlantic super tunnel first detailed thirty years ago [1964? If so, about the first time the word "macro-engineering" appeared in the technology news media.] will attract more backers and supporters". Zey makes no sense sociologically or financially. Geographos asserts a 4,100 kilometer-long "Trans-North Atlantic Ocean Tube-Tunnel" is make-believe reaching its apogee.

Its 21st Century incarnation is a neutrally buoyant steel tube, anchored by strong cables to the seafloor at a depth of 45 to 100 meters below the North Atlantic Ocean's surface, from which air is extracted to create a working vacuum in which supersonic magnetically-levitated trains could move at speeds of 6,400 kilometers per hour. At a 2012 USD cost of 105 to 209 billions, becoming, if built, a fantastically lucrative terrorist lodge target of opportunity! Disregarding possible natural climate change-induced shifts in the Gulf Stream's flow in the North Atlantic Ocean or the shift that might occur were C.L. Riker's Jetty projecting from Newfoundland, Canada into the North Atlantic Ocean were constructed, the seafloor geology along the tube-tunnel's mapped route (**Figure 9**) selected by Davidson and his colleagues is becoming better known. Greenland is covered by moving glaciers that are undergoing drastic geographical changes. Sub-surface, geothermal, warmth caused by intensive volcanism, would be a real macro-problem in Iceland—the only place on Earth's landscape where fissures formed by spreading mid-oceanic ridges, are likely to erupt on a titanic scale. The haze of sulfuric aerosols belched by violent new volcanic eruptions has shut-down, for a period of time, all 21st Century air traffic between North America and Europe. That's a plus in the evaluation process-event for the tube-tunnel. Still, overall, Geographos thinks that the extremely long tube-tunnel of the kind promoted by Davidson and associates passing aircraft-like fuselage would cost the Earth—a real pipedream! Macro-Imagineering/Macro-Engineering's professionals must strongly resist such fixed link fixations and other such juvenilia. Deep-draft calved icebergs traversing the Davis Strait in the hundreds annually pose too great a threat with their deep keels to be ignored so flagrantly.

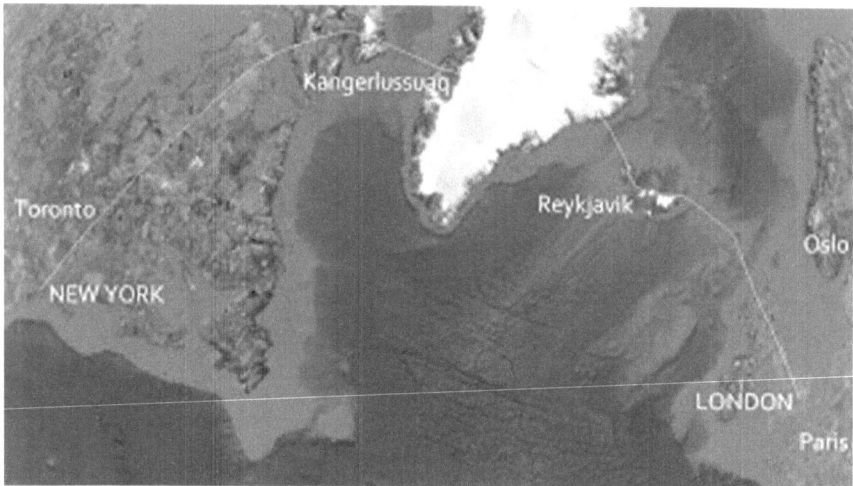

Figure 9. One variant routing of several for the F.P. Davidson "Trans North Atlantic Ocean Tube-Tunnel" rapid transit Macro-Imagineering concept. Of course, from London people could board the English Channel Tunnel trains and in New York City it should be possible to board super-fast trains of "Planetran" (**Figure 10.**) as planned in 1977 by Robert Salter or a regular-style fast train headed towards a tunnel beneath the Bering Strait. George Revill's **Railway** (2012), a thematic and discursive book about the iconic modern cultural meaning of the synergistyic technology's intended function, gives his readers a broad perspective that is endearing and useful, especially in the context of speculated 21st Century Macro-Imagianeering transportation R&D. (Image: Google Images.)

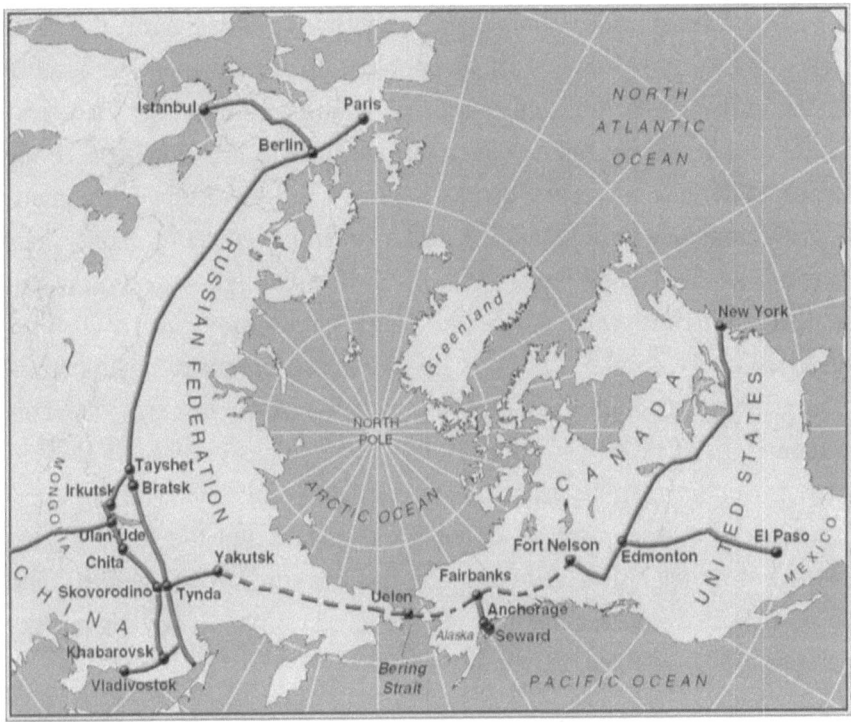

Figure 10. A non-"Planetran" system of railway linkages, postulating a fixed link crossing under or over the Bering Strait. The routing illustrated in **Figure 9**, above, completes a Northern Hemisphere fast transportation system for people and packable things. (Image: Google Images.)

Whilst "Virtual Reality" machines could "send" riders—not aboard high-speed magnetically-levitated trains—to London, Venice or Venus, Rodoman-ALPS won't have the technical capability to really guard any submarine facility unless it were tied via Earth-orbiting satellite into tele-robotic underwater roving guards via "blue-green" laser beams which, because of their wavelength penetrate many meters of seawater. A global, computerized, interconnected network of intelligence, both human and artificial, could lead to a thorough exploration and exploitation of Earth's ocean. Japanese robotics experts have developed a jelly-fish shredding sea-going robot to rid harbors plagued by those pesky creatures! Robots with mobility and "hands" imitate people, who are only persons because they handily wield tools since becoming human, or is that vice-versa (Radman, 2013)? On the world-ocean's surface, 2014's 100,000+ tonne cruise liners may end at the breaker's yards in Alang, India or the mud-flats of Bangladesh where they are torn apart impoverished persons, since "Virtual Reality" could make sightseeing through high-resolution television monitors both inexpensive and comfortable—no lines, no customs searches, no terminal pickpockets, no frantic parents corralling hyper-excited children, no confiscatory money exchange rate or lost luggage! Some persons will always seek physical and psychological pleasures derived from traveling. A more peaceful world might result if "Virtual Reality" "passengers" never physical visited sociologically volatile regional sea-ports. Future television and computer linkages in "Virtual Reality" and could make adventures possible for the handicapped as well as healthy persons.

Jarvis Island, located in the South Pacific Ocean (as one of the Line Islands), uninhabited since 1942 mainly because of its world-ocean seawater circulation-imposed aridity (Frierson et al., 2013), should be designated as a leading prospective Space Elevator construction site. (**Figure 11.**) Jarvis Island is about 41 kilometers south of the Equator. Although not perfectly positioned for a heavenly funicular, Jarvis Island consists of only 65 hectares; it is a low-elevation, basin-shaped coral island with a maximum elevation above local sea-level of only 7 meters—meaning that it could be badly eroded by a higher-than-2014 global sea-level; a seven kilometer-long seawall could defend most of the island from direct wave action. Jarvis Island has a 323,100 square kilometer Exclusive Economic Zone (EEZ) under the 1982 concluded United Nations Organization Law of the Sea Treaty. The atoll has been a USA Possession since 13 May 1936, and is now a part of the National Wildlife Refuge System (administered by the Department of the

Interior's US Fish and Wildlife Service). The island's present-day EEZ extends into the Northern Hemisphere. No major natural hazards are likely to affect the island. Future hypersonic aerospace planes, such as SKYLON, might land on Jarvis Island, something the pre-World War II "China Clipper" seaplanes [M-130's put together by the Glenn L. Martin Company] never did. The Pacific Ocean is almost 50% of the world-ocean's area, and its seawater washes the coastline of about 50 human ecosystems-states in humanity's Earth-Noosphere.

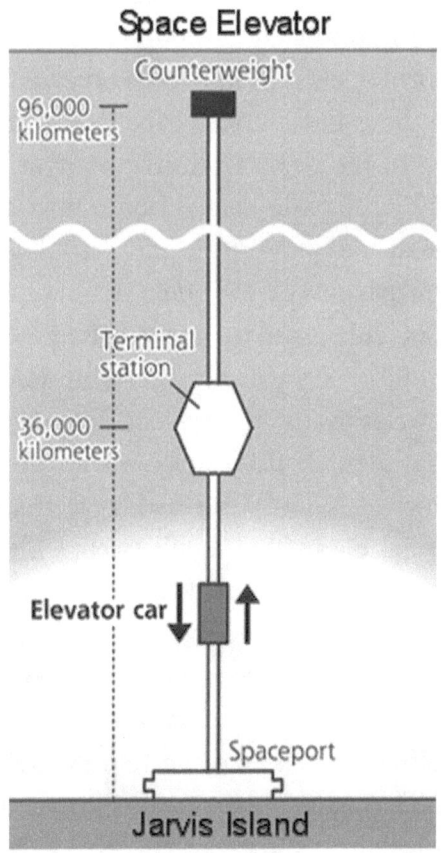

Figure 11. Jarvis Island Space Elevator Base Station in the Central Pacific Ocean. (Image: Google Image, modified by RBC.)

David Noel's maritime, freshwater-collecting and storing "lily-pads", should be fabricated of a gas permeable, non-opaque plastic membrane so as not to shade the Central Pacific Ocean's topmost seawater layer. Indeed, the euphotic zone—the top 100 to 200 meters of the world-ocean's surface—could be enlarged in volume by dangling fiber-optic cables below the floating Noelian reservoir membrane, which would convey natural sunlight (some of which might even be focused on the membranes by Earth-orbiting sunlight reflectors) to deep parts of the covered seas. Noelian collectors could even become sea-farms (mari-culture enterprises) illuminated 24-hours-a-day. Ocean productivity would escalate enormously, in part because the volume of the world-ocean exceeds ten times the volume of the landscapes above sea-level (Mora et al., 2013). In the ocean's belt-like Equatorial Zone, seawater could be enriched by the periodic addition of processed Astythrome sewage, towed to pad sites and other appropriate places in blimp-like floating dracones (capacious plastic pods, Stauber Bags) with minimal capacities of 250,000 cubic meters. Dracone barges are flexible fabric tubes manufactured to carry petroleum, freshwater and other liquids less dense than seawater in which they float. [The density of seawater is 3% more than freshwater.] At this time, dracones are still used to sup-ply many Greek-owned islands, crowded by tourists, with sufficient freshwater. Turkey has planned to supply Crete with freshwater via an immersed pipeline (Finkl and Cathcart, 2011) identical to the trans-atlantic aqueduct investigated by Geographos and closest imaginative colleagues (Badescu et al., 2010). Several David Noel-styled "lily-pads", closely associated with complementary sea-going floating industrial complexes, the Self-contained Oceanic Resource Bases (SORB) that were first detailed during 1970 (Green, 1970), can dangle fiber-optic cables that carry sunlight to the darkest depths of the world-ocean, at-tracting fish and the prey they feed on (Kirk, 2006). Commercial fish-ing has an interesting effect on the world-ocean's mineral iron particle content, and particulate iron is a crucial component for marine life and for carbon storage; the estimated annual fish catch, removed from the world-ocean to human dinner tables, represents 0.5% to 2% of the iron particle content of the upper 4,000 meters of the world-ocean. Erosion of Earth's landscape creates wind-blown mineral dust that is atmospherically transported and deposited onto the ocean; the world-ocean's iron content is so limited, that the miniscule contribution of iron-rich meteorites to Earth's mineralogical inventory is relevant to the sustainability of our world's fisheries!

Evaporation from the ocean's surface can be speeded by the application of effective technologies. The physicist Howard A. Wilcox's **Hot-House Earth** (1975) offered unadulterated Scare Science, threatening that by 2170 the activities of humans would be generating heat at a rate almost equal to 10% of the Sun's energy input to our air and that, by 2230, all seawater in the world-ocean would be close to its boiling temperature (about 100^0 C). Wilcox's scenario, human extinction by way of heat stress, is "Thermogeddon" on steroids (Chaisson, 2008; Kidder and Worsley, 2012); for Aliens, Earth will become the newest global civilization gone the way of "Atlantis"! Even if Wilcox's warning were true, people will have vanished from the face of this planet long before the ocean comes to a boil. Nothing of this global, approximate magnitude has been prognosticated or alleged as a media-hyped macro-problem by radical Greens of extreme anthropogenic "Global Warming"—yet! It is true, however, that ozone and aerosols can accumulate in still air conditions to dangerous densities (air stagnation events, smog) and, as well, that anthropogenic heating of Earth's air is already disrupting the "normal" atmospheric circulation pattern, including a variance of the altitude of the troposphere-stratosphere boundary. Nearly 10% of the ocean has been absorbed deep into the Earth-crust and mantle since our planet's formation billions of years ago and, with the elapse of one billion years more, the ocean will lose about 27% of its extant volume to further geological tectonic plate sub-duction (Bounama et al., 2001). The world's ocean will disappear for-ever, converted to plasma, when the maturing Sun expands to tis Red Giant Stage four or five billion years hence. **Figure 12**, below shows Earth stripped of its ocean, rivers and lakes—waterless from over-heating by Sun alone. Howard A. Wilcox's 1975 screed was opposed in 1976 by Lowell Ponte's *The Cooling: Has the Next Ice Age Already Begun? Can We Survive it?* (Peterson et al., 2008). The 1970s "global cooling" conjecture is evidently related to "global dimming", which is caused by airborne aerosol pollution such as that often imaged by satellites above India and China. Some climate change mitigators have even proposed an anthropogenic increase of sunlight-reflecting aero-sols in Earth's air to counteract human-caused "Global Warming". Such an effort might involve endless monetary costs—the activity could never cease with "perpetual" anthropogenic carbon dioxide gas emissions—and, ultimately, there would be significant human health costs incurred because people have to breathe air! [Terraformers have suggested the uninhabited Mars be forced into a new, more conducive to human life, clement climate regime by deliberate massive injection of aerosols using powerful nuclear explosives or directed meteorites. Geographos believes those plans would be ethically okay and suffi-cient since, unlike Earth's air, such atmospheric macro-engineering there would have no health costs because the future settlers are mostly indoors or wearing pressure suits when outdoors.]

Figure 11. Imaginative artistic rendering of our Earth cooked almost to a crisp by the expanding Sun—far in the unfathomable future! The only tide is that of the whole Earth's body. Returning humans, or Alien tourists, would have to worry some about earthquakes and whatever might fall from the sky! (Image: Google Images.)

The meteorologist who, during the 1930s, proposed the primary mechanism for precipitation's formation in clouds, Tor Bergeron (1891-1977), also proposed—circa 1960—that industrialized humans should deliberately heat a certain ocean region in order to energize a regionalized hydrological cycle for "naturally" watering North Africa's desert and savanna landscapes (Liljequist, 1980/81). Hubert Horace Lamb (1913-1997) incompletely described Bergeron's 1960 Macro-Imagineering concept to moisten the Sahel and the Sahara in *Climate: Present, Past and Future* (1977), mentioning only that Bergeron wished to inject water vapor into air masses formed above the North and South Atlantic Ocean, which normally move inland over Africa and which increase normal rainfall, or break drought spells in that arid demi-continental Anthrome. Clouds are a non-polluting way to transport freshwater—in a manner of speaking, clouds are global Nature's dracones—and cheaply too. Lamb concluded about Bergeron's macroproject concept that is was sound and could "… make a very significant contribution to food production" (Lamb, Volume II, page 662). Bergeron's water vapor would be generated at a region south of Ghana's Three Points Cape, where zero latitude intersects zero longitude (at zero altitude) in the Gulf of Guinea. **Figure 12.**

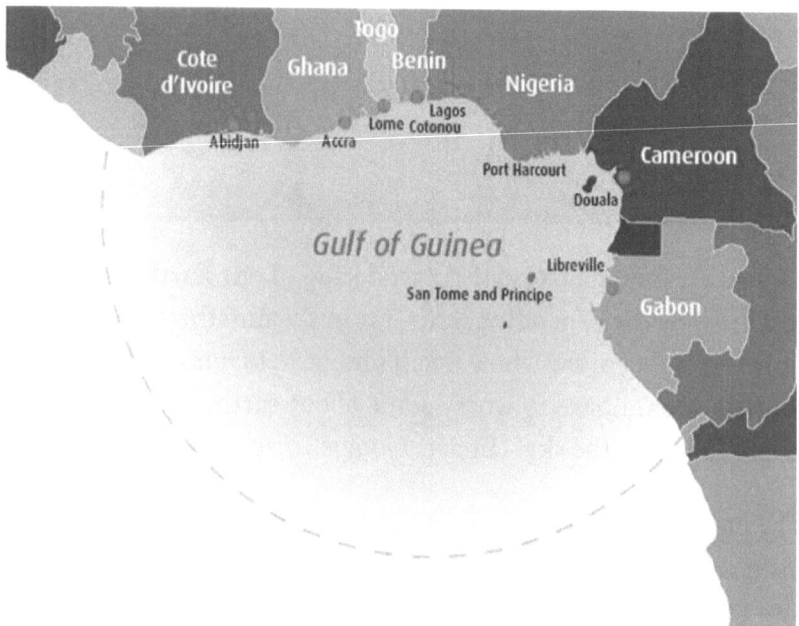

Figure 12. The approximate source-region seawater, shown here in yellow, where the late climatologist Tor Bergeron hoped to anthropogenically create massive quantities of aerial water vapor. During 1434, when seaman Gil Eanes, financed and commanded by Henry

the Navigator of Portugal, rounded the then-dreaded Cape Bojador, he forever eradicated the ancient Greek myths of boiling seawater and fierce monsters once said to be present near Earth's Equator. (Image: Google Images.)

During 1965, Tor Bergeron had identified two energy sources, one of which was nuclear power, to heat (or even boil) that ocean region's surface seawater layer. No discussion of the disposition of the boiled seafood likely to pop up from the murky ocean depths was ever mentioned; cooked, or at least killed sea creatures (plant and animal) might be harvested quickly as a rich fertilizer to be applied to nutrient-depleted soils farmed by North Africans! Today's aquaculture experts might devise means to fence a region of artificially heated open ocean seawater with sonics and other high-technologies. It is known that sea mammals are adversely affected by shipping noise so that sonic beam-forming devices will require expert adjustment to specific marine locales and specific marine creature sound tolerances. In order to increase the aerial watering of the landscape the Gulf of Guinea would have to be warmed about 3^0 C (Cook, 2006). Climatologists have concluded that the Coupled Model Intercomparison Project phase 5 computer models "...demand further work to achieve a reasonable realism" concerning the present and future of the West African Monsoon (Roehring et al., 2013). So, nothing factual is yet settled about Bergeron's scheme. However, a start has been completed during 2014 when the imagined greening of the entire Sahara with desalinated water to produce massive amounts of biomass megaproject was examined by pro-Greenism geoscientists (Bowring et al., 2014). Their conclusion: "Overall, in response to irrigating the *entire* Sahara/ Sahel region at the same rate, the western Sahara near the coast had the highest EROI [Energy Return on Investment] of 6:1 after including the additional precipitation feedback." Also, they found that increased warming and rainfall climate factor changes would happen offshore, over the Gulf of Guinea. Further significant and valuable climatological research is being done by the same group.

21st Century technology R&D will force changes upon the all our world's Noosystems. Noosystem traditions (cultural mind-sets), along with technical progress, will alter with the passing of time. Indeed, out Earth-world is a macro-problem of great complexity. Every ecosystem-nation should have an interest in organizing and funding the augmented Rodoman-ALPS, especially if it was assembled as a super-structure funded by a planet-wide geo-hazard abatement tax. Such facility would provide a global state of consciousness and a permanent visibility that assures the automatic functioning of all subscribing Noosystems. Nicholas John Spykman (1893-1943) expressed the thought, in **America's Strategy in World Politics** (1942), such purchase plans to be one of the "...great variety of techniques designed to wind friends and influence people" (Spykman, page 12). The first fairly complete global ethnographic atlas was compiled by 1826. Carl Stratz de Haag (1858-1924) suggested the possibility of *Homo sapiens* divided into "Progressive" and "Static" groupings. Humanity now knows the impossibility of de Haag's groups. And, very probably by 2050, artificial life will need to be included in any accurate future ethnographic atlas compiled to illustrate the distributions of intelligent persons and Terra-creatures! Harold Sprout (1901-1980) argued that population size will be the eventual determinant of geopolitical status on the grounds that the technological skills are the primary requirement for economic and military strength and that the capacity to acquire those civilian and military skill sets is evenly distributed in our world's populace. The ecosystem-country with the largest population, it logically follows, with be the most influential (Sprout, 1963). However, logic is not reality just as the map is not the world.

Saudi Arabia—that toponym conjures a vision of a desert landscape studded with oil derricks tended by a few people carried to their job-sites and gamboling pleasure palaces by Rolls Royce automobiles! Saudi Arabia is, in fact, under-populated for the infrastructure it must maintain; workers from other nation-ecosystems fill the employment gaps. Saudi Arabia's government should consider the rapid development and deployment of self-replicating machines, which could be field-tested safely in that sparsely-populated arid landscape. Such automatons could be seen (in geopolitical terms) as "force multipli-ers". Visualize an extensive deployment of various sized machines systematically rearranging the countryside's materials by tenting Saudi Arabia's huge Empty Quarter! [A similar geographical scale Sahara tenting macroproject has been outlined mathematically by R.B. Cathcart and Badescu (2004).] If the Empty Quarter were shaded by mirrors, then an amelioration of the local climate regime could take place. In southern Spain, for example, 27,000 hectares of poor soils were covered by glass-houses growing tomatoes and other commercial crops. The effect of the sunshine reflecting glass was to cool the climate as well as protect the plants from temperature extremes (Campra and Millstein, 2013). **Figure 13.**

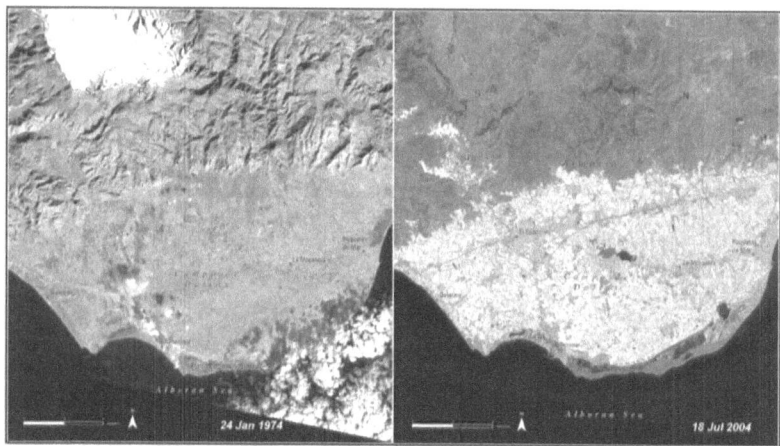

Figure 13. The Province of Almeria, in southeast Spain, from pre-glass-house agriculture to intensive glass-house farming. (Image: www.grid.unep.ch.)

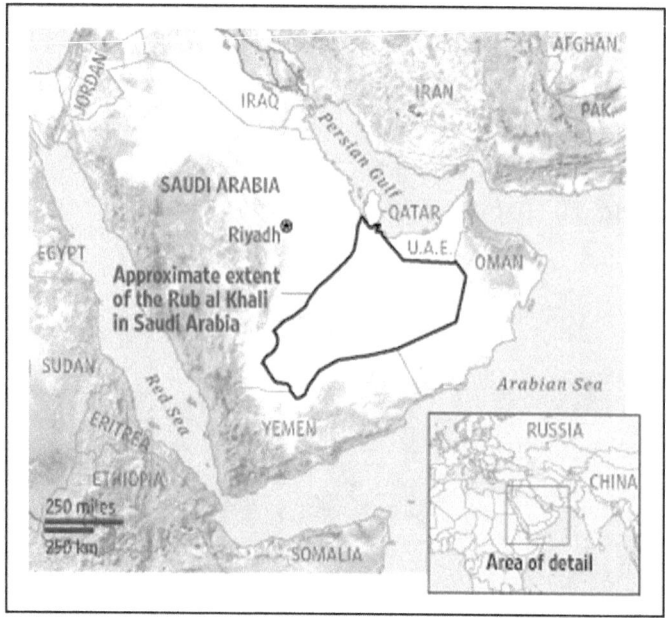

Figure 14. Also known as the "Empty Quarter" is Earth's largest sand desert and its Mars-like silicates are coated with iron oxide that colors

the sand dunes red, orange and purple. The emptiness of a landscape devoid of cities and transportation infrastructure is pervasive and shocking.

The result of such effort would, most reasonably, be best protected by an augmented Rodoman-ALPS since the Wabar Meteorite site clearly shos the effect on the landscape of a powerful space debris impact. The first architectural "duty" of the tenting must be to suppress sand-storms that obscure the vision of ship navigators plying the Strait of Hormuz. People who wanderingly "occupy" the Rub' al Khali, the Bedouin, are nomadic. The Empty Quarter, known to the Saudis as the Rub' al Khali, is situated strategically, lying between the Persian Gulf and the Red Sea (Youssef and Maerz, 2013). (**Figure 14.**) The narrow entrance-exists of the two gulfs can be dammed to produce immense amounts of hydropower (Schuiling et al., 2005; Schuiling et al., 2007). The nearby Arabian Sea's biological productivity is increasing markedly as the landmass of Europe and Asia warms during the 21st Century. Myriad automated devices—devoid of Freud's Id and Super-Ego—operating in the remote region may have the effect of profoundly altering the traditional Middle East's arena of power struggles. Under the Constitution of the USA and its derivative laws, a human is a natural person, but a corporation licensed by statute is an artificial person. Perhaps the tented Empty Quarter installation comprised of robots ought to be called a "satrap", with the domain of their real-world operations to be known as a "satrapy"? Many of our Earth-noosystem's largest economic units are corporations, not country-ecosystems. The biggest international corporations are high-technology, multi-national, with access to (and development rights in) subsidiaries dealing with the newest forms of industrialization. Often, they are staffed with inter-metropolitan elite, using networked computers and televised Board and R&D Meetings!

Such groups may not have a strong allegiance to any single ecosystem-nation. A rather feasible tenting of the Sahara has already been outlined by macro-engineers affiliated with the Candida Oancea Institute in Bucharest, Romania's capital city. It is not an overstretching of our imagination to apprehend the Empty Quarter Tenting Macroproject as a spin-off of a 1960s Japan-originated architectural specialism, Metabolism (Koolhaas and Obrist, 2011). [Since 1969, the in-house planning team Geographos practices a Transcendental Earth Metabolism Macro-Imagineering/Macro-Engineering—its acronym is TEMM—producing non-timorous medicinal macroprojects for humanly inhabitable Solar System celestial bodies.] Motivated by their desire to escape economic austerity and hideous urban blight caused by wartime bombing, ideally, Japan's undaunted post-World War II Metabolist community desired to imagine, plan and construct a futuristic city so flexible in its connections that its parts could grow, transform themselves and die whilst the whole urban-bloc continued to function (live) usefully. A large volume, airy tent would never suffer the macro-problems of sick humans inherent with many modern edifices—that is, the "tight building" or "sick building syndrome"—so fully revealed by Michelle Murphy's *Sick Building Syndrome and the Problem of Uncertainty* (2006). Looking ahead to the 22nd Century, might not Muslims from tented Saudi Arabia enjoy living in a newly Richard L.S. Taylor-style terraformed "Mars Worldhouse" entity or a Martyn John Fogg-style outer space situated parasol-cover for a currently life-repulsive Venus, where naturally arid landscapes will probably always abound even after terraforming? Considering its huge oil and natural gas deposits, Saudi Arabia alone should be able to fund all molecular Nanotechnology R&D necessary to undertake the Empty Quarter Tenting Macroproject. Geographos forecasts that, in future, adventurous Saudis can look forward to urban sprawl, somewhere!

"Geography" is one of the most ancient disciplines (if not the oldest) with continuity of name, its subjective content having changed during recorded human history. 21st Century Geography can be seen existing as a unified, theoretical and predictive Science-Art, surpassing older organizations of the obvious content. All geographers who focus on humanity's Anthropocenic industrial activities find their central intellectual macro-problem in options of where macroprojects should be properly located in Earth's biosphere—called "Geopolitics"—and in other Solar System planets, which ought to be labeled "Geoscience", until geographers arrive there in person and/or via tele-present mobile robots. Macro-Imagineers/Macro-Engineers are basically geographers who, specifically, are and must be lateral thinkers dealing with long-term biosphere change scenarios and such trained persons must be concerned with the future of Earth's human ecumene. An example of this solution-seeking endeavor is that urban planning artwork, *Ecumenopolis: The Inevitable City of the Future* by the late C.A. Doxiadis. Could his 1974 book be the first global vision to initiate large-scale global Nature planning for energy production and human settlement?

Doxiadis's vision can be compared geographically with that of Herman Sorgel (1885-1952), a German architect, who loathed technical quick-fire oral delivery, but who admired macro-projective imaginative thinking as well as real-world building! During the first half of the 20th Century, he, along with some of the most famous architects of the 1920s and 1930s—persons such as Peter Behrens (1868-1940) and Erich Mendelsohn (1887-1953)—designed a reclamation macroproject called "Atlantropa", which encompassed the Mediterranean Sea Basin and the most of northern and central Africa! Sorgel devised his macroproject during the late-1920s for climatic regime ameliorations over the Sahara and the Mediterranean Sea Basin. It is well to anticipate real future climage changes, and currently a scarey super-computer model suggests that the southernmost parts of Europe (Iberian Peninsula, the southern part of France, Italy and the Balkans) will become drier, with stream and river minimum flows reduced by up to 40% and times of freshwater supply deficiency may increase by possibly 80% (Forzieri et al., 2014).

Since the years Sorgel published his macroproject plan for Atlantropa's creation, amazing distortions of the anthropo-geomorphological concept come into existence, with even revered scientists having subsequently made some stupid mistakes in their references to Atlantropa. For instance, Glenn Theodore Seaborg (1912-1999), recipient of one-half of the Nobel Prize for Chemistry in 1951, and his co-author William Roger Corliss (1926-2011), wrote perhaps the last full-length book about the now defunct USA "Plowshare Program" (1958-1975)—the civilian use of plutonium-based "dirty" nuclear explosives for peacetime landscape and seafloor transformations. Seaborg and Corliss' book, **Man and Atom**, an otherwise fine exposition, contains fundamental oceanographic errors when it described Herman Sorgel's Atlantropa realization plan: "...nuclear blasts could help fill in the Strait of Gibraltar, a feat which, according to its proponents, would cause the Mediterranean Sea to rise a bit and freshen to the point where the Sahara could be irrigated. Of course, the advantages of a verdant Sahara would have to be weighed against the loss of Venice and other sea-level cities. We repeat these proposals primarily to stimulate thinking about the pros and cons of planetary engineering" (Seaborg and Corliss, 1971, page 194). They must have meant "thinking" stimulated by reverse Psychology; finding accurate geo-historical data is difficult enough work without having still more confusion generated by paradoxical Psychology!

The Seaborg-Corliss gross factual misstatement was repeated when David D. Caron began his article in a professional Ecology journal with a reprise of it, citing R. Sylves' **The Nuclear Oracles** (1987, pages

193). Nevertheless, their most important point is that it was then, and remains during 2014, technically possible for humans to replace macro-geomorphic event-processes with macroprojects such as Herman Sorgel's mapped nucleus for "World Peace" Geographos finds very exciting. Humans, with their erect posture and rotating neck see the planet and the firmament in a way no other animal does; hominization was inseparable from technologization—both events mark the commencement of human beings. Perhaps "World Peace" (Leidner et al., 2013) is synonymous for technical "Progress" made possible by Macro-Imagineering applied to Earth's fundament (and other, accessible Solar System fundaments)? Humankind's common agricultural pursuits changed from the 1950s, from extension to intensification through farming's industrialization; abandonment of some landscapes for lack of need, helped to reduce the quantity of carbon dioxide gas entering the Earth-atmosphere, certainly a beneficial effect from a Green viewpoint (Shevliakova et al., 2013); in contrast, a certain amount of carbon dioxide gas is absorbed by infrastructure development (Muller et al., 2013). Greens of a slightly different hue, promote the forestation of vast tracts of undeveloped, often ignored, landscapes. Gene Keyes and Scott Seymour (January 1975), for example, hoped that "superordinate projects", applied to wastelands, will promote both "World Peace" as well as positive human attitudes towards all landscapes and the world-ocean. "Superordinate goal", a phrase neologized circa 1966 by the social psychologist Muzafer Sherif (1906-1988), labels an technology progressive objective that cannot be attained by a single noosystem, but which is compelling to several noosystems within a defined region, landscape, island or oceanic realm. "Superordinate goals and interactions in themselves may not produce cooperation, although with any such interactions, at least the most obvious negative preconceptions about other nations[-ecosystems] may be removed" (Keyes and Seymour, 1975, page2).

References Cited

Badescu, V., Isvoranu, D. and Cathcart, R.B. (2010) *Transatlantic Freshwater Aqueduct.* Water Resources Management 24: 1645-1675.

Balcerak, E. (2013) *Anthropogenic aerosols increasing over India.* EOS: Transactions American Geophysical Union 94: 465.

Battersby, S. (4 May 2013) *High and Dry.* New Scientist 218: 36-39.

Bounama, C. et al., (2001) *The Fate of Earth's Ocean.* Hydrology and Earth System Sciences 5: 569-575.

Bowring, S.P.K. et al. (2014) *Applying the concept of "energy return on investment" to desert greening of the Sahara/Sahel using a global climate model.* Earth System Dynamics 5: 43-53.

Boyle, C. (2013) *You Saw the Whole of the Moon: The Role of Imagination in the Perceptual Construction of the Moon.* Leonardo 46: 246-252.

Bracken, P. (April 1982) *The Next 20 Years in Latin America.* The Futurist 16: 8.

Brier, B. (2013) *Egypt-Omania: Our Three Thousand Year Obsession with the Land of the Pharaohs.* (New York: Palgrave Macmillan) 229 pages.

Calcott, B. (2013) *Engineering: Biologists borrow more than words.* Nature 502: 170.

Campra, P. and Millstein, D. (2013) *Mesoscale Climatic Simulation of Surface Air Temperature Cooling by Highly Reflective Greenhouses in SE Spain.* Environmentla Science & Technology 47: 12284.

Cathcart, R.B. and Badescu, V. (2004) *Architectural Ecology: A Tentative Sahara Restoration.* The International Journal of Environmental Studies 61: 145-160.

Chaisson, E.J. (2008) *Long-Term Global Heating From Energy Usage.* EOS: Transactions, American Geophysical Society 89: 253-260.

Charlier, R.H. (1991) *Water for the Desert—A Viewpoint.* The International Journal of Environmental Studies 39: 11-35.

Charlier, R.H. and Finkl, C.W. (2009) *Ocean Energy: Tide and Tidal Power.* (The Netherlands: Springer) 261 pages.

Chanson, H. (2012) *Tidal Bores, Aegir, Eagre, Mascaret, Porococa.* (New Jersey: World Scientific) 201 pages.

Clark, R.H. (2007) *Elements of Tidal-Electric Engineering.* (New York: Wiley-Interscience) 280 pages.

Cook, K.H. and Vizy, E.K. (2006) *Coupled Model Simulations of the West African Monsoon System: Twentieth- and Twenty-First-Century Simulations.* Journal of Climate 19: 3681-3703.

Cook, B. et al. (2012) *Will Amazonia Dry Out? Magnitude and Causes of Change from IPCC Climate Model Projections.* Earth Interactions 16: 1-27.

Damjanov, K. (2013) *Lunar Cemetery: Global Heterotopia and the Biopolitics of Death.* Leonardo 46: 195-162.

Davidson, F.P. and Huot, J-C. (September 1989) *Management trends for major projects.* Project Appraisl 4: 133-142.

Davies, R.J. et al. (2011) *Probabilistic longevity estimate for the LUSI mud volcano, East Java*. Journal of the Geological Society, London 168: 1-7.

Dolgopolova, E.N. (2013) *The conditions for tidal bore formation and its effect on the transport of saline water at river mouths*. Water Resources 40: 16-30.

Donohue, R.J. et al. (2013) *Impact of CO2 fertilization on maximum foliage cover across the globe's warm, arid environments*. Geophysical Research Letters 40: 3031-3035.

Eltahir, E.A.B. et al. (2004) *A see-saw oscillation between the Amazon and Congo basins*. Geophysical Research Letters 31: L23201.

Fiedler, J.W. and Conrad, C.P. (2010) *Spatial variability of sea level rise due to water impoundment behind dams*. Geophysical Research Letters 37: L12603.

Finkl, C.W. and Cathcart, R.B. (2011) *The "Morning Glory" Project: A Papua New Guinea-Queensland Australia Undersea Freshwater Pipeline*. Journal of Coastal Research 27: 607-618.

Forzieri, G. et al. (2014) *Ensemble projections of future streamflow droughts in Europe*. Hydrology and Earth System Sciences 18: 85.

Foster, R.G. and Roenneberg, T. (2008) *Human Responses to Geophysical Daily, Annual and Lunar Cycles*. Current Biology 18: R784-R794.

Freud, S. *Civilization and Its Discontents*, Section III, pages 778-779 **IN** Great Books of the Western World (1952).

Frierson, D.M.W. et al. (2013) *Contribution of ocean overturning circulation to tropical rainfall peak in the Northern Hemisphere*. Nature Geoscience 6: 940-944.

Goncalves, J. et al. (2013) *Quantifying the modern recharge of the "fossil" Sahara aquifers.* Geophysical Research Letters 40: ???.

Gray, S.L. and Harrison, R.G. (2012) *Disgnosing eclipse-induced wind changes.* Proceedings of the Royal Society A: Mathematical, Physical and Engineering Science 468: 1839-1850.

Green, J. (1970) *Concept for a Self-contained oceanic resource base.* Marine Technology Society Journal 4: 88-101.

Hardisty, J. (2009) *The Analysis of Tidal Stream Power.* (New Jersey: Wiley-Blackwell) 321 pages.

Heckenberger, M.J. (October 2009) *Lost Cities of the Amazon.* Scientific American 301: 64-71.

Hemmer, H. (1993) *Partial Irrigation of the Sahara Desert.* Speculations in Science and Technology 16: 65-68.

Hemming J. (2008) *Tree of Rivers: The Story of the Amazon.* (New York: Thames & Hudson) 368 pages.

Hernandez-Leon, S. (2010) *Carbon sequestration and zooplankton lunar cycles: Could we be missing a major component of the biological pump?* Limnology and Oceanography 55: 2503-2512.

Kerr, R.A. (2013) *Humans Fueled Global Warming Millennia Ago.* Science 342: 918.

Keyes, G. and Seymour, S. (January 1975) *The Sahara Forest and Other Superordinate Goals.* Peace Research Reviews VI (#3): 1-62.

Kidder, D.L. and Worsley, T.R. (February 2012) *A human-induced hothouse climate?* GSA Today 22: 3-11.

Kirk, J.T.O. (2006) *Light field around a point of light source in the ocean.* Journal of Geophysics 111: C07008.

Koolhaas, R. and Obrist, H.U. (2011) *Project Japan: Metabolism Talks.* (FRG: Koln) 719 pages.

Kramers, J.D. et al. (2013) *Earth and Planetary Science Letters* 382: 21-31.

Landau, G.D. (1980) *The Treaty for Amazonian Cooperation: A Bold New Instrument for Development.* Georgia Journal of International and Comparative Law 10: 463-489.

Liljequist, G.H. (1980/81) *Tor Bergeron.* Pure and Applied Geophysics 119: 409-442.

Lunniss, R. and Baber, J. (2013) *Immersed Tunnels.* (Florida: CRC Press) 536 pages.

Marchal, C. (1982) *The Moon-Day Project.* Acta Astronautica 9: 391-395.

McGinn, R.E. (1978) *The Problem of Scale in Human Life: A Framework for Analysis.* Research in Philosophy and Technology 1: 39-52.

Meyers, S.S. et al. (2013) *Human health impacts of ecosystem alteration.* Proceedings of the National Academy of Sciences 110: 18753-18760.

Mitchell, J.G. (March 1979) *The man who would dam the Amazon.* Audubon 81: 64-81.

Mora, C. et al. (2013) *Biotic and Human Vulnerability to Projected Changes in Ocean Biogeochemistry over the 21st Century.* PLOS Biology 11: e1001682.

Morgan, C. (April 1994) *Terraforming with Nanotechnology.* Journal of the British Interplanetary Society 47: 311-318.

Muller, D.B. et al. (2013) *Carbon Emissions of Infrastructure Development.* Environmental Science & Technology 47: 11739-11746.

Mumme, S.P. (1986) *"Engineering Diplomacy": The Evolving Role of the International Boundary and Water Commission in the U.S.-Mexico Water Management.* Journal of Borderland Studies 1: 75.

Neave, N. et al. (2011) *Male dance moves that catch a woman's eye.* Biology Letters 7: 221-24.

Oberst, J. et al., (2012) *The present-day flux of large meteoroids on the lunar surface—A synthesis of models and observational techniques.* Planetary and Space Science 74: 179-193.

Panero, R. and Candella, B.J. (1968) *Proceedings of the 4th American Water Resources Conference.* Pages 208-217.

Panero, R. (September 1969) *A Dam Across the Amazon.* Science Journal 5A: 56-60.

Panero, R. (19 July 1972) *The Amazon—A Catalytic Approach to Development.* Hudson Institute HI-1664-RR.

Patterson, T.C. et al. (September 2006) *The Myth of the 1970s global cooling scientific consensus.* Bulletin of the American Meteorological Society, pages 1325-1337.

Pauwels, E. (2013) *Communication: Mind the metaphor.* Nature 500: 532-425.

Radman, Z. (Ed.) (2013) *The Hand, an Organ of the Mind: What the Manual Tells the Mental.* (Cambridge: MIT Press) 464 pages.

Rock, I. and Palmer, S. (December 1990) *The Legacy of Gestalt Psychology.* Scientific American 263: 84-90.

Roehrig, R. et al. (2013) *The Present and Future of the West African Monsoon: A Process-oriented assessment of CMIP5 simulations along the AMMA transect.* Journal of Climate 26: 6471-6505.

Salmon, M. (1982) *Operational Considerations on the Moon-Day Project.* Acta Astronautica 9: 515-523.

Scanlon, B.R. et al. (2012) *Groundwater depletion and sustainability of irrigation in the US High Plains and Central Valley.* Proceedings of the National Academy of Sciences 109: 9320-9325.

Schuiling, R.D. et al. (2005) *The Hormuz Strait Dam Macroproject—21st Century Electricity Development Infrastructure Node (EDIN)?* Marine Georesources and Geotechnology 23: 25-37.

Schuiling, R.D. et al. (2007) *Power from closing the Red Sea: economic and ecological costs and benefits following isolation of the Red Sea.* International Journal of Global Environmental Issues 7: 341-361.

Shevliakova, E. et al. (2013) *Historical warming reduced due to enhanced land carbon uptake.* Proceedings of the National Academy of Sciences 110: 16730-16735.

Spennemann, D.H.R. (2004) *The ethics of treading on Neil Armstrong's footprints.* Space Policy 20: 279-290.

Spennemann, D.H.R. (2007) *Extreme cultural tourism from Antarctica to the Moon.* Annals of Tourism Research 34: 898-918.

Spiegel, D.S. and Madhusudhan, N. (10 September 2012) *Jupiter will become a hot Jupiter: Consequences of post-main-sequence stellar evolution on gas giant planets.* The Astrophysical Journal 756: 132.

Sprout, H. (January 1963) *Geopolitical Hypotheses in Technological Perspective.* World Politics 15: 187-212.

Stenke, A. et al., (2013) *Climate and chemistry effects of a regional scale nuclear conflict [between India and Pakistan].* Atmospheric Chemistry and Physics 13: 9713-9729.

Tollefson, J. (2013) *Footprints in the forest.* Nature 502: 160-162.

Tucker, P. (May-June 2011) *Solar Power From the Moon.* The Futurist 45: 34-38.

Walker, J.A. (1983) *Dream-work and Art-work.* Leonardo 16: 109-114.

Wang, Y. et al. (2014) *Asian pollution climatically modulates mid-latitude cyclones following hierarchical modelling and observational analysis.* Nature Communications. DOI:10.1038/ncomms4098.

Wild, M. (January 2012) *Enlightening Global Dimming and Brightening.* Bulletin of the American Meteorological Society 93: 27-37.

Williams, E. (2014) *Moon: Nature and Culture.* (London: Reaktion Books) 224 pages.

Youssef, A.M. and Maerz, N.H. (2013) *Overview of some geological hazards in the Saudi Arabia.* Environmental Earth Sciences 70: 3115-3130.

CHAPTER 8

MEDITERRANEAN SEA BASIN ENCLOSURE AND REDEVELOPMENT

Fortunately, the USA-manufactured thermonuclear weapons that were accidently misplaced (after two military aircraft had collided and disintegrated in the air) beneath the Mediterranean Sea's cerulean seawater surface near the coastal town of Palomares, Spain, from 17 January until 7 April 1966, were utterly fail-safe gravity bombs (Maydew, 1997). Of the several bombs that fell from the aerial collision zone, only one remained for long on the seafloor, about 15 kilometers east of the Palomares Fault (Weijermars, 1987). In 1972, Arthur Rockwell Miller (1915-2005), Co-chairman of the USA's Delegation to the International Commission for Scientific Exploration of the Mediterranean Sea, foolishly terrified his audience when spoke an exaggeration of awesome geographical scope: "If the sill of Gibraltar [Strait] were removed, the heat budget of the Mediterranean Sea cold be changed, tidal fluctuations might inundate coastal towns, and coasts would be eroded and shifted. One might even speculate that had the nuclear bomb lost off of Palomares exploded and opened the sill of Gibraltar, the ecology of the entire Mediterranean area including the Black Sea would have been affected by a changed regime" (Miller, 1972, page 195). Miller's astoundingly supple statement is, no personal antagonism implied or inferable, unadulterated Green rubbish! [Written in 1986, *Jangada de pedra*, a novel by the 1998 winner of the Nobel Prize in Literature, Jose de Sousa Saramago (1922-2001) proposes the Iberian Peninsula floats away from Europe out into the Atlantic Ocean, thus calling into question its European identity; Earth's geopolitical bureaucrats are forced by this fictive event to deal with traumatic effects whilst the five main characters travel band together as they move about aboard the "Stone Raft". Was Saramago inspired by news media-reported technical nonsense?] Approximately 400 kilometers separated the less-than-ten-Megatonne explosive yield bomb's temporary sea-bottom resting place in the Alboran Sea from the far-distant crustal saddle, which is located at a depth of 320 meters in the Strait of Gibraltar, forming the selected seawater circulatory boundary between the Mediterranean Sea and the North Atlantic Ocean's Gulf of Cadiz, which is situated west of the Rock of Gibraltar. However, most of A.R. Miller's inartistic Absurdism [exaggeration] could take place, to some degree before 2100, if and when the global and local sea-level rises by about one meter. For instance, tidal regimes will be greatly altered everywhere, even the Bay of Algeciras west of the the Rock of Gibraltar (Sammartino et al., 2013).

Immediate fixation of macro-imagineers on the Strait of Gibraltar is centered on diverse superordinate goal proposals to span it with a bridge or a bored tunnel. Completed hydroelectric dam macroproject proposals focused on the Persian Gulf's Strait of Hormuz and the Red Sea's Bab-el-Mandeb, raises the chance—but only very slightly—that a 1930s Depression-era macroproject could promote a 21st Century emplacement of a hydraulic barrier of some kind throttling all seawater flows through the Strait of Gibraltar. By the start of the 21st Century, for sure, the world-public's interest in the pre-World War II scope of Macro-Imagineering/Macro-Engineering as directly related to the Mediterranean Sea Basin, developable as a unitary region, was profoundly renewed. An expected future rise of global and local sea-level, which inevitably will influence the Mediterranean Sea Basin's natural 13,000 kilometer-long shoreline, has instilled reasonable trepidation in Italy's citizenry that is evidenced by the reported public debate since May 2004 about the appropriateness or inappropriateness of shielding Venice against seasonal high-tide and storm surge flooding with an expensive lift-gate equipped permanent dam. Worldwide, approximately ten million persons or more live below present-day sea-level, and as global and local sea-level rises, more may be fated to do so also, for instance, in the Mediterranean Sea Basin. (**Figure 1A and Figure 1B.**).

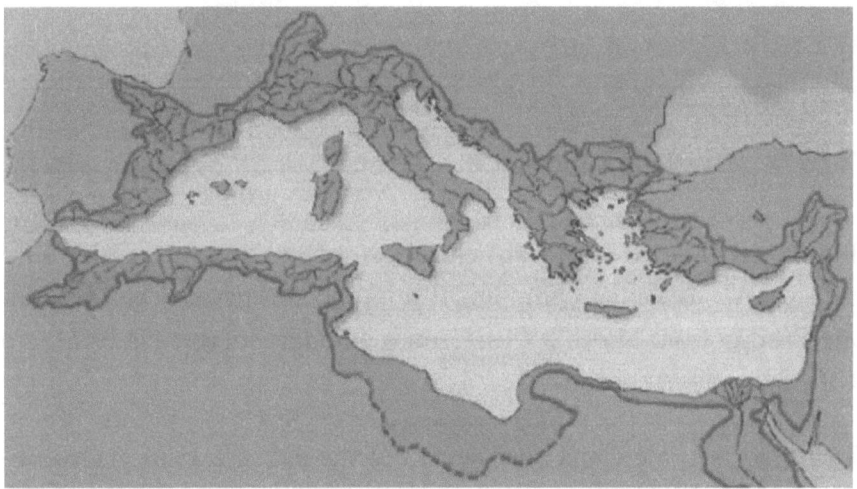

Figure 1A. The hydrologic landscape watershed of the Mediterranean Sea without taking into account the basin of the Black Sea or the upper Nile River. (Image: courtesy JMH).

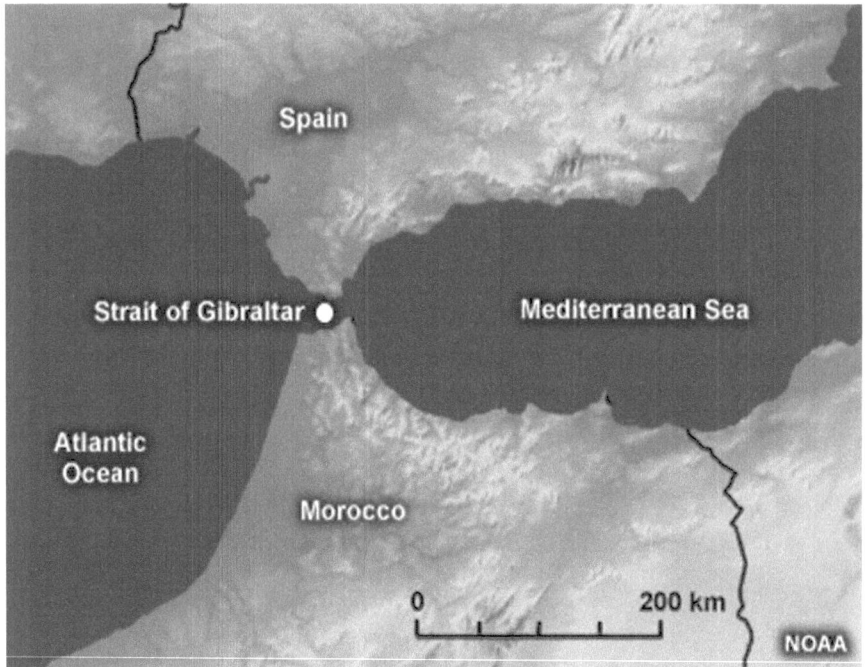

Figure 1B. The all-important entrance/exit at the Strait of Gibraltar for seawater as well as commercial shipping and naval fleets, separating Europe from North Africa. (Image: Google Images via NOAA, a US Government agency.)

Technology's historians have reviewed the old, pre-1930 macroproject proposal known as the "Atlantropa Project", but with no superordinate goal to revive it, merely to note its correct place in the century-long unfulfilled experience of Europe's economic and social integration (van Vleuten and Kaijser, 2006; Badenoch and Fickers, 2010). These historians had available two superb late-20th Century German-language histories of the "Atlantropa Project": (I) Alexander Gall's *Das Atlantropa-Projekt—Die Geschichte einer gescheiterten-Vision: Herman Sorgel und die Absenkung des Mittelmeers* (1998, 187 pages) and (II) Wolfgang Voigt's *ATLANTROPA: Weltbauen am Mittelmeer. Ein Architektentraum der Moderne* (1998, 144 pages). The ongoing shifts in the Mediterranean Sea Basin's human populace demography may foster new, favoring regional public attitudes on the increasing practicality and imperative necessity of a revamped "Atlantropa Project". During April-September 2003, the Deutsches Museum, Federal Republic of Germany, exhibited *Klima: das Experiment mit dem Planeten Erde.* An hour-long television dramatized documentary, laden with colorful three-dimensional special film effects, "recreations", of Atlantropa appeared in Munich, Germany during November 2005. By displaying the inspiring architectural drawings of a consummate draughtsman of the envisioned futuristic facilities planned for the Gibraltar Strait Dam site as well as the empoldered Mediterranean Sea Basin's new harbors, the world-public was made aware of a possibility of new Macro-Imagineering, even new Macro-Engineering, goal planning! These fearless synthesizers are professionals with a strong distaste for the extant, piece-meal approach to our Earth-Noosphere's geophysical and social realities.

During 1929, in Berlin, the World Power Conference promoted the founding of the International Commission on Large Dams in order to foster progress in the design and operation of big barriers. The German dirigible **Graf Zeppelin**, LZ-127, completed a circumnavigation of the world in August 1929 causing public excitement just like the American Space Shuttle's did six decades later. It was an achievement of great consequence because it solidified some European physical teleconnections with overseas colonies Herman Sorgel (1885-1952), a German national, during 1929, first detailed with an arched-in-plan concrete gravity dam to adjustably control all seawater entering the Mediterranean Sea from the North Atlantic Ocean (Spiering, 2002). (**Figure 2.**)

Figure 2. Herman Sorgel, October 1928. (Image: MJH.)

Sorgel had decided, during Christmastime 1927, to willfully dedicate himself to the actual materialization of Atlantropa; Sorgel died during Christmastime of 1952, after the 18 April 1951 signing of the Treaty of Paris, which created the European Coal and Steel Community. The dignitaries who added their signatures to the Treaty did so with the idea that it would be impossible to wage another Europe-wide war because the materials needed to sustain such conflict (coal and steel) would be under the control of a supra-national ruling bureaucracy; in other words, it was naively presumed then that merging Europe's ecosystem-nations should forestall a re-emergence of national animosities, and bring much greater prosperity to Europe. With two powerhouses generating 50,000 Megawatts, the Strait of Gibraltar Dam (**Figures 3-4.**) was the main elemental infrastructure (Schipper and Schot, 2011) intended to unleash chiefly European owned, operated and occupied farm and city constructions during the reclamation of the anthropogenically exposed continental-shelf that would soon appear because of any closure of the Mediterranean Sea can cause a natural evaporative reduction in sea-level of the sea within its Basin (Christensen, 2012). [Geoscientists know of much revealing geological evidence supporting a contention for a future natural Basin closing process-event at Gibraltar Strait that, after millions of years might also close the gap between the Old World and the New World (Duarte et al., 2013)!]

Figure 3. Vivid visualization of the Sorgelian-styled Gibraltar Strait Dam. The light-blue color denotes shallow water in the approach to

the northern powerhouse in Spain as well as, on the right-side, the new Mediterranean Sea 200 meters lower than the North Atlantic Ocean (right-side of image.) The 26 kilometer-long concrete Gibraltar Strait Dam is truly gigantic! The lighter shade/hue of blue is meant to denote a shallower Mediterranean Sea. Other important works details are not shown in this impressionistic image. Notice the spillway/powerhouse is located to minimize any seabed erosion that could destabilize this heavy-weight Sorgelian macro-structure. A recent reconstruction of the landscape adjacent to the Rock of Gibraltar, when the Mediterranean Sea was 80-120 meters shallower than presently, indicates the landscape, because of the cold climate regime, once was a savannah (Rodriguez-Vidal et al., 2013). Some suspect that should climate regimes change [warm] then the Iberian Peninsula might become a source for plant species populating Germany (Bergmann et al., 2010).

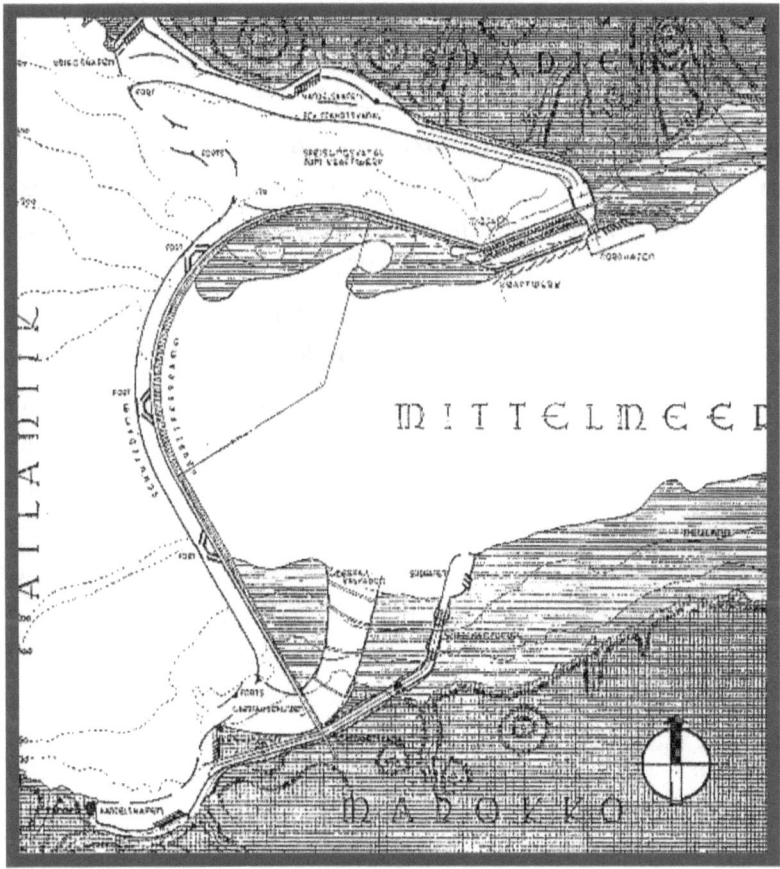

Figure 4. A more detailed drawing of Herman Sorgel's planned dam, illustrating its elaborate architecture and safety precautions. (Image: Sorgel's book modified by RBC.)

Herman Sorgel's plan required a concrete dam to blockade the water connection with the Black Sea and, as well, a series of sea-locks to control the Suez Canal, keeping the Red Sea's seawater from flowing rapidly and uncontrollably into the evaporating, closed-off Mediterranean Sea. One benefit, unforeseen by Sorgel, is that the Suez Canal becomes "immune", because of the installation of strongly-barricaded sea-locks akin to anti-storm surge barriers, to tsunamis generated in the Red Sea and the Mediterranean Sea (Finkl, Pelinovsky and Cathcart, 2012). The impresario Herman Sorgel imagined and proselytized for the rapid creation of additional European colonies in Africa; Greater Europe was to extend from the North Pole to southern Africa, fronting the Indian Ocean. Others, later, have announced similar Techno-Art visions. For example, Henry John Leir (1900-1998), in his *La Grande Compagnie de Colonisation: Documents of a New Plan* (1937) presented a fictional macroproject plan for a culturally united Europe geographically astride a reduced in seawater volume Mediterranean Sea and situated in a world wherein enlightened industrialists consistently helped humanity achieve lasting world peace. Until the late-1950s, Europeans continually pursued a geopolitical and geo-economic union with northern Africa (Muller, 2000). Herman Sorgel, **Figure 5**, first vetted, and absolved of wrongdoing, by an Allied de-Nazification Council, like many Germans had to re-adjust to the postwar international and national social environment (Schroeder, 2012).

Figure 5. Herman Sorgel, date unknown, but very likely photographed after World War II, perhaps during the late-1940s or early-1950s (Image: Google Images.)

By 1950, the human population of the Mediterranean Sea Basin numbered 170 million Europeans (73%) and 63 million North Africans (27%); during 2014 it was estimated that, by 2025, there could be about 305 million Europeans (44%) and 381 million North Africans (56%) living and working within the Basin's borders. For western Europe, "...the Marshall Plan heralded an era of unsurpassed prosperity for [westernmost] Europe: a twenty year period between 1953 and 1973, paralleling America's Golden Age, with no significant economic downturns and 4.8% annual growth rates, more than twice as high as any other point in history" (Magid, 2012, page 5). Especially during the period 1964-1973, in a time of acute labor shortage, Europe became a region of legal and illegal immigration (Therborn, 1987;

Giaccaria, 2011). World War II's casualties in Europe, Muslim immigrants to France from Algeria and Germany's 1950s guest-worker program, together with high birthrates amongst European Muslims and low birthrate amongst traditional Europeans are the chief causes of the remarkable demographic shift. It is now possible that, by 2030, nearly 25% of the people of the European Union will be older than 65 years of age. [Reportedly European automobile purchases slackened noticeably during 2013 and one of the reasons given by manufacturers is that Europeans are getting older and wish to buy other things with their discretionary incomes.] The African Union was proclaimed on 1 March 2001 and on 29 October 2004 Europe's ecosystem-nation leaders signed a European Union Constitution. A demographic shift in Europe seems to presage an epoch—occurring, perhaps, sometime circa 2015-2050—that will alter Europe's still distinctive culture: post-World War II Europe has been colonized by Muslims mostly from North Africa! During the next few years, Muslims in southern Europe may comprise 40% of the available labor manpower. Africa's current population is about 1.1 billion persons and, by 2100, could be 4.2 billion persons. This means that an almost forgotten macroproject such as the Gibraltar Strait Dam and its associated Basin infrastructures may find future acceptance with voting citizens of southern Europe and northern Africa. Forecast warmer and dryer future climate regimes in the Mediterranean Sea Basin are likely to be the initial stimulation for a re-assessment of the old, but unforgotten, macroproject proposal because desertification in the region is a security issue. After 2050, North Africa's population may exceed southern Europe's by nearly one hundred million persons!

The Netherlands owns infrastructure valued during 2014 at approximately USD 3.0 trillion—more than the monetary worth of the annual USA-European Union commercial-trade relationship—that has been set into place to protect the people of that noble ecosystem-country from unwanted storm surge and local sea-level rise incursions. There appears to be a 1% chance that at one meter rise of global sea-level will come to pass during the 21st Century. Assuming a cost of 2014 USD 1.0 million per linear kilometer, a total seawall protection for the Mediterranean Sea Basin's shoreline from an incursive future permanent seawater inundation might cost almost 2014 USD 13 trillions! Kenneth Jinghwa Hsu's *The Mediterranean Was A Desert: A Voyage of the Glomar CHALLENGER* (1983) popularized the confirmed 1972 geoscience theory that the desiccation of a closed-off Mediterranean Sea—during the Messinian Salinity Crisis, named after an evaporate discovered on Sicily, when the Strait of Gibraltar did not exist—occurred, on and off, during a period of several hundred thousand years aabout 6.3 million years ago when virtually all stagnant seawater evaporated, leaving an arid saltpan spotted with briny puddles (Garcia-Castellanos and Villasenor, 2011; Gibert et al., 2013). At 4.8 kilometers below 2014's sea-level, there would have been 1.7 times the normal air pressure impinging the valley floor and this implies that a scorching wind blowing therein would have been about $32\text{-}47^0$ C hotter than at today's Anthropocene sea-level. The highly saline evaporates coating the valley floor precluded most plant and animal life forms—perhaps some extremophiles survived in brine pools. So, the Mediterranean Desert would have been perhaps the harshest desert ever existing on Earth's surface during Geologic Time (Schneck et al., 2010). Eventually, the Mediterranean Desert was submerged by refilling with seawater, mostly flowing into it from the North Atlantic Ocean, about 5.3 million years ago.

Today's familiar Strait of Gibraltar is a shallow (286 meters deep) and narrow channel (12.9 kilometers wide at its narrowest), the Mediterranean Sea's only natural connection with the North Atlantic Ocean (Tanhua et al., 2013; Goffredo and Dubinsky, 2014). [An unnatural connection, for small boats only, is the freshwater canal dedicated in 1680, the Canal du Midi, planned by the French macro-engineer Pierre-Paul Riquet (Mukerji, 2009).] The land-dominated Mediterranean Sea's area (about 2.5 followed by twelve zeros square meters) amounts to less than 0.82% of the world-ocean's area and less than 0.32% of the world-ocean's seawater volume. [The Arctic Ocean is Earth's only other sea that is so land-dominated.] A region ranging from zero elevation [sea-level] to 200 meters below current sea-level—that is, the continental shelf—accounts for 30% of the seafloor area. Supposing the Strait of Gibraltar to be gated, the present-day rate of sea-level reduction due to natural aerial evaporation could be about 0.5 meter annually. Uncovering the Basin's continental shelf, however, will directly affect all container transshipment seaports—commercial shipping harbors with supporting hinterlands that are rich in industrial and agricultural production and consumption—such as Port Said, Damietta, Marsaxllok, Gioia Tauro and Algeceras. A local sea-level rise will deepen all harbors, but damage existing harbor infrastructures. Commercial shipping-scale trade within the Mediterranean Sea Basin, in 2014 served by more than 305 seaports including Barcelona, Marseilles, Genoa, Piraeus and Izmir, will also be greatly affected by any future changes of harbor navigation depths (Pelling and Blackburn, 2013).

First advanced as a macro-planned control of the seawater flow of the North Atlantic Ocean's surface layer entering the Mediterranean Sea for the purpose of hydropower generation and grand-scale continental-shelf reclamation, after World War II the Atlantropa Project was publicly touted by two technology and macroproject historians, separated by a demographic generation, as a "macroproject of the future". Walter Harry Green Armytage (born 1915), during 1961, and Ervan G. Garrison, during 1991, voiced their professional convictions in popular books that a Strait of Gibraltar Dam generating electricity was a desirable, potentially a key piece of world civilization's sustaining infrastructures. [In other words, not one of Dirk van Laak's infamous "White Elephant" macroproject examples described in *Weite Elefanten: Anspruch und Scheitern technischer Grossprojekte im 20. Jahrhundert* (Stuttgart, FRG: Deutsche Verlags-Anstalt, 1999, 304 pages)!] Without endorsement, Henry Petroski recounted Atlantropa's intellectual history and its obvious geographical impacts if built in *Pushing the Limits: New Adventures in Engineering* (2004). On 16 February 1976, "The Convention for the Protection of the Mediterranean Sea Against Pollution"—commonly referred to as the "Barcelona Convention"—was adopted. It is interesting to realize that the international marine shipping steel containers lost overboard from ships traversing the Mediterranean Sea are equivalent to the lost clay amphorae of Greek and Roman times. Some of our ancestors, both European and African, were just as polluting of the environment as some of our peoples are today. In fact, one might see both items as emblems of Europe-Africa human heritage!

In marked contrast to Armytage and Garrison, Stephen Henry Schneider (1945-2010), a Green climatologist, editor of the journal *Climatic Change* from 1975, condemned Herman Sorgel's reclamation macroproject in his ill-considered summarization (Schneider, 1996). Schneider misunderstood and inaccurately described the Atlantropa Project—he badly jumbled all the germane geographical facts. A Herman Sorgel-planned Gibraltar Strait Dam would never cause the Mediterranean Sea's level to rise, raising the ridiculous possibility of enlarging northern Africa's disappearing Lake Chad via a "Second Nile River" dug to convey plant-poisonous saline water overflow away from the Mediterranean Sea! Schneider also claimed that Sorgel's dam could degrade northern Europe's climate regimes because of the stoppage of the high-salinity bottom-seawater outflow into the North Atlantic Ocean. Schneider must have consulted poor-quality published references or gotten some ill-timed, false or bad oral advice. Anyway, he could have had an accurate rendition of the macroproject plan if he had perused only Igor Adabashev's 1966 textbook, *Global Engineering*! Herman Sorgel's early-1930s Depression era plan for the actual physical depression of the Mediterranean Sea's seawater level encompassed the creation of the world's largest man-made freshwater lake in the Sahara, a reclamation-enlargement of shrinking Lake Chad. (**Figure 6.**) Herman Sorgel's "Atlantropa" probably must be appraised, during the 21st Century, as the second largest industrial effort in human history—the impending effort by macro-engineering to extract excess carbon dioxide gas from Earth's air on a huge geographical and economic scale, undoubtedly must be considered the Number One industry of the Anthropocene and the 21st Century's Earth-Noosphere! On 26 September 2013, the Oxford Conference on Negative Emission Technologies commenced (in the United Kingdom) this new macro-project in a spirit of Green-ish resignation—not as bad as Spenglerian depression, but not far from it either (Marshall, 2013).

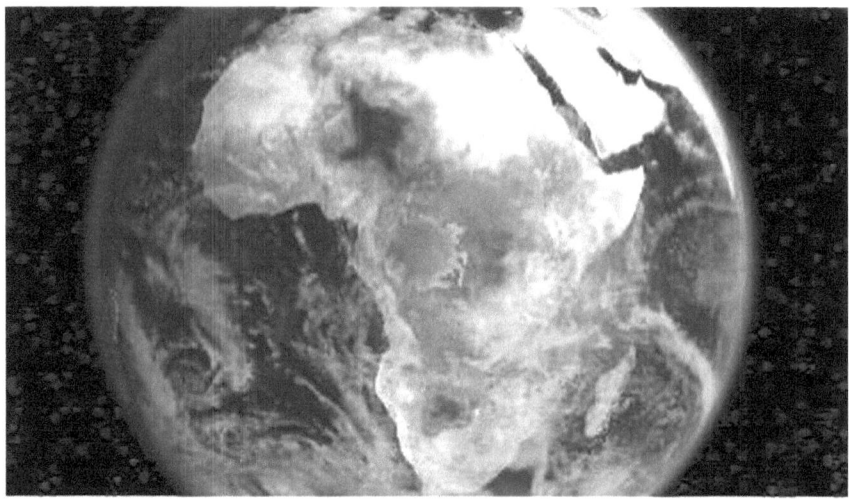

Figure 6. Three enlarged freshwater lakes—please ignore the irrelevant-to-our-story freshwater recreation of ancient Lake Makgadikgadi of 10,000 years ago of southernmost Africa, west of the

island-nation of Madagascar. The Congo Lake in central Africa, and the refreshed Lake Chad, in northern Africa were macro-imagined by Herman Sorgel first! (Image: Google Images.)

If the Congo River, carrying 1,320 to 1,775 cubic kilometers of freshwater runoff to the South Atlantic Ocean yearly (Materia et al., 2012), were dammed at Boyoma Falls [formerly Stanley Falls], then it would impound upstream a very large lake that Sorgel dubbed the "Congo Sea". Central Africa contains the second largest and the least degraded area of contiguous Tropic Zone rain-forest in the world; still, about 40% of Basin of the Congo River is savanna (Verhegghen et al., 2012). A tributary of the Congo River, the Ubangi River, could then flow unimpeded northwest, joining the Chari River. The diverted freshwater, augmenting the Chari's natural runoff, would then finally be deposited onto the nowadays mostly dry lakebed of Lake Chad (Lemoalle, et al., 2012). Subsequently, Lake Chad could be gradually increased in volume to, ultimately, form Sorgel's Chad Sea. After Sorgel's time, geoscientists had discovered that, before the Anthropocene, during the Holocene, an equivalent massive natural pooling of freshwater had existed as "Lake Mega-Chad" (Bouchette et al., 2010; Contoux et al., 2013)—so, in essence, Sorgel's Chad Sea could be viewed strictly as a restoration macroproject! [On 26 September 2012 Chad became the 28th Party to the 21 May 1997 United Nations Organization Watercourses Convention (Resolution 51/229), adopted by the United Nations Organization General Assembly.] Herman Sorgel's version of the Congo Sea and the Chad Sea might inundate 10% of Africa. A "Second Nile River" freshwater artery, Herman Sorgel asserted, could be induced to flow gravitationally northwards, across the mostly uninhabited central Sahara, creating an irrigated Anthrome resembling the crowded narrow Nile River Valley in Egypt. Very like the Second Nile River would become a Mediterranean Sea nutrient enrichment source, just like the original Nile River (Nixon, 2004; Oczkowski et al., 2009), especially after vast adjacent swathes of landscape are systematically, permanently settled and worked beneath the Cathcart-Badescu Sahara Tent. [See: Chapter 2 and Cathcart and Badescu, 2004).] Without creating Sorgel's Congo Sea, a run-of-river intake, powerhouse and turbines as well as a spillway, costing 2013 USD 58 billions, called "Grand Inga Cascades", harnessing the Congo River above and below the seven-level Inga Rapids falling 60 meters over a distance of 100 kilometers, could generate 40,000 Megawatts; that electrical power could be employed to rapidly industrialize Central and Southern Africa (Showers, 2009)! Grand Inga Cascades, as a run-of-river operation won't create major discontinuities in the riverbed morphology or sediment character—no major accumulation of sediment upstream (sediment trap effect) nor downstream bed erosion.

By late-20th Century, Robert Glenn Johnson (born 1922), **Figure 7.** Suspecting that the increasing salinity of the seawater exiting, as an underflow, the Mediterranean Sea Basin at the Strait of Gibraltar, might be the future cause of a new Earth Ice Age commencing in the early- to mid-21st Century, proffered a controversial Macro-Imagineering proposal to study the concept of a porous strait barrier, a permeable rubble-mound dike, emplaced in the fluid connection between the North Atlantic Ocean and the Mediterranean Sea (Johnson, 1997). Briefly, his dike-weir is a 300 meter-high rubble mound structure design with a low 70 meter-wide crest; some sections of it could behave like statically stable submerged low-crested breakwaters with their crests above sea-level whilst other sections act like statically stable submerged breakwaters. All the rocky materials rest at a thirty degree angle of repose. As the structural behavior of Egypt's pyramids clearly demonstrate, even snuggly-fitted stacked cubic rocks can slip as a result of strong local seismic activity. J. Jaeger, a Swiss macro-engineer, shortly before World War II planned to span the English Channel with two parallel dikes between which ships, powered by electricity like trolley cars on city streets, could safely traverse the channel. **Figure 8**, below, shows an architectural cross-section of his massive rock-fill emplacement.

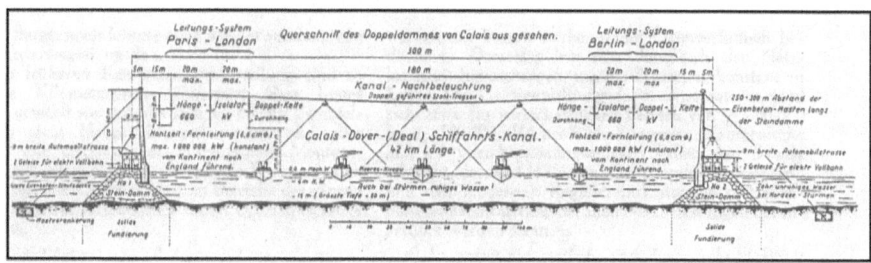

Figure 8. Between the dikes would be calm seawater illuminated at night by outdoor "street" lighting! Railway tracks and a roadway for motor-cars are also provided. (Image: RBC collection.)

It is worth noting, also, that an Italian macro-engineer, Nino del Bosco, considered a crossing of the Messina Strait by emplacement of a rock-fill dike (300 meters wide at its base, 130 meters high with 10 meters projecting above local sea-level) atop which would be electrified trolley railway lines and a super-highway. It was del Bosco's belief that a rock-fill structure could most easily and cheaply be repaired if it were damaged by strong earthquake motions, a major structural challenge that builders of the Messina Strait Bridge will have to face one 21st Century day (Brancaleoni et al., 2010).

Figure 7. Robert Glenn Johnson, date unknown but probably the 1960s. An innovative Geoscience thinker and widely respected academician. (Image: Google Images.)

Johnson's artificial reef-like rubble pile would throttle the seawater flow, especially that which leaves the Mediterranean Sea because his mounded anthropogenic barrier would be thickest at its base, but it was really intended to slow only the outflow of high-salinity seawater, in order to prevent ice-sheet formation in northeastern Canada. His anti-New Ice Age macroproject rests entirely on the proposition that Egypt's Aswan High Dam (closed 1965) has caused the measured increased salinization of seawater leaving the Mediterranean Sea (Skliris and Lascaratos, 2004); Johnson's anthropogenic submarine ridge would become a worthless technological fix if the Aswan High Dam were breached! Were that rock-fill gravity dam suddenly broken, as in Michael Heim's terrifying 1972 novel *Aswan!*, the release of the reservoir's entire contents would elevate temporarily the Mediterranean

Sea's fluid level by about 6.6 centimeters. Robert G. Johnson's theory of the New Ice Age is carefully sketched in *Secrets of the Ice Ages: The Role of the Mediterranean Sea in Climate History* (2002). The public controversy initiated by him remains geo-scientifically unresolved (Bryden and Webb, 1998). Both Johnson's and Sorgel's barrages would alter (de-tune) the Mediterranean Sea's tides, in some instances (such as the Aegean Sea) possibly doubling the amplitudes of its semi-diurnal tides. Such changes must also affect all planning for "The Bridge Over the Adriatic"—actually an extended viaduct—connecting Zadar, Croatia and Ancona, Italy (Romanis,2008). **Figure 9.** [Currently, the longest sea-bridge in the world, the Jiaozhou Bay Bridge spanning 42.4 kilometers, was opened 30 June 2011 by China.]

Figure 9. A fanciful architectural drawing of the trans-Adriatic Bridge offered by Girogio de Romanis. (Image: Google Images.)

In addition, both barriers would probably terminate the atmospheric carbon dioxide gas drawdown currently performed by the Mediterranean Sea naturally. Warmer seawater trapped within the Mediterranean Sea likely would reduce the number of major warm-core hurricane-like cyclonic storms affecting the Sea to as few as one per year (Walsh et al., 2014). It is doubtful that the Gibraltar Strait Dam, fixed in an oceanic constriction [chokepoint], could imitate China's Great Wall on its landscape as a notable impediment to life's gene flow (Su, 2003). Still, if the Mediterranean Sea's salinity exceeded 6%, then all life in the enclosed body of water would die—except, perhaps, for some archaebacateria.

Commercial hydropower dams first began operating in the United Kingdom and the USA from 1880-1882 (Lagendijk, 2008). Germany became and early leader in the construction of big dams and in the reclamation of vast tracts of wasteland, according to David Blackburn's **Water, Landscape, and the Making of Modern Germany** (2006). Commencing circa 1929, and ceasing by circa 1952, the German architect Herman Sorgel proposed construction of the world's most powerful hydroelectric dam, which he estimated capable of generating 50,000 Megawatts at the Strait of Gibraltar Dam, when the Mediterranean Sea enclosed by it had been lowered by 200 meters. That singular facility could be as long as 45 kilometers, weigh almost 104 billion metric tonnes, or about equal to 19,000 Great Pyramids of Cheops! [The weight is approximately five times humanity's annual global redistribution of Earth materials—meaning, if all materials worked yearly were directed to accumulate within the Strait of Gibraltar, then it would take little more than five years of effort to emplace Sorgel's imagined edifice!] Nowadays, perhaps, the widespread use of automated and tele-controlled construction and mining robots would reduce the need for recruitment of trained and acclimatized human laborers. In terms of material volume, Sorgel's edifice would equal the emplaced bulk of 960 Afsluitdijks [in The Netherlands, closing the Zuider Zee in 1932] or 450 Three Gorges Dams [China] or 120 Panama Canals! Members of the Candida Oancea Institute in Bucharest, Romania, estimate the cost of building Sorgel's monumental installation would be approximately an investment of 2013 USD

116 billions, about the early 21st Century property tax valuation of New York City's Manhattan Borough. Off-setting this cost is the average annual avoided oil equivalent saving that will amount, annualized, to tens of billions of USA dollars.

"Atlantropa" means, literally, a macroproject plan that is a suggestion in favor of a "turning towards the North Atlantic Ocean" for hydroelectricity generation. Nowadays, all electricity manufactured at Herman Sorgel's enormously imaginative, some would say megalomaniacal personal monument (Gibraltar Strait Dam), could be efficiently transmitted to consumers within the forthcoming Mediterranean Transmission Super Grid maybe via super-conducting cables (Hawsey, 2005). [This facet was ignored by artist Peter Fend (born 1950) in *Ocean Earth: 1980-Today* (1999) (Joselit and Harrison, 2008) because of his affinity-addiction for locally-produced Green energy. The concept of long-distance power transmission was barely alluded to by any of the contributors to *New Geographies 05: The Mediterranean* (2013), edited by Antonio Petrov.] A 400-kilovolt electrical interconnection, completed by 1999, between Spain and Morocco was the first electric cable linkage established between Europe and Africa. The DESERTEC initiative represents a major advance in the transition towards renewable energy sources because it creates web-like transmission linkages transferring plentiful solar energy from the Middle East and North Africa for distribution in Europe (Clery, 2010; Strahan, 2009; Pearce, 2009; Trieb and Muller-Steinhagen, 2007). Since its first years of revelation, and for a variety of onerous geopolitical reasons, DESERTEC has recently faded in popularity with Europeans, amongst both the public and amongst macro-imagineers/macro-engineers (Smil, 2011; Powell, 2012). DESERTEC could be subjected to damaging climate change event-processes (Patt et al., 2013). By 2014, some Greens were drifting towards locating solar power facilities serving the needs of

Europe placed aboard gigantic, Giga-Float-like, rafts anchored in the Mediterranean Sea as near to the sunniest regions as possible! Such a macroproject would entail an increase in Europe's naval presence in the Mediterranean Sea, but that could be helpful to stem the tide of illegal human migrants originating mainly in Africa. Such Giga-Float power generators might be studded with plasmonic solar-cells that look exactly like an adored children's toy: the LEGO multi-piece sets! (**Figure 10.**) The peak electricity load of the current Mediterranean Power Pool is more than 278 Gigawatts; electrical power can be transferred over a region consisting of three of Earth's Time Zones!

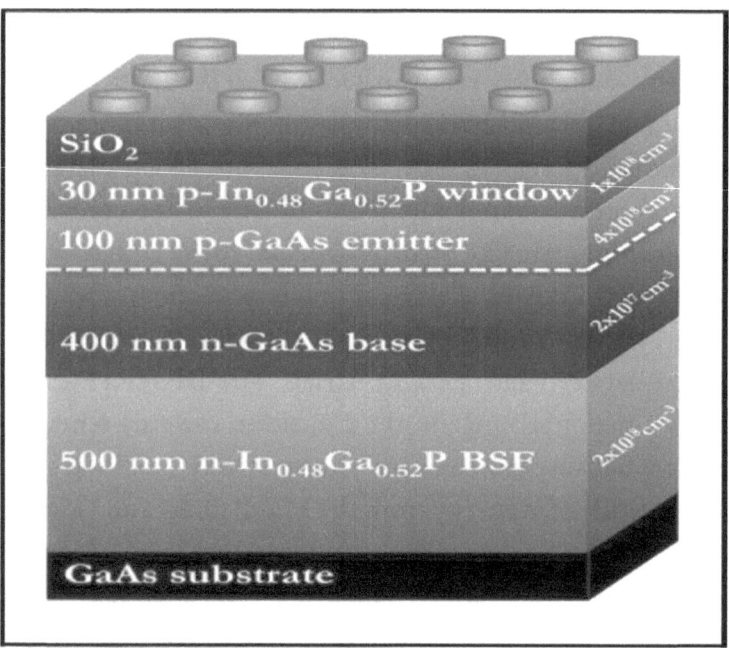

Figure 10. The rows of aluminum studs at the top of a plasmonic solar-cell help solar-cell aggregations extract more energy from sunlight than if the cells were flat-topped. (Image: http://www.science.daily.com/releases/2013/10/131018084452.htm. Altered and enhanced by JMH.)

Last century, R.B. Cathcart opted for an adapted Sorgel-like Gibraltar Strait Dam, affixed to the sea-bottom by a very strong reinforced concrete shear-key, that might operate successfully—in an economic sense—with only a 50 meter reduction of the Mediterranean Sea's sea-water-level, uncovering a new area of landscape amounting to about 8% of the Mediterranean Sea's present-day surface area. However, apparently neither R.B. Cathcart (1998), nor H. Sorgel evidently, thought about or made useful effort to derive a calculated optimal efficiency sea-level technical specification! Unlike dams blocking river valleys, Sorgel's edifice would never suffer sapped hydroelectric power production caused by drought since the installation's head-pool is Earth's world-ocean!

About 10,000 years ago, the Mediterranean Sea was 50 meters lower than during 2014, without 125,000,000 cubic kilometers of seawater that is present now. All other factors remaining the same, if the seawater surface was under an air column that was 50 meters thicker than it is during 2014, it is probable that the air's temperature at the air-sea boundary would be about 0.3^0 C warmer. This estimate ignores climate change trending to warming. Adding the likely temperature increase due to Green-predicted anthropogenic global warming—ranging from say, 2.1 to 4.4^0 C—plus the "atmosphere thickening factor" of 0.3^0 C means that seawater evaporation may be significantly enhanced. In other words, a 50 meter reduction might be accomplished in less than a century, possibly by 2080. [To put this estimate into a bigger context for a proper perspective, Earth's Polar Zones are approximately 0.5^0 C warmer during a full Moon than during a new Moon because the tidal pull alters wind flows sufficiently to transfer heat from low-latitudes to high-latitudes and, thus, to measurably affect average air temperatures in those frigid zones (Shaffer, 1997).] There might be, incidentally, an increase in the number and severity of Mediterranean Sea Basin hurricanes resulting from seawater warming (Emanuel, 2005; Romero and Emanuel, 2013). The 6.3 kilometer-long Corinth Canal, completed in 1893, and only 8 meters deep navigationally, will in about sixteen years fall dry, becoming like the trans-isthmian "Diolkos" route, a ship-slipway, of 700 BC (Werner, 1997)! Still, some 21st Century commercial Roll On/Roll Off maritime shippers might find a dry Corinth Canal quite useful as a 21 meter-wide portage ship-railway or hovercraft guide-way—the inventor Christopher Cockerell (1911-1999) tested the first practical hovercraft in 1959, as mentioned previously (Sambracos, 2003).

Many physical and biological changes would inevitably occur upon completion of a Sorgelian Gibraltar Strait Dam—for example, freshwater artesian springs that are undersea springs nowadays may become valuable in the future as sources of freshwater for new coastal settlements and farmland irrigations. Fernando Gomez examined some of the seawater chemistry consequences of a Mediterranean Sea enclosure (Gomez, 2003). Its seawater will warm and become saltier, causing many extant species of its rather sparse life component to decline in total biomass or become extinct. In brief, Atlantropa's exacting builders will instigate an anthropogenic marine salinization. Any future climate changes will affect the storm tracks and storm characteristics that govern the Basin's hydrological cycle (Mariotti and Struglia, 2002; Camilleri, 2012) and that event-process may affect the North Atlantic Ocean's thermohaline circulation were either Robert G. Johnson's underwater piled rock ridge or Herman Sorgel's concrete barrier built. Once emplaced, Sorgel's Gibraltar Strait Dam will give the remaining Mediterranean Sea somewhat the character of an "aquarium". Aquatic biological invasion of the Mediterranean Sea is a fact of life, and preliminary open-ocean fertilization experiments have been done elsewhere to stimulate a part of the surface seawater layer. It is possible this technique could be applied to the isolated Mediterranean Sea in future. To remedy local pollution problems, some new species of seaweed may be introduced (Fei, 2004). Isostatic rebound-related earthquake activity and reduced hydrostatic pressure caused by regional isostatic Earth-crust rebound might destabilize gas hydrates present in the sea-bed. Some extant coastal infrastructures will, undoubtedly, benefit greatly from a 50 meter reduction of Mediterranean Sea altitude. For future human Atlantropans, not only will its Macro-Engineering be experimental—colossal (Grigsby, 2012) seawater barrier building, harbor and city reconstructions—but the macro-management of the various completed, operational Basin-wide megaprojects will also be profoundly experimental! However, macro-engineers should see their contacts anywhere in the Basin not as impure destructive touching but, rather, as the induced next natural existence-state steps for all materials (fluid, solid, gaseous, or agreeable/noxious Green persons) affected; soon Synthetic Biology may extend *Homo sapiens'* touch to the extension of the living realm's natural and artificial life-forms—even outwards to other Solar System places, via outer space-transiting intelligent vehicles. Positivist, reasonable Greens might take the standpoint that such an exit constitutes, in Macro-Imagineering/Macro-Engineering's terminology, a reduction of the Earth-biosphere's live loading!

Atlantropa embodies an idea of a combination of civilian and military geopolitical "regime". Regimes are most wanted and/or needed in geo-political situations where great interdependence prevails or will prevail (Keohane, 1982). An operating Mediterranean Sea Basin Authority should seek to use the seawater and all its contents, including that on, in or under the sea-bottom, as a source of international revenue, and such monies should be expended to ameliorate any hardship environments imposed by harsh climatic regime changes over Atlantropa's politically demarcated ecosystems. Linked concepts provide a neutral basis for international monetary transfers and geophysical justification for special development public-works by macro-engineers attending projects in those least-developed, peaceful country-ecosystems sharing the Mediterranean Sea's newest shoreline.

Collapse of the Sorgelian Gibraltar Strait Dam, from any imaginable cause, would foster catastrophic dam-break seawater wave propagation with tsunami-like characteristics as well as terribly deleterious effects upon the Basin's new and old shorelines. A dam-break disaster of such imagined magnitude has never before been physically or computationally modeled, but a first-step start has been made in another oceanographic investigation context (Speich, 1996). In effect, its collapse would restore the Earth's Main Watershed as defined in 1887 by Aleksei Andreevich von Tillo (1839-1899) (von Tillo, 1887). Geographos estimated that the leading seawater wave from such a rapid Gibraltar Strait Dam's crumbling might reach a velocity of 150 kilometers per hour! Sensitive supercomputer modeling applied to the particular dam-break macro-problem ought to be helpful simulations for persons trying to fully assess the nature and totality of the hydraulic situation. Geographos estimates that 17,680 cubic meters per second would fall into the Mediterranean Sea Basin as an irresible torrent at a broken Gibraltar Strait Dam. Such an infrastructure fracture would be an emergency and restoration macro-management headache; in terms of public relations, it could be a long-term geoscience, Geopolitics and Macro-Engineering setback! Since the induced reduction in seawater volume of the Mediterranean Sea will raise the world-ocean's level by around 33 centimeters, the quick re-establishment of a long-time aerial water vapor export region as a single inrushing seawater flow restoration would adversely affect the navigational level of all world seaports (Madron, 2011; Portman, 2012)! And, still another, possible, catastrophe impends: what if a really large superbolide—perhaps with a diameter greatern than 50 meters—hit the Mediterranean Sea's surface?

Rene Magritte (1898-1967) painted a disquieting visual paradox, one of his most Surreal "La Fleche de Zenon" ["Zeno's Arrow"] which illustrates a huge gray boulder suspended over a frothy and aquamarine-colored ocean like some stop-motion photograph of an incoming asteroid! If an asteroid fell into the central, lowered Mediterranean Sea, it would generate a tsunami wave which would radiate outwards in all directions and, in part, move rapidly towards the Strait of Gibraltar where it would smack the air-exposed backside ["downstream face"] of any Gibraltar Strait Dam! Can highly-stressed static structures tolerate such sea-wave impacts?

The world-ocean is the surface temperature boundary for the Earth-atmosphere over 71% of our planet's surface; a segment of the ocean, the Mediterranean Sea, is considered the "Mare Nostrum" of Anthropocenic geoscience (Krijgsman, 2002; Bergamasco and Malanotte-Rizzoli, 2010). A modern historiographer, Allan Megill, instructs truth-seeking historian with the answer to the question, "What was actually the case?" in the recorded past is "recounting". Historical explanation, he wisely opined, is dependent on a professional "recounting", which he likened to an event-process "...like the winning of [2,300 square kilometers of] land from... [The Netherlands'] Zuider Zee" (Megill, 1989, page 648). A lower sea-level Mediterranean Sea would be a maritime archaeological treasury. An in-depth recounting the Atlantropa Project, promoted from 1927 until 1952 in Germany by Herman Sorgel, deserves a thorough re-assessment in the event that impending global climate change challenges Space Age humans everywhere—including even those possibly visiting and/or contemplating Mars' future terraformation to macro-engineer our Earth (Hempshell, 2005).

"Technology" is employed loosely to denote the development and use of hand and other kinds of sophisticated tools in mobilized human

pursuits, yet Geographos is more than mindful that any definition is problematic because the world's meaning changes at a rate as
rapid as that at which what it signifies alters our Noosphere, commerce, weltanschauung, ecosystem-nations and Solar System planets
and moons. In humans, technology creates a sense of wonder, helping to solve the problem of boredom, absence of meaning for lives,
and loss of personal motivation; it can serve as a casting formwork
for future actions. In terms of boredom alieviation, visiting skiers at
Zermatt, Switzerland schuss on artificially manufactured and artisti-
cally sculpted snow, a theatrical stage spraying a man-made landscape
into existence via computer controlled sprinklers! Likewise, rigorous
definition of technological "expertise" is also elusive, but Geographos
herein employs the term to denote the idea of specialized knowledge
of Science, Art and progressing Technology points to an unbounded
period of human development at least in the Earth-biosphere; the real
danger to humans are unreasonable and ignorant (because of lacking or improper schooling) people; *Homo sapiens*' civilization cannot macro-engineer Earth without "designing" humanity. "Technical
Fix"—really, "Techno-Art"—is the full-story of all extinct and extant
human noosystems! Planetary stylization resembles what oil and wa-
tercolor painters do to a blank canvas: elite techno-artists apply the
crust of Technology (instead of paint) to a possibly currently unused
Earth-surface.

It seems obvious to all those who, in a balanced manner, are truly informed about, and intently concerned with, the urgent, over-
arching macro-problem of humanity's long-term survival and
prosperity that adherents, activists, advocates and the actionists
of Macro-Imagineering/Macro-Engineering's elite must properly
maintain the Earth-biosphere, or any applied Terraforming effort
elsewhere in our Solar System will lack its planet-stylization model.
Global Nature cannot "self-correct" entirely against climate change,

and the scientific and technology communities have both been over-estimating the impact of plants and animals, even soil micro-organisms, and exposed rocks in how they absorb carbon dioxide gas and ultimately impact global warming, whatever its source. Humans are global Nature's "trump card", capable of reversing the effects of such air warming. Oceanographers have mapped ship-dumped clinker as recognizable rock units of the North Atlantic Ocean's seafloor, deposited by coal-fueled vessels during a short historical period of steamship navigation—from the early 1800s to circa 1940 (Kidd and Huggett, 1981, page 102): "A surprising discovery from our [sea-bottom] sampling is that clinker from coal-burning ships can, even in north[eastern] areas, be more abundant on abyssal seafloors than any debris material deposited by ice-rafting or other geologic agents". The base for our Anthropocene efforts in the Mediterranean Sea Basin, probably, ought to be the obvious Earth-surface changes scaled against the distribution of nuclear- and thermonuclear weapon test generated radioactive decay particle deposits, or by something or some means that is similar (Garcia-Orellana, 2009; Roether et al., 2013). Informed geoscientists know *Homo sapiens* is the only species that is equipped with large-scale Earth feature changing Techno-Art, which is certainly selectively and randomly pressuring all other living, biosphere-confined organisms. Vegetation and animal life on our world's landscapes and even the creatures living in the deepest places of the ocean—especially after 23 January 1960 when the still-used bathyscaphe *Trieste* descended into the Pacific Ocean's depths—are subject to our discretion. All life—including that dwelling within our guts and bloodstreams—is potential future artifact. Because of the (retired) Space Shuttle and the International Space Station, humans are adapting, via gravitational physiology, to variances of gravity during short aerospace-plane flights, extended journeys in outer space, and the gravity of our Moon. As the technically progressing pilot-passengers of R. Buckminster Fuller's "Spaceship Earth", increasingly capable of voluntarily exiting the presumed vehicle via smaller aero- and spacecraft, *Homo sapiens* must teach its species and its future teammates, mindkind (*Machine sapiens?*), to properly operate this Space Age planet before people can ever hope to successfully occupy and operate for an extending Anthropocenic period a "Spaceship Mars" or a "Spaceship Venus" residence-factory.

To reiterate, there is a chance, approximately 1%, the world-ocean will increase in volume, its seawater stock becoming higher by about one meter by circa 2100. One of the contributory reasons is that, because dammed freshwater reservoirs of our landscape fill with sediment, the storage capacity of reservoirs declines with time, thus causing more runoff to reach the world-ocean (Wisser et al., 2013). Many of our planet's major coastal metropolises nowadays monitored from Earth-orbiting satellites (Taubenbock et al., 2012; Dorrian and Pousin, 2013) will be gradually affected adversely; within the Mediterranean Sea, the port cities of Alexandria (Egypt), Naples (Italy) are at great potential geo-hazard risk. The Mediterranean Sea-Black Sea basins comprise a region where much can be done to mitigate a local sea-level rise because the basins abut technologically developed ecosystem-nations and there is a common availability of high-technology R&D. "More importantly" and "most importantly"—these adverbial phrases of degree cannot be made to fall trippingly from the English-language speaker's tongue and they smack of unseemly conceit. If degrees of importance must be compared an answer lies in "of greater importance" or "of paramount importance". In a regime of choice, where unfettered use, or use with some permanent physical limitations, of the Mediterranean Sea Basin-Black Sea Basin high-seas is weighed, Geographos suggests the reader think of paramount importance that the second option be adopted in order to abruptly counter-act a possible future doubling of the air's carbon dioxide gas content. How might it be possible, using Macro-Engineering techniques of climate control, to cool the planet's air economically and quickly?

A "Checker-board Macroproject"—alternation of black/white surfaces—installed upon the Mediterranean Sea and Black Sea surfaces would tend to promote convection if the albedo of that fluid surface was diminished in some sub-regions (so that less solar energy were reflected back to the sky above), then the conversion of energy at the surface would raise the temperature at its surface (and, thus, increase the radiation and evaporation) from the darkened regions, as well as heating the superincumbent air by conduction and convection. By changing the thermal conductivity or heat capacity of the topmost seawater layer of those liquid bodies, or both, or its seawater content available for evaporation, the surface layer would be affected, and thus, the outgoing radiation. Seawater reflects about 5% of impinging sunlight under calm sea states. [Geographos wonders what a mirrored artificial horizontal layer beneath the seawater's surface might do to the seawater trapped or situated between the air-seawater's interface and the mirrored fabric's topmost layer: could such a system also be used in the Gulf of Guinea east of Africa north of the Equator?] A self-replicating plate-like hydro-robot mono-layered Mediterranean Sea-Black Sea—approximately 0.008% of the Earth's ocean—could be harnessed to roughly "fine tune" that closed-off segment's sea-level and its internal oceanic hydro-climates [underwater "weather"] whilst functioning as a blanket beneath a modified Mediterranean Climate type! To offset global warming stimulated by a doubling of the air's carbon dioxide gas content, about 10% of the world-ocean, primarily in the Tropic Zone and Temperate Zone, must be colored white, like sea-ice floes in the Polar Zones. The strategic importance of the Mediterranean Sea-Black Sea is that the region is an enclosed body of seawater close to large Astythromes in Europe and the traditional Middle East where air-cooling could be most needed quickly.

Obviously, this "Checker-board Macroproject" makes any reduction of the Mediterranean Sea via Herman Sorgel's methods very, very counter-productive if the hinted climate regime control scheme were needed in a hurry!

A large patch of solar-powered white-colored bobbing hydro-robots having an anti-beaching capability will be effectively for the job required in the time allotted. Just like Heinrich Hemmer's transatlantic freshwater pipeline, each of the hydro-robots must be coated to prevent corrosion and to prevent reflectivity-degrading organisms preferentially attaching themselves to the hydro-robot hulls (Aliani and Molcard, 2003; Occhipinti-Ambrogi and Galil, 2010; Seebens et al., 2013). Barring painting all sea-going ships with highly reflective white paint, white-colored hydro-robots are doable. At the Equator approximately seven square kilometers of white-color, flat-topped hydro-robots could, theoretically, reflect to outer space about one Gigawatt; at a mean latitude an area of 7.7 square kilometers would suffice to accomplish that amount of energy reflection. In other words, a square-shaped installation of many hydro-robots (a square with each side 2.8 kilometers long) accomplishes an enormously useful task! Teflon has been marketed only since 1960. J.B.S. Haldane (1892-1964), in *Daedalus, or Science and the Future* (1923), foresaw Earth's ocean turned the color purple after a cultivated alga fertilized by manufactured chemicals is introduced to increase *Homo sapiens*-harvested fish stocks! [Was Joseph Ward Moore, 1903-1978, so impressed by Haldane's remark that he was stimulated to write his science-fiction novel *Greener Than You Think* (1947)?] Floating artificial plants, as proposed by Edward Forrest Moore (1925-2003), could detoxify polluted seawater in harbors (Moore, 1956). Only 3% of the sunlight striking a green plant is actually stored as chemical fuel within the organism; by contrast, some miniscule non-living devices that convert sunlight into usable energy have theoretical maximum efficiencies greater than

35%. Maintenance workers—*Homo sapiens* (humans) and/or *Machine sapiens* (mindkind) can be housed and headquartered safely on floating bases (Kunz, 1995).

Actually, the reflective mutli-robot sea covering mimic's the effect of the Sahara Tent Greenbelt proposed by R.B. Cathcart and Viorel Badescu, but in a smaller geographical context. Qualitatively, examining images of Earth's surface from outer space or fly-by spacecraft, anthropic activities are turning the planet into a more absorptive macro-object and, as a direct result, darker in the infrared and brighter in the visible wavelengths. The Sahara Tent Greenbelt proposed importation of Amazon River freshwater to help in the enormous improvement effort to green the desert. Later, during 2009, a Swiss writer, Jean-Edouard Buchter, proposed in **Saharia: les nouveaux paysans** (Paris: Editions du Madrier, 107 pages) and during 2012 with **Reverdir le Sahara** (Paris: de 'Aire, 134 pages) to water unsettled Saharan tracts with freshwater to create and vast green-belt of wealth and comfort for humans by tapping the freshwater that is current wasted by its melding with the world-ocean at river mouths. [Buchter was awarded, on 21 May 1974, US Patent 3,811,382 for a "Process for packing and leveling railway tracks and device for performing same", indicating his macro-engineering expertise is not insignificant! Human passage through the ancient Sahara is well established—three large ancient rivers [Irharhar River ending near the Algeria-Tunisia chotts region, the Sahabi River of northern Libya and the Kufrah River. The "Chotts" may become seawater-flooded, making those below sea-level basins into inland extensions of the Mediterranean Sea during the 21st Century], now buried by desert dunes served as viable routes for human migration across the Sahara to the Mediterranean Sea Basin's plusher landscapes (Coulthard et al., 2013).

At the same time, the greatest and largest landscape source of air-dimming dust [mineral aerosols], the Sahara of North Africa, remains virtually unaltered by humans; about 50% of the Earth-atmosphere's dust content emanates from northern Africa, especially the all but life-less hot landscape of Chad's east-west elongated 134,000 square kilo-meter dry lake basin, the Bodele Depression in the Sahara (Field et al., 2010). [Above the Sahara, the reader must visualize the view of the fictional film passengers aboard the1950s-era C-82 airplane who sur-vive its crash landing caused by a violent dust-sand storm—*The Flight of the Phoenix* (1965) dramatized, more or less realistically, the awful predicament of those parched survivors stranded in an extremely un-forgiving, waterless arid place. **Figure 11.**]

Figure 11. A film frame extracted from the Hollywood motion picture of the C-82 inflight above the "Sahara". (Image: Google Images.)

Bodele Depression is a part of the dried out Ice Age time Mega-Chad, becoming so about 1,000 years ago. The Depression is the source-region for about 20% of the aerosol deposition on the Amazon River Basin, helping to fertilize plant growth therein. These long-standing geographical facts can be altered radically, according to calculations done at the Candid Oancea Institute, principally by Viorel Badescu, during 2003-2004 (Cathcart and Badescu, 2004). Badescu and

Cathcart proposed a geographically gigantic Techno-Art encrustation macroproject, the pneumatic tenting of 3.5 million square kilometers of desert—about 50% of the Sahara's area! Their plan may seem wildly optimistic, yet a plan was offered in the early 20th Century to construct a series of shiny, foil-like cloth reflectors at least 5,200 square kilometers in area to signal Martians of our presence (Flowers, 1921). The Sahara Tent Greenbelt will be composed of, roughly, 700,000 almost-identical big white-colored inflated buildings. Their stated desire was to quickly change the Sahara's albedo and to afford agriculturalists, even some nomads, with potential greenhouses made to grow by harnessing freshwater pumped from known freshwater aquifers underlying the Sahara Tent Greenbelt. It was foreseen that the entire facility could be macro-managed via the Internet and all ecosystem-nations will maintain their current territories. Electrical power, in addition to solar power naturally, can be furnished to residents of the Sahara Tent Greenbelt by utilizing a unique hydrologic opportunity afforded by the Qattara Depression, where seawater from the Mediterranean Sea may be directed to pass through turbines evaporating in the Qattara Depression. Shortly after completion of the Sahara Tent Greenbelt, the air will become 50% cleaner so that regions immediately surrounding North Africa, namely southern Europe, will benefit from this macroproject with better air quality! Computer simulation of the effect of the desert's subsequent albedo that is unequivocally attributable to the Sahara Tent Greenbelt will be effective in countering global warming's regional effects by lowering local air temperatures, through its shading and sunlight reflectance. The spacious and capacious compartmentalized building constructed, in part, of fireproof fabric and air ought to be conceived as an enclosed botanical and economic industrial park.

Someday, perhaps, the poisoned rocks and soils (Certini et al., 2013) instigated by France's 17 nuclear weapon tests—four were atmospheric, exploded near Reggan, and 13 were subterranean, triggered in specially-mined tunnels near Ekker—all in southern Algeria (from 13 February until 1960) can become outdoor museums uncovered by the Sahara Tent Greenbelt? Such spoiled ground can become reminders of the days when the Sahara was a wilderness/wasteland through which humans trudged or aimlessly wandered or trekked determinedly on well-worn trails (Connelly, 2012-2013; Chianese, 2013; Xue and Shukla, 1996). The Sahara's present-day arid conditions began about 6,000 years ago; it is now possible to reverse this condition technologically. On the Mediterranean Sea Basin's northern climate boundary, it is very interesting to note that a mountain pass in Switzerland at an elevation of 2,756 meters, the Schnidejoch, situated between Sion in the south and Lenk in the north, on the south face of the Bernese Alps, was buried by ice and snow all year and, therefore, was never considered a practicable land route joining these towns—that is, until and exceptionally hot summertime during 2003 opened it to hardy hikers, some of whom were archaeologists. The numerous carbon-datable relics they discovered strewn on the landscape showed that Schnidejoch was closed to littering travelers from 4300 to3700 BC and 1500 to 200 BC. In other words, air temperatures were not warm enough to sublimate [melt] the ice and most of the snow in the mountain pass short-cut during those periods well before the carbon dioxide gas producing Industrial Revolution of 1784 (Hafner, 2012)!

Geographos has devised a safer and significantly less costly means for a truly 21st Century Atlantropa to come into concrete reality—but without such great masses of real concrete required as in the Herman Sorgel-envisioned dam structure, or Geographos' own reduced utility seawater barrier structure planned version from earlier times: terracing the Mediterranean Sea-Black Sea basins by a single one- or two meter tall Gibraltar Strait Dam, fabricated with inexpensive textiles, that excludes the effect of a global/local one-meter sea-level rise! In other words, the Mediterranean Sea's water level would remain as it is during 2014, and the fabric hydro-electric dam proclaimed will project above its level, but hold back a higher sea-level North Atlantic Ocean (Pugno, Cathcart and Bolonkin, 2011). If overtopped by the North Atlantic Ocean, it would appear as a 13 kilometer-wide anthropogenic waterfall, enormous by any standard of Earthly comparison (Hudson, 2012). By 2000, flexible barriers consisting of vertically tensioned membranes spanning the entire seawater depth were being first reported in the technical literature of coastal macro-engineering (Lo, 2000; Cullen, 2005). The plan by Geographos, Bolonkin and Pugno follows up on these advances. Geographos here cites USA Patent 3,785,158 issued to Andrew Noel Schofield (born 1930) on 15 January 1974 for his "Hydraulic Engineering Installations". Schofield opted to place a seawater-impermeable sheet on the seabed with a liquid pumpable drain below the sheet. In this way, a pressure differential (weight of seawater above pressing on the plastic sheet) seals the sheet, and the vertical fabric dam attached to it as an anchorage, securely to the seafloor and the sides of the submarine canyon that the Gibraltar Strait Textile Barrage (GSTB) blockades. Its necessarily shore-housed seawater pumps could be solar-powered or powered by plentiful, cheap hydroelectricity!

Space Techno-Art's proponents favor the construction of various symbolic artifacts, in low and high Earth-orbit outer space. With plastic sheeting and textile envelopes, Air Techno-Art's advocates opt to exploit all the potentialities of compressed air or naturally generated wind (Cathcart and Cirkovic, 2006). Land Techno-Art's fans reshape the Earth's rocks and soils into unique sub-aerial landscapes of some artistic merit appreciated by its aficionados. Geographos is inspired by the outdoor artworks of Robert Smithson (1938-1973)— *Spiral Jetty* (actually built in the State of Utah's Great Salt Lake during 1970) and his undefined *Proposal for a Monument on the Red Sea* (1966). Christo's planned *Mastaba*, a 150 meter-high, flat-topped pile of 410,000 multi-color steel oil barrels planned for the barren desert landscape at Al Gharbia, 160 kilometers from Abu Dubai city, resembles Smithson's monument, each appear to attempt to surpass the ancient pyramids of Egypt. *Spiral Jetty* required the expenditure of expensive fuels to power bulldozers and dump trucks, but it is composed of inexpensive, actually cost-free materials, just immediately available dirt and rocks at the worksite/art installation site. Christo installed eleven Flamingo-pink floating plastic mats covering approximately 600,000 square meters of a coastal lagoon's seawater in Biscayne Bay near Miami, State of Florida, for his temporary *Surrounded Islands Project* of 1983. During 1969, Peter Hutchinson (born 1930) and Dennis Oppenheiem (1938-2011) installed artworks beneath the ocean's surface off Tobago, West Indies. Oppenheim's *Arc* and *Threaded Calabash* are two examples. [The latter resembles, in miniature, ANTARES, the largest high-energy astrophysical neutrino telescope in the Northern Hemisphere, the instrument's volume is approximately 0.02 cubic kilometers.] Their installation is misidentified as "Oceanographic Art" since it is essentially decorative. However, Ocean Techno-Art is a form of seawater sculpting by aquatic terracing focused on the 71% of Earth's surface that is the world's oceoan; the modern-day originator of spatially large-scale intentional Ocean Techno-Art is the German architect Frei Otto, who first contemplated the concept's scope in 1953. Ocean Techno-Art has a commercially viable aspect that mere Oceanographic Art lacks and must, as a macroproject application, be of great interest to 21st Century Macro-Imagineers/Macro-Engineers!

About 5,000 years ago, the Earth's atmospheric methane gas content began to increase. Methane's generation source was human cultivation of rice in flatland fields. About 2,000 years ago, humans had begun to cultivate rice in watery paddies carved from hillsides, like those picturesque stair-like terraces to be found today in central Bali,

Indonesia (Scarborough, 2003) and at Ifugao in the Philippines. [21st Century technology and neglect brought about by social change has made such spectacular terraces uneconomic to maintain except as outdoor native cultural museum exhibits.] Methane is a greenhouse gas and the anthropogenic contribution to the air causes some global warming, possibly also influencing the Earth's predicted global sea-level rise (Ruddiman, 2005). Perhaps one-half of all seven billion or more humans alive in 2014 eat rice as a main staple. Microorganisms living in anoxic rice-field soils contribute between 10% and 25% of the yearly global methane emissions; by 2030, there maybe five billion human rice consumers in the Earth-biosphere (Groenigen et al., 2012). Artificial wetlands introduced to shaped hillsides have in the past, and continue during the present-time, to contribute to the world-ocean's sea-level instability. It is possible future genetic manipulations will produce plants with different albedos and/or more favorably responsive to carbon dioxide gas increases since such modest changes in plant photosynthetic prowess would affect Earth's climage-change process-events (Jansson et al., 2010; Ridgwell et al., 2009;Singarayer et al., 2009). Huge methane releases from hydrate dissociation triggered by depressurization following submarine slumping of seafloor sediments have taken place under the Mediterranean Sea (Maslin et al., 2004). Defrosting of the Northern Hemisphere's Arctic Zone permafrost during the 21st Century is expected to release large amounts of methane gas, also adding the air's burden of that greenhouse gas. An anthropogenically maintained existing sea-level within the Basin of a fabric-dam isolated Mediterranean Sea, Geographos believes, won't prompt unnatural methane releases whatsoever.

Thousands of freshwater reservoirs on Earth's landscapes, storing at least 6,500 cubic kilometers of freshwater, have had the effect to the present time of the Anthropocene of retarding the world-ocean's natural and/or unnatural tendency to measurable rise worldwide (Ngo-Duc et al., 2005) but, these reservoirs are undergoing a measurable decline the freshwater-holding capacity due to trapped sedimentation (Wisser et al., 2013). The Netherlands' famous Ijsselmeer did not directly terrace the North Sea, but simply replaced a segment of seawater with an equal volume of freshwater, according to Robert J. Hoeksema's *Designed for Dry Feet: Flood Protection and Land Reclamation in the Netherlands* (2006) and France's La Rance Tidal Barrage merely harnesses the daily tides. The barrage does affect the tidal regimes to a certain degree, according to Robert H. Clark (1921-2007) in *Elements of Tidal-Electric Engineering* (2007). Absent all reservoirs on land, the ocean would be higher in elevation than it nowadays is. Evidently, and prospectively, it is *Homo sapiens'* intention to occupy the landscape and the ocean with various kinds of infrastructures; per contra, humans can still only use the Earth's air.

By 2100, the world-ocean could be one-meter higher relative to its 2014 level. Earth's coasts will be impacted directly. Human activity to make and earn a living will "globalize" the Mediterranean Sea—its seawater, organic and inorganic contents and shores. All ecosystem-nations that are not landlocked should now be planning to endure a future 0.5-1.0 meter rise of sea-level that will affect their coasts. States within the Mediterranean Sea Basin-Black Sea Basin have several old and potentially expensive macroproject plans to accommodate to a global sea-level elevation. However, recent R&D and new products derived from advanced materials technologies—particularly, technical textiles exhibiting high-performance, purely functional, and precisely woven or non-woven fabrics but, additionally, even atomically interfacial materials used in micro-electronics—offer the prospect of cheap regional sea-level rise control and storm surge mitigation macroprojects (Cathcart, 2006)! Geographos is often astonished when contemplating what human-developed technologies can do to rearrange our world. John H. Lienhard's entertaining and enlightening attempt to reconcile large-scale technologies with the chain-linked inventors, *How Invention Begins: Echoes of Old Voices in the Rise of New Machines* (2006), begins in Chapter One with the sad imagined story of a male human Ice Age hunter-gatherer, "Otzi", who died in the freezing-cold snowdrifts of the Atztaler Alps circa 5,300 years ago; owing to local air warming which is, in part, caused by the effects of Europe's industrialization. If a rise in average air temperature during the 21st Century does occur, then "...the Alps would become almost completely ice-free" (Zemp et al., 2006). In other words, more deceased persons, additional "Otzi" bodies, may then be revealed, but few new curiosities will ever again become ice-entombed in that specific high-altitude ice-field Anthrome.

Commercial shipping industry interests and tour group organizers and related industries are the most obvious pro-amelioration macro-project constituencies examine the environmental impact of future sea-level rise within the Basin in addition to the ecologically sustainable fisheries, recreational boating/yachting, harbor maintenance and the militaries. The Mediterranean Sea Basin, renowned for its beautiful beaches and healthful climates, suffers extensive boundary layer air pollution originating mostly from other regions of the world because the Basin lies at the "crossroads" of global wind currents and receives inordinate amounts of industrial pollutants. Ship stack emissions contribute substantially to air pollution over the summertime Basin, which also modifies the Basin's cloud properties—more sky glow from shoreline cities, sometimes exceeding the light levels produced by the Moon in all its phases—as do innumerable jet-aircraft contrails. Literally, clouds are present were no clouds existed before because of human machines! Ship tracks and airplane contrails modifies the region's cloud properties affecting the Basin's climate radiation budget through sulfate forcing. Ancient Roman harbors have been submerged by sea-level rise, and even those were often massively dredged to maintain their viability as entrepot ports (Marriner and Morhange, 2006). A Basin-wide sea-level higher than 2014's by one meter, could also lead to higher wintertime storm surges affecting the Basin's coasts and its costly-to-repair-or-replace infrastructures.

Peak evaporation of the Mediterranean Sea, driven by energy release from seawater, takes places during the wintertime. The Mediterranean Sea measurably warmed during the 20th Century and the potential effects on the pelagic food webs, as well as the Basin's future climate regimes, is barely known. Future volcanic landscape eruptions may shroud the Basin with ash-clouds catastrophically; Santorini might again, as it did during 2011-2012 on a modest scale of activity, rouse from its apparent long-term dormancy and, as during the ancient past, cause great havoc in the Eastern Basin of the Mediterranean Sea, causing heavy ash-falls on Egypt and Turkey as well as the coastal places between those states (Friedrich et al., 2006); future seafloor volcanic eruptions could create more land—much as the eroded-from-human-sight island named Ferdinandea during July 1831 (**Figure 12.**), a submarine volcano located for about a year's time, set between Sicily and Tunisia (Dean, 1980; Hamilton, 2012).

Figure 12. Ferdinandea is now a reef, once a wave-eroded island, now reduced to a volcanic rock and rubble stub under about six meters of seawater. (Image: Google Images.)

Another place where new land might abruptly appear through volcanic action is at the Marsili Ridge seamount north of Sicily (Ventura et al., 2013). If large ash plumes are coincident with volcanic eruptions then the same general effects as occurred during the 2010 Eyjiafiallajokull eruption in Iceland will prevail—that is, air travel could be curtailed in the Basin and even shipping traffic may have to be rerouted to avoid bad air heavily loaded with corrosive aerosols. Geothermal heating can certainly warm seawater (Mashayek et al., 2013) and such heating can affect the Mediterranean Sea by overturning circulation. Because a warmer surface seawater layer is already present due to accumulated climate regime changes, during the 21st Century it is, therefore, possible that hurricanes occurring in the Basin (Emanuel, 2005) may become slightly more intense storms, with higher boundary layer wind speeds and increased vertical mixing of different seawater masses.

For the moment, ignoring the existence of the anthropogenic Suez Canal, opened in 1869 and creating the first artificial seawater exchange between the Mediterranean Sea and the Red Sea, the Mediterranean Sea-Black Sea basins connect naturally with the North Atlantic Ocean through the Strait of Gibraltar. **Figure 13.** Long-span bridges are becoming lengthier because macro-engineers have the assistance of computers capable of fully calculating the forces impinging such open-air structures and because new ways of isolating suspension bridge towers bases from earthquakes have been developed. Macro-engineers must meet managerial macro-problems such as financing and resource accumulation. Technical challenges remain, of course, but new materials like carbon fibers embedded in composites are being research and developed as practical solutions to most anticipated macro-problems.

Figure 13. Satellite-made and transmitted image of the Spain-Portugal and North Africa divided by the Strait of Gibraltar, a physiographic seawater-filled gap. (Image: Google Images.)

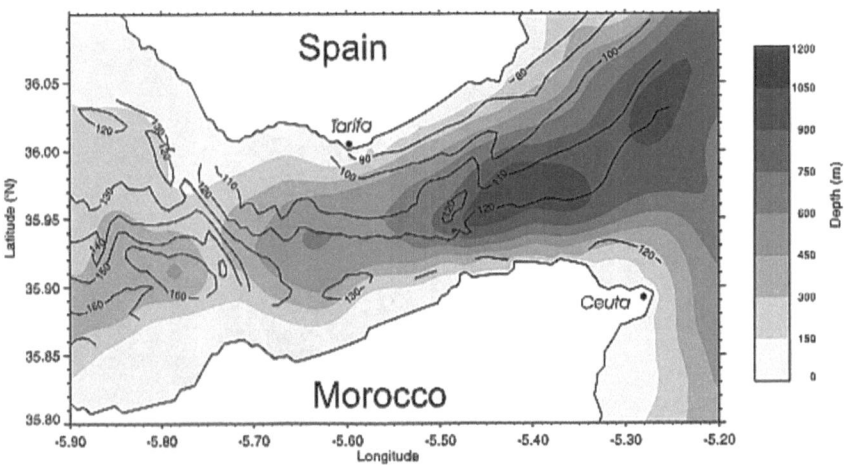

Figure 14. Bathymetric chart, non-navigational in purpose, of Gibraltar Strait. (Image: Google Images.)

Three-dimensional, computer-generated bathymetric charts (**Figure 14**) and high-resolution geologic sections are already available that generally reveal the Strait of Gibraltar's geologic framework. International geo-scientific studies of the Strait's geologic framework have been done to prepare for its bridging or tunneling during the 21st Century. Additional substantive pre-planning site studies may be required if the Gibraltar Strait Textile Barrage (GSTB) macroproject plan is adopted for implementation by a "Confederation of Mediterranean Sea Basin-Black Sea Basin States". With an land:ocean area ratio of 4.4, the Mediterranean Sea Basin-Black Sea Basin is the most landscape-dominated of all the sub-divisions of the world-ocean; the landscape are connected to the region by hydrology is dominated by Africa (65%), Europe (28%) and Asia (7%). The hypothesized Confederation might be named "Atlantropa" as Herman Sorgel wished! The Confederation will be an institutional organization in which the policies of every member ecosystem-country will be, at least in part, influenced by the preferences of voters/taxpayer from other member ecosystem-nations (Cremer and Palfrey, 1999); such a group, the Confederation, will indisputably govern the as-built GSTB and its associated electric power generation and transmission facilities.

Devised by R.B. Cathcart with vital bonus technical R&D help from Alexander A. Bolonkin (both are USA-based Science and Technology research and development operators) and the structural engineer Nicola M. Pugno (based in Italy), the Gibraltar Strait Textile Barrage will be draped across the gap between Spain and Morocco on a general alignment between Tarifa on Spain's southern coast and close to Ksar e' Sghir on Morocco's shore. [**Figure 13** indicates Tarifa, Spain's absolute geographical position and the southern terminal site for the strung GSTB will be near 35.90 North latitude by -5.47 East longitude in Morocco.] Geographos was first inspired to think about the GSTB by seeing the Techno-artist Christo's famous **Valley Curtain, Rifle, Colorado, 1970-1972** outdoor artwork. **Figure 15.**

Figure 15. Installed at Grand Hogback, Rifle, State of Colorado (USA), the translucent orange-colored **Valley Curtain** is made of Anthropocene technical materials—plastic [18,586 square meters of rayon polyamide] and alloy steel metal cables—spanning 417 meters to a height of 112 meters. Kevlar thread has a self-supporting length of about 200 kilometers—almost four times that of drawn steel wire! It connected briefly the slopes of a Rocky Mountain valley above a roadway continuously open to vehicular traffic. The suspending cable is catenary. About 3.2 kilometers upstream of Christo's installation is the rolled earth-fill Rifle Gap Dam built 1964-1967 which endured, on 10 September 1969 an underground nuclear explosion done by Project RULISON just 29.7 kilometers distant. (Image: Google Images.)

The Gibraltar Strait Textile Barrage may not be a definitive solution to a perceived macro-management problem. "State-of-the-Techno-Art" is always provisional—but Geographos does maintain that it probably is the best macroproject plan that can be devised considering all of the technologies available. The GSTB, if built as intended, will be a submarine artwork imitative of Christo's aerial Techno-Art, macro-engineered by Ernest C. Harris (1915-1998), **Valley Curtain**. Deployed properly, and configured appropriately, because of seawater currents, the GSTB will bow eastwards like a sailing ship's spinnaker (from the selected 20 kilometer-long alignment work-site landscape and world-ocean segment); hydraulic billowing will also be caused by a difference in sea-level on a two-sided, sea-bottom anchored membrane as well as seasonal wind pressure acting on the GSTB. Macro-Imagineers/Macro-Engineers of the GSTB macroproject might well draw on the recorded installation experience with heavy wire nets, floatation systems and their moorings derived from World War II anti-submarine installations in strategic harbors and that documented experienced available from the 72 kilometer-long "barrier" of anti-submarine steel indicator nets used during World War I in the 1915-1919 Otranto Strait Barrage. One of the main cost factors governing the GSTB's monetary cost will be the seafloor cut-off wall—made to minimize seawater seepage beneath the GSTB. A submarine cut-off trench need not be dug across the Strait of Gibraltar, nor an uninterrupted underwater grout curtain installed, to ensure proper functioning of the completed Gibraltar Strait Textile Barrage.

From its western sea approach, the GSTB will have the characteristic of Architecture's visual deception, often featured in an English Garden or an exotic animal zoo landscape architect's "ha-ha" (also known as "sunken fence") in that—absent any warning light-buoys and radar reflectors—ship navigators will visually misapprehend the true nature of the sea-route ahead! Those mariners, such as private-sector fisherman and yachtsmen piloting their boats on an easterly course from the North Atlantic Ocean without benefit of up-to-date navigational charts (paper or electronic), will have no inkling via normal optical clues that a one-meter drop in sea-level obstructs the entrance to the Mediterranean Sea. Similarly, mariners without radar readouts or displays using the eastern approach will visually espy a one meter-high tensioned fabric GSTB wall, which if fabricated of aquamarine-color textile might be almost invisible until closely sighted! It would appear as if Mediterranean Sea seawater were moving upwards towards the North Atlantic Ocean; indeed, a most baffling situation and confusing visual! Ignorance is not bliss, especially on the high-seas.

At least 50,000 vessels of all types currently pass through the Strait of Gibraltar annually and the two officially separated ship traffic lanes are considered to be military chokepoints; at least one, and possibly two, Frei Otto-style tensioned fabric ship-locks will be required to accommodate post-construction GSTB ship transits. A ship-lock failure, once in a great while is possible. However, even structural failure of a blocking one meter-high GSTB, extended 20 kilometers across the waterway, won't necessarily be a disaster for the postulated Confederation; more or less, it will be similar to a strong storm surge event with a constant 106,000 cubic meter per second incoming "surge"—with the same appearance as a large, noisy tidal bore—rippling eastwards towards Greece and Turkey, even the Black Sea beyond, to eventually inundate (by one meter) some unprotected parts of the Basin's strand. [From which direction it came is unreported, but a tsunami in ancient times may have swirled through the Bosphorus (Pickard, 1987).] It is possible that precautionary seawalls, permanent or temporarily deployable, might be justified, built just to cope with this accidental "extreme" short-term inundation event-process. Certainly, coastal infrastructures on the landscape will be impacted most directly and, subsequently, in need of repair and some replacement; submarine infrastructures—such as the underwater acoustic detector of high-energy neutrinos, ANTARES, in the Mediterranean Sea off the coast of Toulon, France—should be immune to damage of any kind. Once the broken sea-lock gates are restored, natural evaporation from the Mediterranean Sea will re-establish the status quo ante.

Other than collisions caused by errant ships and the cycling pressure changes of small intra-Basin tides, the most significant prospective structural integrity maintenance threat facing the Gibraltar Strait Textile Barrage are tsunami generated with the Basin or in the North Atlantic Ocean. The small tsunami resulting from a Black Sea 70 meter-diameter rocky asteroid impact won't pose much of a threat, but the social chaos caused in the immediate region surrounding the afflicted Black Sea could undermine the Confederation's stability (Schuiling et al., 2007). The tsunami caused by the 1 November 1755 Lisbon, Portugal seismic episode caused a maximum 11 meter-high run-up seawater wave at Tarifa, Spain and 2.5-5 meter-high wave at Tangier, Morocco; the Rock of Gibraltar's Harbor was hit by a maximum wave run-up of 2 meters above normal local sea-level (Mendes-Victor et al., 2009). The incident tsunami interacted violently in harbors with seaport infrastructures as well as floating boats and anchored/docked shipping. A 70 meter-diameter asteroid impact west of the Strait of Gibraltar, in the North Atlantic Ocean, or inside the Basin onto the Mediterranean Sea, could be catastrophic. A one meter-high tsunami hitting the air-exposed eastern face of the GSTB—then behaving as a flexible hydrostatic seawall—will likely exert a momentary pressure of 10,000 kilograms per square meter, or about 200,000,000 kilograms overall. Tsunami momentum effects must be very carefully considered in great detail. When a tsunami meets suddenly a horizontally semi-slack fabric barrier, the vertical barrier absorbs some of the momentum from the tsunami's collision and transfers it to, for example, the GSTB if it can move, making it extremely taut; after the snap loading, some of the momentum is bounced back to the ocean in various compass directions. If a single GSTB is deemed insufficient to withstand the brief strong forces applied in this planning scenario, then another paralleling GSTB can be installed beforehand since each is not as costly monetarily as a single Gibraltar Strait Bridge made of exotic materials (Meier, 1987). Elongation of the GSTBs super-ropes under dynamic loading will dissipate some tsunami-deposited energy.

In the particular case of the Gibraltar Strait Textile Barrage, there is a quite interesting newly apprehended hydraulic effect observed in India by B.R. Rao after the memorable major earthquake in Indonesia that took place, and tragically ended so many human lives, on 26 December 2004 (Rao, 2005). Pending further computations, Rao's observation is only partially comprehended by Alexander Bolonkin and R.B. Cathcart, the originators and chief proponents of the GSTB macroproject. Any North Atlantic Ocean tsunami over-topping the Gibraltar Strait Textile Barrage will encounter a sudden one meter hydraulic descent immediately after falling over the GSTB. Is it possible, Geographos asks, a sudden descent of fast flowing seawater at the GSTB impedance in a one meter-high waterfall will effectively attenuate a potentially devastating tsunami's subsequent run-up on 13,000 kilometers of Mediterranean Sea Basin coast? The acute observation, and quick reporting of it, by B.R. Rao of the remarkable effect of precipitous topographical drop on tsunami propagation hydrodynamics may be as instigative subsequently as the contemplations of Benjamin Franklin (1706-1790) on oil's spread and calming effect on seawater (Mertens, 2006).

When a tsunami slams into the GSTB, part of the wave's energy is transmitted through the GSTB, part is reflected from the GSTB and another part is absorbed in the various materials of which the GSTB is constructed. Tsunami overtopping of the impermeable tensioned fabric Techno-Art dam with zero freeboard (on its westernmost face) will cause a North Atlantic Ocean seawater hydraulic flow (from supercritical to sub-critical) that induces vibrations in the GSTB (near-critical seawater flow-induced vibrations, vortex shedding and suction on the air-exposed "downstream" eastern face of the dam-artwork). In effect, the Mediterranean Sea close to the intact, quivering/resonating GSTB will become a seawater stilling basin that will dissipate the kinetic energy of the overtopping seawater flow. A one meter overtopping results in a temporary 106,000 cubic meter per second seawater flowage, and a two meter overtopping results in a 300,000 cubic meter per second seawater waterfall. The optimum value of the drop height at the GSTB must still be determined by elaborate mathematical exercise and by physical model testing since macro-engineers will wish to stabilize the GSTB's geographical position and not cause any dangerous structural damage the GSTB. Unfortunately, R&D reports on tensioned vertical membranes spanning water depths of limited extent are rare in the appropriate literature scientific literature. The total area of the seawater-retaining fabric drape comprising the GSTB is about 200 square kilometers, but only approximately 20,000 square meters of the drape will actually be fully exposed to the air and material-degrading sunlight on its eastern face whilst under continuous imposed one meter seawater head (hydraulic pressure) on its submerged western face. Of course, the GSTB must "survive" the daily tidal regime present in the Strait of Gibraltar (Sanchez-Romain et al., 2013) as a matter of course.

Cables, super-ropes and membranes are the essential components of Frei Otto's architecture proposals. Application of advanced technical textiles and super-ropes composed mainly of Kevlar—or, soon, low-cost carbon nanotubes—could permit safe emplacement and implementation of a pontoon bridge spanning the Strait of Gibraltar. Membrane materials also continue to become ever more useful. Several years ago, laboratory chemists in Japan, Germany and the USA devised an artificial spider thread that is stronger than steel. Whilst not yet brought to market, further development is aimed at the manufacture of the most durable fabric every made (Negishi, 2013). Durability submerged in seawater and air is yet to be investigated: many fabrics used in existing tensioned membrane structures exposed to Tropic Zone conditions do not fare well in terms of wear survival under outdoor working conditions (Wang, et al., 2013)., so the GSTB's builders will welcome, indeed relish, any materials science laboratory-led improvements in the qualities of employed membrane fabrics! Effectively, the fabric used in the GSTB drape is a geo-synthetic material. The 2010 International Commission on Large Dams (ICOLD) Bulletin on *Geomembrane Sealing Systems for Dams* (ICOLD Bulletin No. 135) supersedes and greatly elaborates two earlier versions published in 1981 and 1991. For more than 50 years geomembranes have been used to provide water-proofing on various types of large dams, most have been installed in the dry, but some success at underwater sealing has also been achieved. A vehicle-carrying floating bridge macroproject, connecting Tarifa, Spain and Point Cires, Morocco, imitating the span used by Xeroxes during 480 BC to support his marching troops and their baggage train as they aggressively crossed the Hellespont, has been proposed by the USA architect Eugene Tsui (**Figure 16**).

Figure 16. American architect Eugene Tsui. GOTO: http://www.tdrinc.com/gibraltar.htm

Figures 17 and 18, below are images of Tsui's drawings for a heavy transport-oriented bridge-city floating between Spain and Morocco in the Strait of Gibraltar. Notice that he has taken into account the bow effect mentioned as a characteristic of the GSTB fixed Strait of Gibraltar crossing proposal by Geographos.

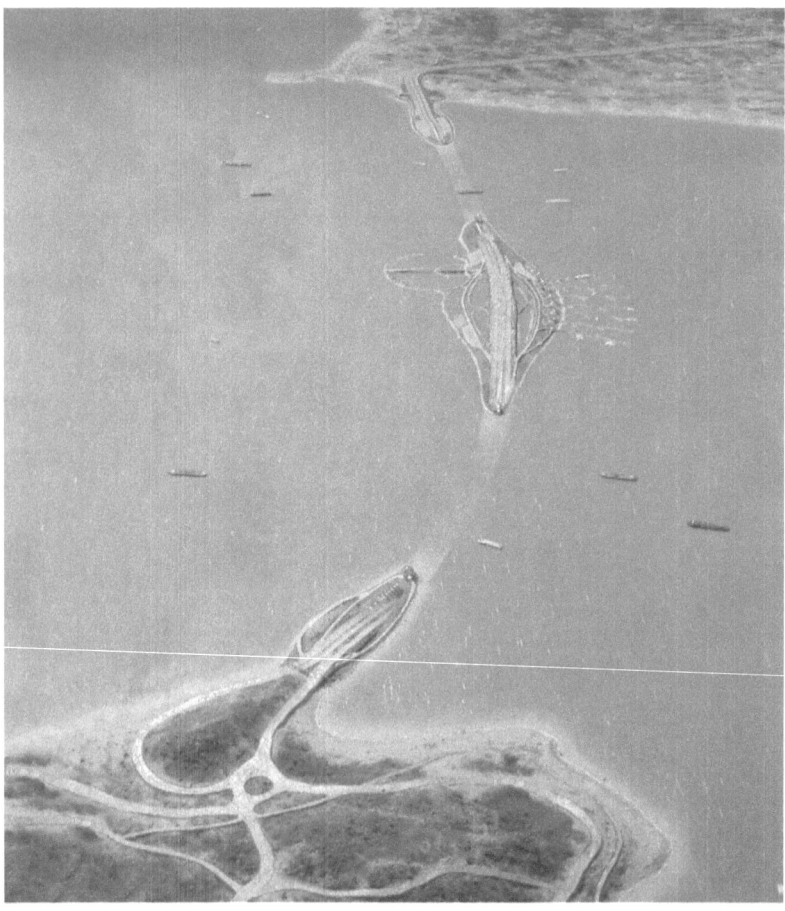

Figure 17. Tsui's conceptualization of a permanent partial filling for the geographical gap between Morocco (bottom of picture) and Spain (top of picture). Left side represents the North Atlantic Ocean and the right side represents the Mediterranan Sea. It resembles a deployed floating causeway with decorations (Russell et al., 2013) (Image: Google Images.)

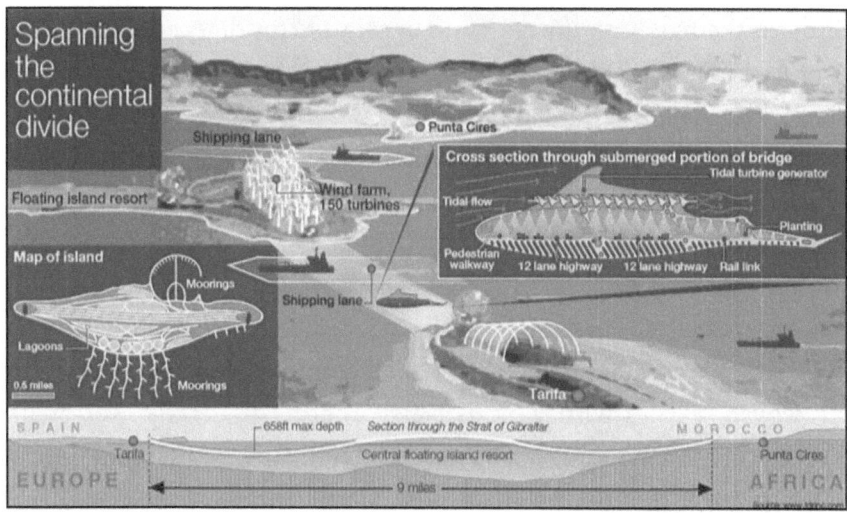

Figure 18. Explanatory infographic about Eugene Tsui's floating bridge-city Macro-Imagineering dream. This drawing's perspective is a view from Tarifa, Spain towards North Africa. The central island might be named "Atlantis" (Ashe, 2012) since the realms of landscape and ocean is never really separated in the story (Mack, 2013). One proposed site for Atlantis is on the planed underwater mountains, islands in former time when the world-ocean was lower during the Ice Age, virtually within the present-day Strait of Gibraltar (Gutscher, 2005). Floating cities are an unresolved conceptual challenge for International Law (Kieth, 1977; Woodliffe, 1978). (Image: Google Images.)

Braided or stranded super-ropes could stabilize a pontoon bridge in a fixed geographical alignment for a long period, especially in a one meter per second eastward flowing seawater surface current refilling the evaporating Mediterranean Sea, which has a yearly seawater deficit of about 0.5 meters.

The ultimate hydrostatic head supported by a textile (woven or non-woven) is the measure of the resistance to the passage of seawater through the material; the standard applicable for determining the resistance to seawater penetration is the hydrostatic test. Several international materials testing standard regulatory organizations, as well as many national standard-setting agencies, generally accept the height of a seawater column, expressed in metric units of distance, as the applicable validation of a test method primarily intended for dense fabrics and flexible films. "Waterproof" and "watertight" are synonyms in this instance. In the past, the resistance to seawater penetration (in ship sails, ship cargo hatch-covers) has been technically achieved by coating woven textiles with various waterproofing materials; watertight textiles can now be achieved by dense weaving of strong fibers. Multi-axial, multi-ply textiles are bonded by a loop system, consisting of one or more yarn layers stretched in parallel; year layers can have different spatial orientations and different yarn densities. The combination of multi-directional fiber layers is known to be capable of distributing extraordinarily high strain forces; multi-axial, multi-ply textiles are dimensionally stable in any direction and exhibit isotropic distribution of stress forces with uniform behavior. Kevlar (29, 49, 149)—since 1971 a Dupont Corporation (USA) trade-name for aromatic polyamides—with tensile strength of more than three Gigapascals, a failure strain of 3% and a material density of 1.4 grams per cubic centimeter is a good example. All extremely strong materials able to perform as unitary form-active structures ought to be investigated for use in the proposed Gibraltar Strait Textile Barrage. Perhaps these innovative uses of exotic materials might be pioneered at the proposed 35 kilometer-long TRANSMAR [Sea Crossing System at the Sea of Marmara] floating super-highway called the "New Waterway" development which may be emplaced south of Istanbul, Turkey. TRANSMAR is designed primarily to carry heavy vehicle traffic, moving it outside of the already highly congested traffic in the vicinity of Istanbul (Alp, 2001). Like the GSTB, will have to cope with tsunamis but, even worse, the confident expectation of major future earthquakes (of magnitude 7 or more) of the North Anatolian Fault Zone, slightly south of Istanbul (Bohnhoff, M. et al., 2013).

The characteristic strength of a structural or film material must have a low probability (5% or less) of not being reached during the period of the material's use in the GSTB and the characteristic load must not have more than a 5% probability of being exceeded during the GSTB's design lifetime. Potentially, embedded fiber-optic electronics—detectors, reporters and automated alarm actuators—ought to be incorporated to monitor in real-time the super-ropes as well as the draped barrage, which must fit tightly to the Gibraltar Strait's sea-bottom and submarine canyon sidewalls to successfully fulfill its macroproject functions, giving instant alerts to immediately responsible shore-based monitory computer and its human supervisors of all developing GSTB structural issues related to the GSTB's safe and efficient performance. Should the GSTB become separated from its two sidewalls and sea-bottom anchorage, swept away by an overwhelming tsunami after full loss of structural integrity, then possibly it might be partly retrievable/salvageable and, if so, its quick post-failure replacement would "re-initialize" the artificial Mediterranean Sea reduction by evaporation in a manner timely. Again, the air of the macro-engineers will be to keep the present-day seawater level of the Mediterranean Sea at less than the rest of the world-ocean will be once global sea-level has made the difference important. A clever collapse design could even optimize recovery of a broken GSTB's components—in other words, the GSTB might be constructed with a design philosophy including the possibility of semi-controlled collapsibility, even pre-planned folding!

The width of the Strait of Gibraltar at the place most likely designated is 20,000 meters and the charted maximum depth is 900 meters, with an average depth of 450 meters. The GSTB will have a seawater surface difference of one meter affecting the whole deformable barrier. If the barrage is supported by pontoons floating on the North Atlantic Ocean, the installation may be utilized as a vehicular highway between Europe and Africa. Sea-going ships arriving and departing the Mediterranean Sea, including via the [newly sea-locked] Suez Canal, will bypass the GSTB by using sturdy fabric ship-locks built at each terminus of the oceanic surface-floating road-barrier. A simple sketch of the Gibraltar Strait Textile Barrage, originally done by Alexander A. Bolonkin and, later, modified by R.B. Cathcart, is provided in **Figure 19**, below.

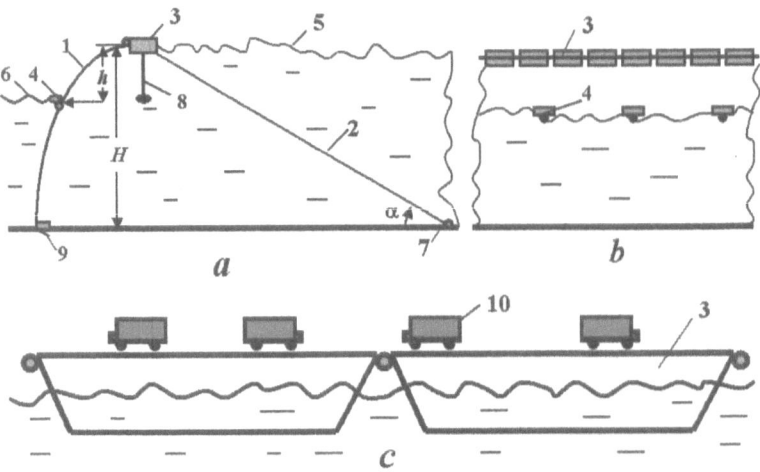

Figure 19. Whatever the flexible, plastic film-like material used an impermeable curtain will be used to blockade all seawater flows other than what is used in the associated sea-locks and to generate some hydropower. Legend: (a) side-view, (b) front-view, (c) pontoon super-highway with cartoon trucks traveling to and from Europe and Africa. Notations: (1) flexible textile dam, (2) support cable, (3) pontoon, (4) hydro-electric turbine, (5) North Atlantic Ocean, (6) Mediterranean Sea, (7) anchor, super-rope spool, motor for support cable spooler, (8)

hydrodynamic stabilizer, (9) anchors and (10) angle a = 30⁰. (Image: Bolonkin and RBC, 2009, Part II, Chapter 4, Figure 1 at page 370.)

Here are some purely Macro-Imagineered statistics for a realized Gibraltar Strait Textile Barrage. Since the GSTB duplicates some of the major transportation-entertainment functions of Eugene Tsui's proposed macroproject, Geographos supposes that monetary costs of emplacement will be approximately the same. The GSTB should be cheap to construct, significantly less expensive than the [English] Channel Tunnel. Geographos estimates the total cost, conservatively, at USD 11.9 billions as of the end of Year 2013. The GSTB could generate at 500 to 800 Megawatts. Constructed of a film with a thickness of 0.05 millimeters, the total weight of the dam's impermeable plastic drape would be around 810 metric tonnes. Adding the 36 millimeter-thick support super-ropes, each separated from the other by a distance of 10 meters, results in a total securing cable weight of 3,240 metric tonnes plus 32 metric tonnes of connecting super-ropes. The total weight of the GSTB installation—without any floating pontoons—should total 4,182 metric tonnes. Pontoons can also be constructed of non-woven film and additional weight will be small. Depending on the strength of the artificial fiber used, with a safety stress of 100 kilograms per square millimeter, the total weight of the installation can be decreased by ten times (up to 418 metric tonnes), making the installation millions of time lighter—and much less expensive to build—than a convention reinforced concrete gravity dam as Herman Sorgel first proposed nearly a century ago. A still sturdier GSTB might carry high-speed trains across the Strait of Gibraltar—Geographos offers the Gedankenexperiment of a Maglev's operation between Casablanca and Paris at a constant acceleration (accelerating halfway and braking the other half at 0.5 the force of Earth's gravity) a fast train would cover the distance in about twenty minutes! "The idea behind this is to save cultural roots without impeding work and business in the most suitable places" (Marchetti, 1994, page 81).

More plainly said, North Africans can shop in Paris and return home to have dinner with the family, and vice-versa!

Summarizing, the Gibraltar Strait Textile Barrage is a unique Sea Techno-Art piece—the combination of a seawater barrier, hydroelectricity generators and an inexpensive holistic anthropogenic solution for major known, and forecasted, multiple environmental macro-problems sure to affect the Mediterranean Sea Basin-Black Sea Basin during the 21st Century. Proof of concept by operation, say by 2029, may allow this very practical technology to be adopted elsewhere. The GSTB is a new kind of hydrostatic seawall capable of successfully resisting unpredictable incident one meter-high tsunami. Its biggest drawback is that the Mediterranean Sea is likely to trend to warmer seawater and this poses the risk of extinction for some sea-life because the forms most affected cannot move northwards to cooler waters. In some sense, then, some sea-life will find itself actually between a figurative "Scylla" and "Charybdis" (Defant, 1940) of another physical kind entirely! Not a swirl of heated seawater but an impassable northern Basin shoreline! In other words, there can be no migration exit for biota trying to flee looming hostile environmental conditions. Joseph-Marie Jacquard (1752-1834) invented the automatic loom and his work with calculating machines eventuated in 2014's computers, according to James Essinger's *Jacquard's Web* (2004). How fitting, then, that commercially available woven textile/braided or stranded super-ropes and super-computers are precisely the two industrial tools most needed to successfully resolve this gedanken-experiment in Macro-Imagineering/Macro-Engineering! During humanity's Anthropocenic period "Globalization, broadly defined, is simply the weaving of the webs that knit the [E]arth into the Ptolemaic 'unit of itself'" (Lewis, 2000, page 625). To express it another way, Dr. Hubertus Strughold's "Dosmozoicum" impends, a topic for another time, another place, another forum.

References Cited

Aliani, S. and Molcard, A. (2003) *Hitch-hiking on floating marine debris: macrobenthic species in the Western Mediterranean Sea.* Hydrobiologia 503: 59-67.

Alp, A.V. (2001) *World's Longest Floating Sea Crossing—Istanbul bypass "TRANSMAR".* Pages 317-331 *IN* Krokeborg, J. (Ed.) *Strait Crossings 2001: Proceedings of the Fourth Symposium on Strait Crossings, Bergen/ Norway/ 2-5 September 2001.* (Tokyo: A.A. Balkema Publishers) 696 pages.

Ashe, G. (2012) *Atlantis.* (New York: Thames & Hudson) 128 Pages.

Badenoch, A. and Fickers, A. (2010) *Materializing Europe: Transnational Infrastructures and the Project of Europe.* (New York: Palgrave Macmillan) 333 pages.

Bergamasco, A. and Malanotte-Rizzoli, P. (2010) *The circulation of the Mediterranean Sea: a historical review of experimental investigations.* Advances in Oceanography and Limnology 1: 11-28.

Bergmann, J. et al. (2013) *The Iberian Peninsula as a potential source for the plant species pool in Germany under projected climate change.* Plant Ecology 207: 191-201.

Bohnhoff, M. et al. (2013) *An earthquake gap south of Istanbul.* Nature Communications, Article 199: DOI: 10.1038/ncomms2999.

Bolonkin, A. and Cathcart, R.B. (2009) *Macro-Projects: Environment and Technology.* (New York: Nova Science Publishers, Inc.) 537 pages.

Bouchette, F. et al. (2010) *Hydrodynamics in Holocene Lake Mega-Chad.* Quaternary Research 73: 226-236.

Brancoleoni, F. et al. (2010) *The Messina Strait Bridge: A Challenge and a dream.* (New York: CRC Press) 324 pages.

Bryden, H.L. and Webb, D.J. (1998) *A perspective on the need to build a dam across the Strait of Gibraltar.* News Letter European Geophysical Society, No. 69, pages 1-2.

Camilleri, D.H. (June 2012) *Tsunami and wind-driven wave forces in the Mediterranean Sea.* Proceedings of the ICE-Maritime Engineering 165: 65-79.

Cathcart, R.B. and Cirkovic, M. (2006) *Extreme Climate Control Membrane Structures: Nth Degree Macro-Engineering.* Chapter 9, pages 151-174 **IN** Badescu, V., Cathcart, R.B. and Schuiling, R.D. (Eds.) Macro-Engineering: A Challenge for the Future. (The Netherlands: Springer)316 pages.

Cathcart, R.B. (1998) *Land Art as global warming or cooling antidote.* Speculations in Science and Technology 21: 76-83.

Cathcart, R.B. and Badescu, V. (2004) *Architectural Ecology: A Tentative Sahara Restoration.* International Journal of Environmental Studies 61: 145-160.

Cathcart, R.B. (2006) *Sethusamudram ship channel macroproject: Anti-tsunami and storm surge textile arrestors protecting Palk Bay (India and Sri Lanka).* Current Science 91: 1474-1476.

Certini, G. et al. (December 2013) *The impact of warfare on the soil environment.* Earth-Surface Reviews 127: 1-15.

Chianese, R.L. (September-October 2013) *Regeneration on Tree Mountain.* American Scientist 101: 350-351.

Christensen, P. (2012) *Dam Nation: Imaging and Imagining the "Middle East" in Herman Sorgel's Atlantropa.* International Journal of Islamic Architecture 1: 325-346.

Clery, D. (2010) *Sending African Sunlight To Europe, Special Delivery.* Science 329: 782-783.

Connelly, J. (Winter 2012-2013) *The Living Lighthouse.* Cabinet, Issue 48, pages76-77.

Contoux, C. et al. (2013) *Megalake Chad impact on climate and vegetation during the late Pliocene and the mid-Holocene.* Climate of the Past 9: 1417-1430.

Coulthad, T.J. et al. (2013) *Were River Flowing across the Sahara During the Last Interglacial? Implications for Human Migration through Africa.* PLOS ONE: 8: e74834.

Cremer, J. and Palfrey, T.R. (1999) *Political Confederation.* American Political Science Review 93: 69-83.

Cullen, Martin N. (2005) *Tension membrane water retaining structures.* WIT Transactions on The Built Environment 79: 427-436.

Dean, D.R. (1980) *Granham Island, Charles Lyell, and the Craters of Elevation Controversy.* Isis 71: 571-588.

Defant, A. (1940) *Scylla und Charybdis und die Gezeitenstroemungen in der Strasse von Messina.* Ann. Hyddr. Marait.Meteor. 5: 145-157.

Dorrian, M. and Pousin, F. (2013) *Seeing from Above: The Aerial View in Visual Culture.* (UK: I.B. Tauris) 320 pages.

Duarte, J.C. et al. (2013) *Are subduction zones invading the Atlantic? Evidence from the southwest Iberia margin.* Geology 41: 839-842.

Emanuel, K. (2005) *Genesis and maintenance of "Mediterranean hurricanes".* Advances in Geosciences 2: 217-220.

Fei, X. (2004) *Solving the coastal eutrophication problem by large scale seaweed cultivation.* Hydrobiologia 512: 145-151.

Field, J.P. et al. (2010) *The ecology of dust.* Frontiers in Ecology and the Environment 8: 423-430.

Finkl, C.W., Pelinovsky, E. and Cathcart, R.B. (July 2012) *A Review of Potential Tsunami Impacts to the Suez Canal.* Journal of Coastal Research 28: 745-759.

Flowers, J.B. (March 1921) *Signaling the Planets with a Cloth Reflector.* Illustrated World, pages 84-85.

Friedrich, W.L. et al. (2006) *Santorini Eruption Radiocarbon Dated to 1627-1600 BC.* Science 312: 548.

Garcia-Castellanos, D. and Villasenor, A. (2011) *Messinian salinity crisis regulated by competing tectonics and erosion at the Gibraltar arc.* Nature 480: 359-363.

Garcia-Orellana, J. et al. (2009) *Distribution of artificial radionuclides in deep sediments of the Mediterranean Sea.* Science of the Total Environment 407: 887-898.

Gebert, L. et al. (2013) *Evidence for an African-Iberian mammal dispersal during the pre-evaporitic Messinian.* Geology 41: 691-694.

Giaccaria, P. (June 2011) *The Mediterranean Alternative.* Progress in Human Geography 35: 345-365.

Goffredo, S. and Dubinskyk, Z. (Eds.) (2014) *The Mediterranean Sea: Its History and present challenges.* (The Netherlands: Springer) 237 pages.

Gomez, F. (2003) *The role of exchanges through the Strait of Gibraltar on the budget of elements in the Western Mediterranean Sea: consequences of human-induced modifications.* Marine Pollution Bulletin 46: 685-694.

Grigsby, D.G. (2012) *Colossal: Engineering the Suez Canal, Statue of Liberty, Eiffel Tower, and Panama Canal.* (New York: Periscope Publishing) 200 pages.

Groenigen, K.J. van et al. (21 October 2012) *Increased greenhouse-gas intentsity of rice production under future atmospheric conditions.* Nature Climate Change. DOI: 10.1038/NCLIMATE1712.

Gutscher, M-A. (2005) *Destruction of Atlantis by a great earthquake and tsunami? A geological analysis of the Spartel Bank hypothesis.* Geology 33: 685-688.

Hafner, A. (2012) *Archeological discoveries in Schnidejoch and at other sites in the European Alps.* Arctic 65, Supplement 1, page 189.

Hamilton, J. (2012) *Volcano: Nature and Culture.* (London: Reaktion Books), Chapter 5, pages105-108.

Hempsell, M. (2005) *Terraforming in context of the evolving space infrastructure.* Journal of the British Interplanetary Society 58: 385-391.

Hudson, B.J. (2012) *Waterfall: Nature and Culture*. (London: Reaktion Books) 248 pages.

Jansson,C. et al. (2009) *Phytosequestration: Carbon Biosequestration by Plants and the Prospects of Genetic Engineering*. BioScience 60: 685-696.

Johnson, R.G. (1997) *Climate Control Requires a Dam at the Strait of Gibraltar*. EOS: Transactions of the American Geophysical Union 78: 277-281.

Joselit, D. and Harrison, R. (Summer 2008) *A Conversation with Peter Fend*. October, No. 125, pages 117-136.

Keohane, R.O. (1982) *The Demand for International Regimes*. International Organization 36: 325-355.

Kidd, R.B. and Huggett, Q.J. (1981) *Rock debris on abyssal plains in the Northeast Atlantic: a comparison of epibenthic sledge hauls and photographic surveys*. Oceanologica Acta 4: 99-108.

Kieth, K.M. (July 1977) *Floating cities: A new challenge for transnational law*. Marine Policy 1: 190-204.

Krijgsman, W. (2002) *The Mediterranean: Mare Nostrum of Earth Sciences*. Earth and Planetary Science Letters 205: 1-12.

Kunz, B.P. (1995) *Open-Ocean, Air-Supported, Stable Platforms*. Sea Technology 36: 47-50.

Lagendijk,V. (2008) *Electrifying Europe: The Power of Europe in the Construction of Electricity Networks*. (Amsterdam: Aksant) 246 pages.

Lemoalle, J. et al. (2012) *Recent changes in Lake Chad: Observations, simulations and management options (1973-2011)*. Global and Planetary Change 80-81: 247-254.

Lewis, M.W. (October 2000) *Global Ignorance: geographical errors by scholars, journalists, intellectuals writing on globalization and world history.* The Geographical Review 90: 603-628.

Lo, Y-M. (2000) *Performance of a flexible membrane wave barrier of a finite vertical extent.* Coastal Engineering Journal 42: 237-251.

Mack, J. (2013) *The Sea: A Cultural History.* (New York: Reaktion Books) 272 pages.

Madron, X. et al. (2011) *Marine ecosystems' response to climatic and anthropogenic forcings in the Mediterranean.* Progress in Oceanography 91: 97-166.

Magid, J. (2012) *The Marshall Plan.* Advances in Historical Studies 1:1-7.

Marchetti, C. (1994) *Anthropological Invariants in Travel Behavior.* Technological Forecasting and Social Change 47: 75-88.

Mariotti, A. and Struglia, M.V. (2002) *The Hydrological Cycle in the Mediterranean Region and the Implications for the Water Budget of the Mediterranean Sea.* Journal of Climate 15: 1674-1690.

Marriner, N. and Morhange, C. (2006) *Geoarchaeological evidence for dredging in Tyre's ancient harbor, Levant.* Quaternary Research 65: 164-171.

Mashayek, A. et al. (2013) *The role of the geothermal heat flux in driving the abyssal ocean circulation.* Geophysical Research Letters 40: 3144-3149.

Marshall, M. (12 October 2013) *Transforming Earth.* New Scientist 220: 10-11.

Maslin, M. et al. (2004) *Linking continental-slope failures and climate change: Testing the clathrate gun hypothesis.* Geology 32: 53-56.

Materia, S. et al., (2012) *The effect of Congo River freshwater discharge on Eastern Equatorial Atlantic climate variability.* Climate Dynamics 39: 2109-2125.

Maydew, R.C. (1997) *America's Lost H-Bomb! Palomares, Spain, 1966.* (Manhattan, Kansas: Sunflower University Press) 160 pages.

Megill, A. (1989) *Recounting the Past: Descriptive, Explanation, and Narrative in Historiography.* American Historical Review 94: 648.

Meire, U. (1987) *Proposal for a carbon fibre reinforced composite bridge across the Strait of Gibraltar at its narrowest site.* Proceedings of the Institute of Mechanical Engineers 201: 73-78.

Mendes-Victor, L.A. et al. (2009) *The 1755 Lisbon Earthquake: Revisited.* (The Netherlands: Springer) 600 pages.

Mertens, J. (January 2006) *Oil on Troubled Waters: Benjamin Franklin and the Honor of Dutch Seamen.* Physics Today 59: 36-41.

Miller, A.R. (Spring 1972) *Ecological Balances in Semi-Enclosed Seas.* Environmental Affairs 2: 195.

Moore, E.F. (October 1956) *Artificial living plants.* Scientific American 195: 118-126.

Mukerji, C. (2009) *Impossible Engineering: Technology and Territoriality on the Canal du Midi.* (Princeton, New Jersey: Princeton University Press)304 pages.

Muller, K. (2000) *The Birth and Death of Eurafrica.* International Journal of Francophone Studies 3: 4-17.

Negishi, M. (9 July 2013) *Stronger Than Steel: The Amazing Spider Web*. The Wall Street Journal CCLXII: B4.

Nixon, S. (2004) *The Artificial Nile*. American Scientist 92: 158.

Occhipinti-Ambrogi, A. and Galil, B. (2010) *Marine alien species as an aspect of global change*. Advances in Oceanography and Limnology 1: 199-218.

Oczkowski, A.J. et al. (2009) *Anthropogenic enhancement of Egypt's Mediterranean fishery*. Proceedings of the National Academy of Sciences 106: 1364-1367.

Patt, A. et al. (2013) *Vulnerability of solar energy infrastructure and output to climate change*. Climatic Change 121: 93-102.

Pearce, F. (24 October 2009) *Sunshine Superpower*. New Scientist 204: 38-41.

Pelling, Mark and Blackburn, S. (Eds.) (2013) *Megacities and the Coast: Risk, Resilience and Transformation*. (London: Routledge) 272 pages.

Pickard, W.F. (1987) *The Symplegades*. Greece & Rome 34: 1-6.

Portman, M.E. et al. (2012) *From the Levant to Gibraltar: A Regional Perspective for Marine Conservation in the Mediterranean Sea*. Ambio 41:: 670-681.

Powell, D. (2012) *Sahara solar plan loses its shine*. Nature 491: 16-17.

Pugno, N.M., Cathcart, R.B. and Bolonkin, A.A. (2011) *Sea Art: The Mediterranean Sea Terrace Proposal*. Pages 1441-1447 **IN** Brunn,

S.D. (Ed.) Engineering the Earth: The Impacts of Megaengineering Projects. (The Netherlands: Springer) 2266 pages.

Rao, B.R. (2005) *Buckingham Canal save people in Andhra Pradesh (India) from tsunami of 26 December 2004.* Current Science 89: 12-13.

Ridgwell, A. et al. (2009) *Tackling Regional Climate Change By Leaf Albedo Bio-geoengineering.* Current Biology 19: 146-150.

Rodriguez-Vidal, J. et al. (2013) *Undrowning a lost world—The Marine Isotope Stage 3 of Gibraltar.* Geomorphology 203: 105-114.

Roether, W. et al. (2013) *The transient distributions of nuclear weapon-generated tritium and its decay product 3He in the Mediterranean Sea, 1952-2011, and their oceanographic potential.* Ocean Science 9: 837-854.

Romanis, G. de (2008) *The Bridge over the Adriatic.* (Milan, Italy: Edizioni L'Arachivolto) 171 pages.

Romero, R. and Emanuel, K. (2013) *Medicane risk in a changing climate.* Journal of Geophysical Research 118: 5992-6001.

Ruddiman, W.F. (March 2005) *How Did Humans First Alter Global Climate.* Scientific American 292: 46-53.

Russell, B. et al. (2013) *Reconceptualization and Optimization of a Rapidly Deployable Floating Causeway.* ASCE Journal of Bridge Engineering. DOI: 10.1061/(ASCE)BE.1943-5592, 0413013.

Sambracos, E. (2003) *Market analysis and pricing policies for sea canals: the case of the Greek Corinth Canal.* Maritime Policy & Management 30: 175-190.

Sammartino, S. et al. (2013) *A numerical model analysis of the tidal flows in the Bay of Algeciras, Strait of Gibraltar.* Continental Shelf Research 72: 34-46.

Sanchez-Romain, A. et al. (2013) *Spatial and temporal variability of tidal flow in the Strait of Gibraltar.* Journal of Marine Systems 98-99: 9-17.

Schipper, F. and Schot, J. (2011) *Infrastructural Europeanism, or the project of building Europe on infrastructures: an introduction.* History and Technology 27: 245-264.

Schneck, R. et al. (2012) *Climate modeling sensitivity experiments for the Messinian Salinity Crisis.* Palaeogeography, Palaeoclimatology, Palaeoecology 286: 149-163.

Schneider, S.H. (1996) *Geoengineering: Could—or Should—We Do It?* Climatic Change 33: 292.

Schroder, S.M. (2012) *To Forget It All and Begin Anew.* (Toronto, Canada: Toronto University Press) 240 pages.

Schuiling, R.D. et al. (2007) *Asteroid Impact in the Black Sea: Death by Drowning or Asphyxiation?* Natural Hazards 40:327.

Seebens, H. et al. (2013) *The risk of marine bioinvasion caused by global shipping.* Ecology Letters 16: 782-790.

Shafer, J.A. (1997) *Polar temperature sensitivity to lunar forcing.* Geophysical Research Letters 24: 29-32.

Showers, K.B. (2009) *Congo River's Grand Inga hydroelectricity scheme: linking environmental history, policy and impact.* Water History 1: 31-58.

Singarayer, J.S. et al., (2009) *Assessing the benefits of crop albedo bio-engineering.* Environmental Research Letters 4: 1-8.

Skliris, N. and Lascaratos, A. (2004) *Impacts of the Nile River damming on the thermohaline circulation and water mass characteristics of the Mediterranean Sea.* Journal of Marine Systems 52: 121-143.

Smil, V. (2011) *Global Energy: The Latest Infatuations.* American Scientist 99: 212-219.

Speich, S. (1996) *A Strait Outflow Circulation Process Study: The Case of the Alboran Sea.* Journal of Physical Oceanography 26: 320-340.

Spiering, M. (2002) *Engineering Europe: European Idea in Interbellum Literature, the Case of Panropa.* Pages 177-200 **IN** Spiering, M, and Wintle, M. (Eds.) *The Idea of Europe Since 1914: The Legacy of the First World War* (Great Britain: Palgrave Macmillan) 201 pages.

Strahan, D. (14 March 2009) *Green Grid.* New Scientist 201: 42-45.

Su, H. (2003) *The Great Wall of China: a physical barrier to gene flow.* Heredity 90: 212-219.

Tanhua, T. et al. (2013) *The Mediterranean Sea system: a review and an introduction to the special issue.* Ocean Science 9: 789-803.

Taubenbock, H. et al. (2012) *Monitoring urbanization in mega cities from space.* Remote Sensing of Environment 117: 162-176.

Therborn, G. (4 September1987) *Migration and Western Europe: The Old World Turning New.* Science 237: 1183-1188.

Trieb, F. and Muller-Steinhagen, H. (2007) *Europe-Middle East-North Africa cooperation for sustainable electricity and water.* Sustainability Science 2: 205-219.

Van Vleuten, E. and Kaijser, A. (2006) *Networking Europe: Transnational Infrastructures and the Shaping of Europe, 1850-2000.* (Sagamore Beach: Science History Publications/USA) 335 pages.

Ventura, G. et al. (2013) *The Marsili Ridge (Southern Tyrrhenian Sea, Italy): An island-arc volcanic complex emplaced on a "relict" back-arc basin.* Earth-Science Reviews 116: 85-94.

Verhegghen, A. et al. (2012) *Mapping Congo Basin vegetation types from 300 m and 1 km multi-sensor time series for carbon stocks and forest areas estimation.* Biogeosciences 9: 5061-5079.

Von Tillo, A. (1887) *Ein Wort uber die Hauptwasserscheide der Erde.* Pettermanns Mitt. 33: 101.

Walsh, K. et al. (2014) *Mediterranean warm-core cyclones in a warmer world.* Climate Dynamics 42: 1053-1066.

Wang, C. et al. (2013) *Defects of Tensioned Membrane Structures (TMS) In Tropics.* ASCE Journal of Performance of Constructed Facilities. DOI:10.1061/(ASCE)CF.1943-5509.0000530.

Weijermars, R. (1987) *The Palomares brittle-ductile Shear Zone of Southern Spain.* Journal of Structural Geology 9: 139-157.

Werner, W. (1997) *The largest ship trackway in ancient times: the Diolkos of the Isthmus of Corinth, Greece, and early attempts to build a canal.* The International Journal of Nautical Archaeology 26: 98-119.

Wisser, D. et al. (2013) *Beyond peak reservoir storage? A global estimate of declining water storage capacity in large reservoirs.* Water Resources Research 49: 5732-5739.

Woodliffe, J.C. (January 1978) *Floating cities: further thoughts on their legal status.* Marine Policy 2: 79-81.

Xue, Y.A. and Shukla, J. (1996) *The Influence of Land Surface Properties on Sahel climate. Part II: Afforestation.* Journal of Climate 9: 3260-3275.

Zemp, M. et al. (2006) *Alpine glaciers to disappear within decades?* Geophysical Research Letters 33: L13504.

CHAPTER 9

MARS-CONFIGURATION OF MOTHERSHIP FUTURE #2?

It is possible, just, that Greens could extend their collective attention and Gaia outlook, to Mars during the 21st Century; after all, for many Greenies the ideal sustainable "human being" is something close to an autonomous, solar-powered Terra-creature! Clive Staples Lewis (1898-1963), certainly no psilosopher, coined the term "planetolatry" to properly label those who exhibit an idolatrous worship of the planets. Geographos considers the Earth to be humanity's first "mothership", a term that combines a once globally natural "Mother Earth" with the technical "Spaceship Earth" assessments. According to its most recent [2012] assessment, **Number of Planets Needed**, the Global Footprint Network (www.footprintnetwork.org) alleges that if everyone on Earth enjoyed the Standard of Living of an average American, then our globe's needful humans must have access 4.16 Planet Earth-biosphere equivalents, or more! Today's poorest ecosystem-nations, once generally the most rural, now have enormous slums in their burgeoning urban regions (Marx, Stoker and Suri, 2013).

So, it is just possible that Mars might become, for future Earthlings, Mothership Number Two! But, will there be slums? Humans started to make their mark on Earth because of the Ice Age—or so it is alleged by Anthropology—and Mars is today, climatically speaking, existing in a super-Ice Age climate condition and state. [Earth's earliest mapped climate zones were based on the distribution of living vegetation and precipitation, absent obvious plant life Mars's climate zones must be differently based; GOTO: http://planetologia.elte.hu.] The Remote Radiation Assessment Detector on the Mars Science Laboratory's "Curiosity" rover began its measurements of the cosmic ray and energetic particle "weather" of Mars on 7 August 2012 during a solar maximum period in our Solar System. Currently, more and more of humanity is urbanized—peoples in Europe and the North America spend 85% of their time indoors, even in sunny Spain and California! To get to Mars, spationauts will have to dwell inside complex and vulnerable portable buildings called "spacecraft" and whenever they do get to, and land, on Mars in large numbers, Geographos believes, they will enter desirable homes—crude, but not slums—that can be deployed and relocated to suit the needs and desires of communities of nomadic peoples, just like humans behaved during the worst of the Earth's most recent Ice Age (Kemp, 2013). In fact, well-housed mobile Mars frontier communities will tread lightly on Mars's rugged surface, traveling inside portable, componentized, shipyard-like Astythromes! Greens yearn for simpler Earthly lifestyles—these will not for a long time be found on Mars (Anker, 2005); human terraformers can only try to configure Mars to suit our species' living requirements. At its surface, the pressure of Mars's atmosphere is equivalent to that of Earth's air at an altitude of 32 kilometers about two and one-half times the maximum altitude commercial airliners normally fly in 2014. Portable Astythromes will house highly-educated and highly-skilled workforces. When the long-period comet C/2013 A1 (Siding Spring) passed Mars on 19 October

2014, its coma [surrounding cloud of various materials and gases flying separate from the guiding comet's core] enveloped Mars and all the man-made satellites circling it (Moorhead et al., 2014). Imagine being on Mars's surface, witnessing that process-event!

Geoscience and Macro-Imagineering are never-ending pursuits of rock and water-related ultimates; no fabled Aztec treasury, storied Incan gold cache, El Dorados, Fountain of Youth or Seven Cities of Cibola ethnic myths stimulating an exploration Geography of Avarice! As of 2014, there are three kinds of imperfections in humankind's knowledge of Mars: (I) intrinsic—human spationauts cannot ever reach the planet's center of mass; (II) unlinked theoretical idealizations—any planet with almost closed material cycles has no almost no isolated systems; (III) ground truth—planners lack still useful knowledge of many operationally relevant planetary constants of non-Earth planets. Future direct examination of Mars can cure two of these imperfections through practical on-worksite data acquisition. From 1966 economists worldwide were speculating on the coming economics of Spaceship Earth, some even dreaded the apparent shift from the "open" world of the past to the "closed" world of the near-term future; some economists even pejoratively labeled this hypothetical economic closure as the "Spaceman's economy"! In fact, what the Anthropocene has produced, via its ongoing Space Age segment, is a strengthening of the human will and drive to make a closed, evidently lifeless Mars "open" whilst, at the same time, keeping Earth "open" to flourishing life of all kinds! Mars's monochromatic, cloudless sky even suggests a unifiable potential world. [Nature will add cloudy skys to Mars, a billion years hence, when our Sun's luminosity increases gradually so that our Earth becomes a Venus-like furnace (Leconte et al., 2013).]

Green fanaticism has taken people generally to the tipping point of silliness that even our most adored pets must be "sustainably" fed

(Swanson et al., 2013)! Between 15,000 BC and 1500 AD, there were remarkable social differences evident in peoples living in the Old World and the New World (Watson, 2012) and very many of these differences have been culturally homogenized during the past five centuries. The once and future Mars must be described in terms of global Nature as it was, as it is in 2014, as it could be before 2100. Since its coinage in 1987, "sustainability" has permeated the literature of Geoscience (Costanza et al., 2011)—to the obvious detriment of intellectual reason and spiritual optimism! There are, literally, several hundred so-called definitions of "sustainability". Truthfully, the term has never been officially adopted, rather it has been infiltrated, added to other monotonous and repetitious phrases, intruded upon jargonized statements and stealthily entered the political vocabulary as feel-good "eco-speak". Essentially, it is a worthless term of the art of deception, Green salesmanship for an impracticality that is human civilization destroying if allowed to mesmerize the world-public! Two years ago, an Australian physician, Bryan Furnass (born 1927), coined a barbarism, "Sustainocene" to label the future Earthly Geologic Time period extending one billion years from 2012. Needless to say, "Sustainocene" is an erroneously constructed word—"sustain" is Latin for "to hold up" and "cene" is from Greek! Furness says that artificial photosynthesis will be the primary energy supply for people living in his "Sustainocene" and the idea is best related by Thomas Alured Faunce (born 1958) (Faunce, 2012). Geographos prefers the retention of "Anthropocene" and, furthermore, to apply it to Mars once people truly settle thereon, especially if they use controlled nuclear reactors and/or explosives!

Acquisition of such valuable data will involve pioneering humans in Macro-Engineering efforts seemingly encompassing a whole planet at the outset; actually, terraformers will demarcate the upper and lower boundaries of Mars's environmentally relevant crust based on the vertical range of their investigational instrumentalities of reconnaissance, their mechanical means of Mars-surface movement and, as well, their ability to mobilize materials and energy from therein (Anker, 2005). Nevertheless, as for the Earth-biosphere, all of Mars's measurements—even those relevant to commerce—commence at the planet's center of mass (Palaia et al., page 395 *IN* Badescu 2009). Measurements of the Mars crust, made from movable platforms conveying air-filled sealed homes, offices and workshops, will survey domains of particular interest. Exploitative macro-projects can then follow at sites deemed worthwhile for resource excavation, ore and fluid processing, industrial and residential use and final waste disposal. [As on Earth, or inside individual spaceships or squadrons of interplanetary spacecraft, material recycling activities will predominate (Minter, 2013).] Every artful product of this far-distant settled human social group will be built on, in, or with Mars resources, except those products that fly and float, or collapse and those last three must start or end with some contact with Mars's pebble, cobble and sandy granular surface materials, as well as its water ice. Virtually all exploration and development of extra-terrestrial places involves navigation in and on loose granular media and little is certain about how artificial, mobile objects behave in and on granular media with gravitational accelerations other than Earth's [normalized at 9.78 meters per second squared]! Humans have some little experience on the Moon [gravity 1.622 meters per second squared] and Mars [3.711 meters per second squared]. Overall, Mars's normalized gravity is assumed to be just 38% of our Earth's. **Figure 1.**

Figure 1. The size of Mars, right-side, and Earth, left-side. The Earth's land area is approximately equal to the whole soil+ice surface of Mars. Whilst Venus has about the same bulk and mass as Earth, it is an infernal "Earth" climatologically. To thin Venus's atmosphere by undertaking a subtractive macroproject, making it equal in mass to Earth's, an expenditure of 2.5 followed by twenty-eight zeros Joules would be required to remove 4.75 followed by 20 zeros Kilograms of carbon dioxide gas. (Image: NASA/JPL.).

Nowadays, it is possible to imitate experimentally all appropriate temperature and pressure regimes that might exist inside Mars, subsequently, tentative conclusions derived from such laboratory experiments will be revised by extensive and intensive geoscience and macro-engineering fieldwork done and many selected places on the Mars-surface. The future human Martian's chief concern is Active Tectonics, which officially codified all results from geo-scientific observations of tectonic movements of parts of the Mars-crust that are occurring, or expected to happen, within a period of vital concern to a particular socialized human ecosystem-organization. A period of 10,000 year or less, since that is a timeframe commonly adopted by prudent Terran macro-engineers to ensure the natural total isolation of industry-generated radioactive wastes from the Earth-biosphere, seems most appropriate and the period selected permits Martian Macro-Imagineering/Macro-Engineering's use of the tectonic "undulation theory" of planet-crust motion during a planet-specific Geologic Time period, a gravitational explanation first proposed by Reinout W. van Bemmelen (1904-1983). For this chapter, Active Tectonics (van Bemmelen, 1967) encompasses all future Mars landscapes and seascapes from the interplay of artificial climate *versus* global natural and terraformer-induced magmatic event-processes and crust tectonics (Anthropogeomorphology) (Haqq-Misra, 2012). In other words, the Mars crust will become a kind of kinetic architecture trod by humans moving about in a tiny room, the anthropomorphic bags known as pressure suits (Roach, 2010); such adventurers could even emplace small works of artistic merit (Philippe, 1996).

Probably, Mars is comprised of a number of concentric zones of different physical and chemical properties; with increasing depth the topmost zone—during the 21st Century to become Homo sapiens' domain—becomes more and more complex, or simpler! Van Bemmelen identified planetary structural zones as "Stockwerke", and envisioned their vertical and horizontal movements following great inputs of energy. Super-bolide impacts, causing extra-Mars energy to be shot into the planet's Stockwerke, can on occasion heat the sub-aerial crust everywhere to its melting point, and energetic rocky ejecta (including rock vapor) can escape to outer space whilst, at the same moment, less energetic ejecta are globally redistributed. Small surface-impacting bolides can affect the future domain of humans, the Stockwerke, by (I) disturbing carbon dioxide reservoirs consisting of dry ice, clathrate, and even liquid carbon dioxide and by (II) rapidly injecting extra-Mars minerals and energy into the Mars-crust. Additionally, vaporized small bolides classed as ice-volatiles—"iceteroids"—can produce rapid changes in Mars's atmosphere composition and condition (Wilde and Quinby-Hunt, 1997). Many researchers suggest that, approximately, 200 new craters form each year on Mars that have diameters of 4 meters or larger and these same experts foresee that a Mars Exploration Mission lasting three years would likely experience a one Megatonne bolide blast event during their visit! [The Comet C/2013 A1 (Siding Spring) passed extremely close to Mars on 19 October 2014—there was no chance that it could have impacted the planet. If it had, the collision event would have been witness by robots rolling about Mars's surface. Had people been there, perhaps in a portable "Las Vegas", then the collision might have been seen like those atomic explosion mushroom clouds LasVegans used to observe during the 1950s!] Robert Alan Mole (1995), writing on the future terraformation of Mars, opted to fill the atmosphere with aerosols and cover its South Polar Cap with dark dust, enough to induce sublimation through solar heating, by exploding thermonuclear warheads equal to ten million one-Megatonne bombs! During the night of 8 December 1951, Tsuneo Saheki (1916-1996), one of Japan's leading planetary astronomers saw a sudden "… small but extremely brilliant spot…" of light located at the eastern end of a feature on Mars known as Tithonius Lacus. Saheki offered many explanations of its rapid appearance and five minute duration, among them was the "unreasonable", yet possible, that it was a "…product of intelligent beings". At the time of the Cuban Missile Crisis of October 1962, a plant physiologist born during 1927, Dr. Frank Salisbury (1962; Edwards, 2012) voiced a decidedly non-botanical thought: "Was this volcanic activity, or are the Martians now engaged in debates about the long-term effects of nuclear fallout?" Truthless rhetoric certainly, yet there is a transient Mars climate effect of large bolide crust impacts.

Undoubtedly comet and super-bolide impacts have already happened, and could recur. If humans equipped with anti-bolide technology were actively present, and had a vital interest to prevent a normal event-process, then such cataclysms may not occur after settlement, ameliorated at reasonable financial cost in the form of a Mars version of Earth's future Rodoman-ALPS. In that case, Mars's crust loses it Nature-imposed curse, afterwards holding the long-term status of a protected and exploited litho-technical system of rock formations influenced—ultimately to be controlled as a catastatic structure (Stockwerke)—by an ever-migratory Homo sapiens. Monitoring of the litho-technical system and, perhaps, prediction of its state in the nearest and distant Geologic Time future will be one of the most important stages in the induced change from ecological expansion (imposed by Martian terraforming industry) to ecological co-existence of biologically established people with Mars's solid and fluid material components. Our extended species' Earth-biosphere activities are remarkable: "The total earth moved in the past 5,000 years would be sufficient to build a 4,000-m-high mountain range, 40 kilometers wide and 100 kilometers long" (Hooke, 2000). Located at a single place, this is enough solid material—thicker and heavier than an Ice Age glacial remnant—to activate a localized (purely isostatic) vertical tectonic landscape movement. Martians may challenge that record for digging and piling!

21st Century Earthlings apprehend that Mars appears inimical to superficial Earth-life forms and, absent a closed cycling of rocks, the planetary crust must be intelligently altered and sensitively adjusted by any long-term future settlers by the bridled use of a truly comprehensive Technology—perhaps molecular Nanotechnology. R.A. Mole chose to use thermonuclear explosives—the highest yield thermonuclear bomb ever tested aerially above the Arctic Polar Zone island of Novaya Zemlya on 20 October 1961 used just 2.7 Kilograms of energetically liberated matter to produce a blast exceeding 50 Megatonnes.

But, transformation of Mars by Mole's technique is complicated; it is also an unnecessarily compromising means to proceed since it creates destabilizing and costly geotechnical macro-problems that must then be carefully considered, and somewhat mitigated where necessary. Violently fracturing the planet's crust is a counter-productive global Mars Nature alterative technique that leaves a residue of radioactivity matter. When Mars's crust first becomes a Stockwerke, it really should be treated like a mixture of rocks and infrastructure. **Figure 2.**

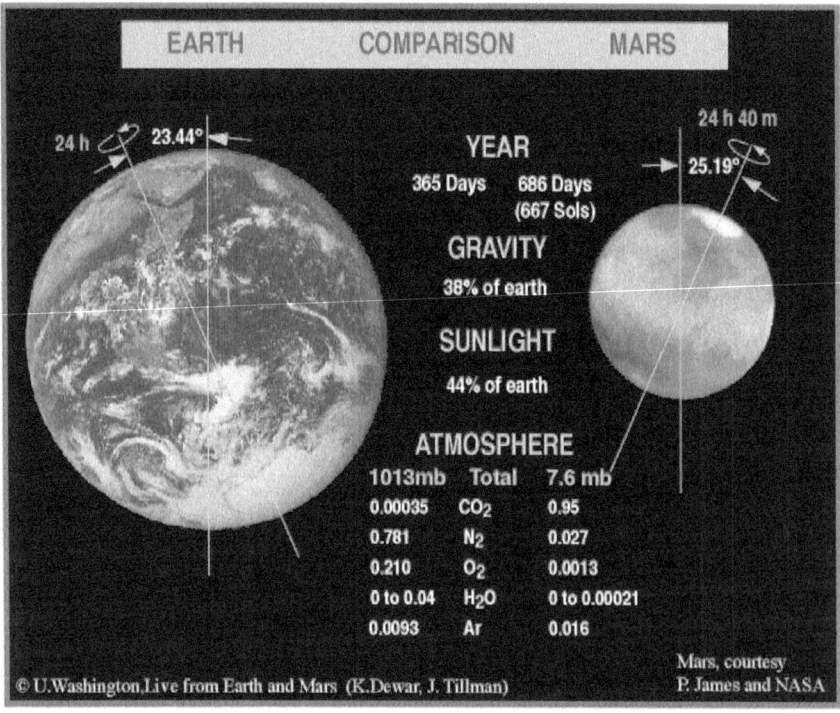

Figure 2. Self-explanatory chart, courtesy of Google Images.

Currently, there are no beachfront, river-front or lakefront properties on Mars. Getting healthy humans to Mars is no easy task since the closest Mars ever comes to Earth exceeds by more than 100 times the distance between the Earth and the Moon! (Image: Goolge Images)

"Infrastructure" is a broad term that refers to the overall system of services and physical facilities (involving transportation systems, freshwater supplies, energy distribution linkages, waste treatment and disposal systems, and telecommunications networks) that are provide through public and privately-financed activities. Egyptian pyramids or the Great Wall of China, perhaps, comes closest to predicting what a planetary-scale effort would entail to transform Mars into a place pleasanter than it currently is. As living, intelligent inhabitants of Earth's biosphere, our ambitious species has endured Techno-Art's many vital breakthroughs and excitingly frequent breakdowns. To construct an unbalanced artificial global Nature (enclosing Mars) requires skills as yet unperfected and untested. Molecular Nanotechnology's idealized machines, imagined by Micro-Imagineering's practitioners who do not yet fathom the full effect of Gremlins, may convey humans to a planet-wide named Geologic Time period and geophysical-state called "The End of Technology and the Beginning of Pure Art"! An Earthly Ballardia—the offshoot of John Heaver Fremlin's conceptualized nightmarish future warehousing of billions of human beings within an Earth-crust enveloping building, a monumental ultra-computer epidermis—in fact, a single shell-shaped geological formation installed and typified by our species of the human life-form—covering Earth's surface at a distance determined by its present-day crust's highest topographic elevation. By capping, Fremlin's building would cause a closed Rock Cycle for Earth; atmospheric heating forced by a near-Red Giant Stage Sun will cause all lateral tectonic movement to naturally cease in 2.5 billion years (Lenton and von Bloh, 2001)— the equal, therefore, of a man-made Ballardia. It seems entirely permissible to think Mars's obvious hazards and ecological inadequacies eventually can be overcome economically; one result of the transformative effort on Mars will be to embed various networked systems of infrastructure within the Mars-crust, and to mix Martian life-support facilities through very broad imbrications with the planet's Nature-given surficial features.

Advocates of "footprints", a novel public relations idea that living standards and personal choices in life-style can be used to measure humanity's impact in the Earth-biosphere is espoused by scientists and others in the popular Science news media. Practically speaking, a fully developed Mars can only be about 25% of the volume of an Earth-biosphere. Today's Science leaders typically shun proponents of both Macro-Imagineering and Macro-Engineering—Macro-Engineering is a practical umbrella term of Art signifying Geo-engineering and Terraforming—because these concepts empower Techno-Art's managers and because they wish to clearly dissociate themselves in the world-public's mind from "oneiric macroprojects". Yet, to forestall high-prices on a human necessity (breathable air, for instance) of life, Mars must be terraformed by oneiric Macro-Imagineering/Macro-Engineering so that its future trekking settlers need not wear pressure suits supplied with bottled air. Completion of such a macroproject—actually an exercise in creative geography and metabolism—will concretize the post-1972 vogue word "environment" and combine it with R.W. van Bemmelen's Stockwerke perspective. Theoretical terraformers attempt by speculative means, prognostic if not prophetic, to outline realistically what Homo sapiens' next efforts ought to be in regard to Mars's development and a world.

Modern architect Remment Lucas Koolhaas, **Figure 3**, revealed one of the biggest philosophical macro-problems nagging 21st Century oneiric Macro-Imagineering/Macro-Engineering: "Beyond a certain scale, architecture acquires the properties of Bigness.... Bigness is ultimate architecture.... [Only] Bigness instigates the *regime of complexity* that mobilizes the full intelligence of [Hyper]-architecture and its related fields.... The absence of a theory of Bigness—what is the maximum architecture can do?—is architecture's most debilitating weakness.... Big mistakes are our only connection to Bigness.... [The] attraction of Bigness is its potential to reconstruct the Whole.... Bigness destroys, but it is also a new beginning.... Bigness...is the one architecture that can survive, even exploit, the now-global condition of the *tabula rasa*: ...it gravitates opportunistically to locations of maximum infrastructural promise, it is, finally, its own *raison d'etre*" (OMA, Koolhaas and Mau, 1995).

Figure 3. Born during 1944 in The Netherlands, "Rem" Koolhaas, famous architect. (Image: Google Images.)

Koolhaas's "Hyper-architecture" is synonymous with "Macro-Imagineering/Macro-Engineering"; but, more importantly, Koolhaasian "Bigness" correctly indicates its near-future spatial planning scale, the economic scope of its prospective cost-benefit analysis/synthesis work, construction and demolition. In a configuring work of Techno-Art, Mars's crust, its future infrastructure, when it becomes a humanly useful mixture of natural and artificial components, a colossal composition that will be a new kind of technically configured global

Nature—Number Two!—it will become a Global Gizmo blending infra-structures and various human life-styles!

By whatever means the new [human and Terra-Creature, **Figure 7**] peripatetic Martians first trudge on the Red Planet's surface—possibly by 2100—they will likely dwell in inflatable composite habitats forti-fied against cosmic radiation with frozen freshwater brought to the site via flexible and readily repositionable tensioned hose-pipelines (Badescu et al., 2009). [The Red Planet resembles a vast section of Central Australia (Rey, 2013).] The Poles of Mars have freshwater ice-caps and even the Equatorial canyon Valles Marineris contains an estimated one million cubic kilometers of fossil ice (Gourronc et al., 2014; Komatsu, 2007), its fossil ice is a leftover from ancient times when large lakes were present in the canyon, sometimes spilling over to cre-ate an elaborate drainage system of rivers on Mars's northern plains (Cabrol and Grin, 2010). Such shifting "igloo"-type buildings, allow-ing room for pedestrians, however, will be subject to ground motion ("Marsquakes"); thus, for inflated, iced-coated buildings there is the requirement that a nearly indestructible bottom seal be present since soil has great porosity (Jones et al., 2012). Still, considering Homo sa-piens' profound ignorance of Mars' Active Tectonics and the risks im-posed by destructive cosmic radiation, it might be safest to lodge and shield all explorers and settlers inside movable (via telescoping me-chanical legs, by hover-crafting and by endless treads). Topographical instability promotes adoption of "Walking City"; highly mobile hous-ing and work exploration-exploitation platforms, with the capability to rapidly self-elevate should conditions warrant. It is a technically sweet solution, responding to any and all threats vividly elaborated by Marsologists.

What might a new Martian, mechanical or organic former Terran, see whilst inside one of these "igloos"? From 16 March until 30 December

2013, at Oberhausen in Germany, the Techno-Artist Christo displayed his sculpture, the "Big Air Package: The Largest Inflated Envelope in History"", with a volume of 177,000 cubic meters, kept inflated at 27 Pascals of constant air pressure by two fans. Some 20,350 square meters of white-colored semi-transparent polyester fabric bag, weighing 5.3 metric tonnes, was inflated with filtered air inside the largest disc-type gas-holder [gasometer] in Europe, visitors entered only by air-locks. The diffuse light present throughout the interior of the bag was quite attractive, well illustrating the potential for such buildings to have a maximum psychological effect with minimal material usage. Suitable materials containing shielding ice would have much the same appearance and effect on human residents attempting to thrive on a Mars undergoing massive alterations at the hand of Homo sapiens.

Ron Herron (1930-1994) designed in 1964 Archigram's "Walking City"; it is possible to build such vehicle-cities (mobile Astythromes) because there are extant actual parts of huge buildings that move at the USA's Cape Kennedy—the recently remodeled Vertical Assembly Building (VAB) on Merritt Island that first served the Advanced Saturn C-5 Launch Complex 39 and now is used to service post-Space Shuttle rockets has a tread-moved launch platform as part of its total architecture. VAB has an interior volume of 3.765 million cubic meters. A platoon of "Walking City" urban regions (Deyong, 2001), with the characteristics of artificial armadillos, are meant to house large populations of world traveler-workers, moving onwards to wherever the Martian economy and civilization dictates a need or want. **Figure 4.**

Figure 4. The VAB moving platform, here playfully adapted with no-slum zone "Las Vegas" set atop the crawler, meant to tease the reader's mind about how future mobile Martian Astythromes might appear... well, sorta! (Image: JMH.)

Moving under their own volition, "Walking City"-type buildings will insert their extractive mechanical equipment into the underlying Mars terrain from which they shall draw raw materials and geothermal energy. Fertilizer can be made on Mars by "Walking City"-style factories, drawing from the present-day atmosphere consisting of about 2.7% nitrogen gas. A drivable "concept vehicle" such as **Figure 4**—is introduced herein to showcase a new technology/style to gauge future customer reaction to a new and radical design which may never be mass-produced on Mars—is clearly meant to satisfy the common human insatiable drive to explore! Regions of the State of Arizona, especially during wintertime, are carpeted with popular mobile homes! "Walking City" is not that far removed from the no-need-for-mobility supposedly imposed on Earth's Macro-Imagineering/ Macro-Engineering by the microchip's mid-20th Century invention

(Kronenburg, 2013). Something similar to the spin-off Geographos proposes, an endless treaded city for use on Mars's landscape has been proposed during 2013 by the architect Manuel Dominguez for use on Earth! **Figure 5.** [GOTO: www.gizmag.com/very-large-structure-moving-city/29779/.] When they are extant, each "Walking City" will recycle its harvested resources as much as practicable, thereby adding few "wild card" elements to Mars's changing atmosphere. However, defining a suitable level of durability—the ability of a building to perform its required function over time in spite of degrading sub-aerial forces that act upon it—for any "Walking City" means first defining the service life of a proto-vehicle. Like Earth-connected cities, a "Walking City" ought to be sturdy, with a long design service lifetime, and not become obsolescent because of a change in Martian requirements or expectations regarding their use. So far, there is no helpful published data permitting a credible durability assessment of Mars's first "Walking City".

Figure 5. An Earthly future mobile city artwork—an undeniable off-spring of "Walking City" envisioned by Manuel Dominquez. Sci-Fi writer Greg Bear, in *Mandala* (1978), featured a movable city walking on elephantine legs, endless tractor treads and wheels! (Image: Google Images.)

Just at The Netherlands is a ecosystem-nation whose citizen-ry fought to reclaim from the ocean much of the land on which it is built, Mars's citizens will be building on a high-economic value landscape—terraformed at a not inconsequential cost—where build-ings are perhaps never to be thought as permanent, but rather some-thing to be changed in response to the emergence of new planet settler needs within a budding civilization. "Walking City" settlements are just one possibility, of course. Just getting people from the Earth-biosphere to a vacant yet "inviting" Mars, practically speaking, im-mediately instigates the "Concorde Fallacy"—the maladaptive effect of sunk economic cost that is generally manifested when macroproject investment in energy, money and effort time has been made. It is a state of human affairs possibly due to "...humans' overgeneratization of the 'Don't Waste' rule of common behavior (Arkes and Ayton, September 1999, page 591): Once something is started, one does not stop!

As the technically progressing "pilots" of Spaceship Earth, according to some Greens a "depleted world", increasingly capable of voluntarily exiting the presumed "homeland" vessel at will, *Homo sapiens* must ed-ucate its group membership to properly operate our present-day planet of full-time occupation before its representatives can ever hope to suc-cessfully homestead and operate (for an extended post-terraformation period) a Spaceship Mars residence-factory. The Earth-biosphere is not the only realistic habitation for humans. Macro-Imagineers/Macro-Engineers allege that Terraforming will give humankind a second home, where global Nature must be altered from a cruel alternative to community life-style into a place of play, another planetary megalopo-lis-bivouac! Most of the published game-plans about getting Mars into a geophysical state suitable for settlement by persons unencumbered by $12 million NASA-ESA-Russian-Chinese-Indian spacesuits ignore the post-makeover Mars's likely sub-aerial crust surface features and, most significantly, the probable industrial uses of those planetary fea-tures such as regenerated seas or oceans and their coastal landscapes (Parker and Currey, 2001; Carroll and Lopes, 2013). **Figure 6.**

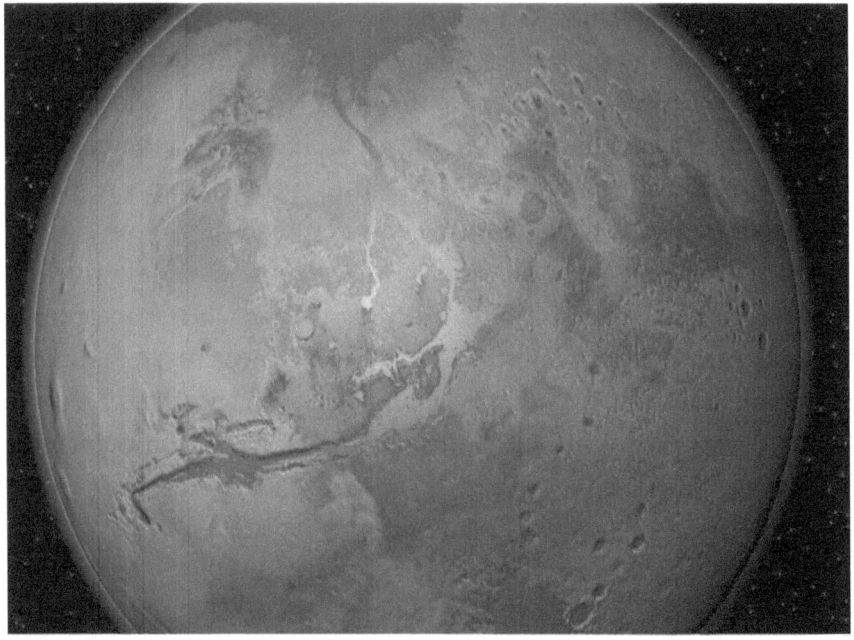

Figure 6. Mars after a lot of expensive terraforming work. (Image: Google Images.)

Climate modelers have imagined Earthly landscapes devoid of vegetation—as a "desert world" like Mars—for the educative purpose of contrasting Earth's obvious present-day biosphere vitality with its prospectively harsher alleged near-term future ecological state (Kleidon et al., 2000). A planet such as Mars that is drier than Earth seems to have a wider stellar habitable zone tolerance, H. Strughold's "ecosphere". Mars may currently be in a hydrologically dormant state and there is potential for hydrological activity to be technically regenerated in the future following large-scale anthropogenic changes of the planet. In other words, terraformers anticipate the re-creation of superficial lakes, seas and oceans empounded on Mars's now arid landscape; ter-raformers boldly assert their restorative Greening reintegration event-process ought to be hastened via hard work by brave persons helped by reliable Terra-creatures! Vociferous and visionless radical Green non-enthusiasts are afflicted with an anti-Macro-Imagineering/Macro-Engineering form of Daltonism since they can, apparently, "see" only a natural Red Mars, not an unnaturally Green Mars! It is a form of ecological brinksmanship. As long as humans can wear only pressure-suits on Mars, galumph from place to place at an optimum walking speed of just 3.4 kilometers per hour, they can leave only sterile shoe-prints like those impressed on our Moon's regolith during 1969-1972 by elite American spationauts, but which are gradually disappearing there owing to solar-induced regolith thermal pulsation of an atmosphere-less Moon's pulverized rock surface. Mars can become an arena for audacious possibilities for architectural design. Specially trained terraformers will wisely draw upon a Macro-Engineering history mostly written, so far, by Earth-bound Geo-engineers who are cognizant that a natural global Earth-atmosphere climate change 2-3 million years ago markedly increased the rate of erosion of our world's landmass (Peizhen et al., 2001); a similar event-process is sure to follow any global climate alteration by humans of Mars's gaseous envelope! Someday rain could again fall on Mars's landscape, not only snow, followed by runoff, pooling and new waterborne sedimentation.

Circa 3.5 billion years ago, Earth and Mars were basically alike; over the aeons since the planets' natural geophysical event-processes diverged and, today, Earth and Mars are fundamentally dissimilar. Today, Mars is a **Dune**-type planet. Frank Herbert (1920-1986), author of **Dune**, is often credited with the emergence of a Green-style "Ecology" as a popular idea in the USA's public mass-media because his 1965 Climate-fiction [Cli-fi]/Sci-Fi novel greatly influenced some of today's leading opinion-shapers. In **Dune**, it is the character "Liet Kynes", son of the Imperial Planetologist, leading the Fremen of Arrakis, who devises a planet-encompassing macroproject plan to moderate its harsh Sahara-like desert condition. To the junior Kynes, "...the planet was merely...a machine being driven by its sun. What was needed was reshaping to fit it to man's needs. His mind went directly to the free-moving human population, the Fremen…. What a tool they could be!" Real Earthlings may soon function as non-fictional Fremen. "Liet Kynes" performed the work of a Geo-engineer (Wood et al., 2013), not a perfect Terraformer. Bold Earthlings are going to try to make a markedly different planet (Mars) Earth-like; they won't try to duplicate Earth [forming-Terra]! ["Terran", the English-language word for a human inhabitant of Earth dates from circa 1953; somewhat oddly, "Martian", used in the sense of a non-human, Earthling-equal, inhabitant of Mars was neologized circa 1998.] The geophysical differences between these two celestial bodies is far too great to induce, by any known or speculated Technogenic means, a smaller mass Mars to be an exact match for Earth. Eco-deviant life conditions, in part caused by a low gravitational field strength, may cause disturbances of physiology or behavior in some organisms transferred from Earth; without and Earth-strength magnetosphere, Mars is exposed to biological cell-damaging electromagnetic radiation (x-rays, gamma-rays) and particulate space radiation (protons and electrons). Thus, Terraforming is the altering of a planet or large moon's environment to suit our tastes—in fact, one might call it "geographical stylization". Humans must safely shed their pressure-suits if they are ever to become true Martians! This is an imaginable prospect. Whilst Earth is a circumstellar planet, in fiction the circumbinary *Star Wars Episode 1—The Phantom Menace* (1997) movie planet "Taatooine", like *Dune*, an all-desert planet, is invoked by geoscientists when they discuss the formation of sandy Saharan land-forms (Lorenz et al., 2013); expertise acquired with the Sorgelian-style of improvement of the Mediterranean Sea Basin, including its Sahara segment, will support human settlement efforts on Mars. "Those who imagine, define, and transform landscapes bring material changes in the biogeophysical environment, which in turn influence social organization, values, understandings, and actions" (Gellert and Lynch, 2003, page 17; See also, Priemus and van Wee, 2013).

Terraforming's professionally-schooled 21st Century Nimrods without nescience will have their jobs cut out for them; such talented off-planet experts will not attack Mars's daunting dry-land climatic condition—macro-problems of definition for conscious global biosphere-creation—without abundant backup from infrastructure-constructing in-planet Geo-engineers. In other words, both elements of a two-pronged professional Macro-Imagineering/Macro-Engineering will undertake, persist and succeed eventually in the geophysical transformation of Mars; isolated as a practical and operational field of endeavor, Macro-Imagineering/Macro-Engineering is a useful means to form integrative conceptual links amongst, at least, planetary Science's arbitrary sub-divisions; consilience means all of these sub-divisions ought to be internally consistent with all known laws of Science.

Free men and women, even boys and girls enjoying liberty, as working pioneers from the natural Earth-biosphere, are going to stabilize Mars's exterior features—perhaps, as Geographos suspects in a strictly supplemental published suggestion, by using a planet-encapsulating Mars Tent (Cathcart, 1998). Dr. Viorel Badescu and R.B. Cathcart, both long-time founding members of the prestigious Candida Oancea Institute of Solar Energy situated in Bucharest, Romania, computer-simulated the regional albedo effect of a white-color pneumatic fabric covering of 3.5 million square kilometers of Earth's Sahara. Out modeling proved there are beneficial effects to "tenting" a vast Earth-surface region analogous to a future Mars Tent, which would be more usefully tinted to absorb impinging solar energy, rather than reflect it. The most effective Greens are those undeterred persons striving to advance humanity, giving birth to "New Earth-worlds" in our possibly unique Solar System. More and more, terraformers will adopt a three-dimensional perspective on planetary system geo-dynamics and, as well, a three-dimensional grasp of real property rights in another Global Commons (Yandle, 1999). One of the postulated Mars Canals, "Agathodaemon", in the gazetteer contrived by Perceival Lowell (1855-1916), after the USA's Mariner 9 flyby mission, was identified to correspond to Valles Marineris, a deep strategically located valley near the planet's Equator in the Southern Hemisphere highlands (Sagan and Fox, 1975). Valles Marineris is a good place for human settlement: its depth means that atmospheric pressure is higher, it would have fewer dust storms and dust devils affecting above-ground structures and it has plenty of freshwater ice. Food and fiber crops needed by a settled colony may be irrigated with freshwater, even brine.

Billions of years ago, Mars had a cold-water ocean covering about 33% of its surface; that ocean even had icebergs! If a single circum-polar ocean (or several widely-distributed circular seas and/or a Valles Marineris Reservoir (VMR) hydropower complex—either naturally or artificially made to exist—then all regolith-submerging bodies of liquid water existing at that planet's surface will have a cumulative area of, roughly speaking, 10% of today's Mars area. Creation of an ocean would reintroduce an air-sea interaction, resulting in advection fogs over a large region of its Northern Hemisphere. Before some part of Mars's crust is loaded with water, the entire planetary surface must be seismically monitored and vetted for potential "Marsquake" hazards, an integrated seismic risk map up-dated constantly. An adaptation of the 1964 "Walking City" idea originated by Archigram ought to be used by future prepared invaders, human Martians, for this seismological survey. Wind strength will be variable, owing to the atmosphere's changing density, during the time of terraforming, but will become steady and predictable after completion of the macroproject. As on Earth, water waves will be generated by winds; wave characteristics are going to be dictated by then extant wind velocity, duration and fetch parameters. Murky water conditions may prevail near shores. High-seas waves are not liable to be much higher than one meter even during the winter season's stormiest weather; the water will be well stirred by its constant movement and it is highly probable there will be season-guaranteed episodes of floe ice.

"Walking City", actually a hovercraft or endless tread vehicle-city, should compile the bathymetry data for a planned "Neo-Oceanus Borealis". ["Oceanus Borealis" is a descriptive term for a speculated Northern Hemisphere ocean emplaced early in Mars's existence.] If all the water ice in the two polar caps of Mars melted, it would be equivalent to a layer of 20-30 meters of water spread over the whole surface of Mars; if the water were constrained only to the Northern Hemisphere's plains, then it would produce, maximally, an ocean about 100 meters deep on average. An artificial global atmospheric warming, causing total glacial sublimation of Polar Zones is equivalent to a natural planetary geological catastrophe, a forceful event-process regenerating an ocean on Mars. Mars's terraformers will view their handiwork on a daily basis, even from multiple planet-orbiting satellite viewpoints; there might be a historical parallel with the events in the Nordic Countries stimulated by land-ocean elevation changes (Nordlund, 2001). Marine transport can occur as a result of event-processes such as wind-driven fluid flows, tidal currents, tsunamis, surf zones along the anthropogenic coastline. All of the shoreline will be well documented historically and topographically mapped. Water mass anoxia on a large scale and, perhaps, even algal blooms are potential macro-problems. If there is no endemic life in a pristine Mars, then Homo sapiens can deftly resolve the "paradox of invasion" (Sax and Brown, 2000) through carefully controlled breeding of all intentionally seeded life-forms, especially microscopic life-forms, in high-security off-planet biological laboratories.

A Valles Marineris Reservoir could harness impounded water's grav-itational potential energy, it could be designed in two main styles: (I) as a single large pool or (II) as a series of cascading pools, will all released water flowing into the Neo-Oceanus Borialis finally. At the mouth (an estuary or delta) of this Nile River or Suez Canal-like topo-graphical feature, there will inevitably be land erosion and water-laid deposition of sediment; a complicated deltaic construction/erosion event-process will be affected by VMR's operator decisions on the timing and quantity of all water releases. Possibly, where the chan-neled water enters the new ocean there might be some eutrophica-tion macro-problems due to the use of plant growth-stimulating soil fertilizers. (Sewerage might even exacerbate that macro-problem.) Either design approach permits VMR creation through the astute utilization of inflated fabric dams, each dam having a length ap-proximately ten times its crest height, according to conservative structural design theory developed since the 1970s. If these dams become inflated with a fluid such as water, then they would amount to a permanent water storage capacity that is never influenced by atmospheric conditions (evaporation). Future Martians will likely be techno-addicts, a global civilization organized around technical in-novations like solar and nuclear energy facilities as well as safe ad-vanced bio-technology R&D. People and Terra-Creatures will alter that place in ways humans alive today cannot yet imagine clearly (**Figure 7**).

Figure 7. Terra-Creatures, left-side, and humans, right-side, create life and civilization on Mars! These are the operational motto of Geographos' macro-imagineers!

Looking farther into the future, it is possible to foresee shipwrecks littering the anthropogenic seafloor! Earth's ocean has, approximately, 50,000 hydrographic chart-indicated shipwrecks; Mars's future ocean may become, over historic time, modestly littered with a few shipwrecks and these will, of course, serve as televised bottom indicators of sediment transport (Gray, 2013). The Neo-Oceanus Borialis will be extraordinarily difficult to navigate in other than purpose-designed and well-constructed ship types because of the ever-present unpredictable bubble eruptions of carbon dioxide gas effervescing from the recently water-inundated seabed. Mars's Chandler Wobble will be excited by fast emplacement of an ocean, so much so that it is likely to be the most drastic change in the rotational parameters of Mars since the liquid water disappeared from its surface eons ago! (The VMR may induce a countervailing effect.) Under circumstances of induced quick global warming and fluvial exhumation, frozen carbon dioxide gas, embedded for eons in Mars's sediments and fractured rock layers can become a gas and leak from the water-loaded crust into the superincumbent Neo-Oceanus Borialis. This poses a very great operational hazard for future seafarers because floating objects can be sunk by sub-sea gas blowouts that suddenly lower water's density. On land, there is also a possibility that large quantities of expelled carbon dioxide gas could overwhelm an unprotected human community just like the ground-hugging deadly event-process of 21 August 1986 at Lake Nyos in Cameroon [Africa]. Since not every drop of Terraformer-freed fresh and geothermally variable mineralized water on Mars will be deposited in the Neo-Oceanus Borialis, some could become captured precipitation ponded in bolide-excavated craters, maars or thermo-nuclear explosive-excavated crater lakes and so come eventually to mimic Lake Nyos.

Mars's landscape is red because of iron, like Australia's great deserts. By time-tabled dumping of iron flakes—possibly by the infall of directed iron bolides derived from interplanetary space—into a body of water coevally seeded with specially-selected and/or genetically-engineered oceanic plankton, a process would commence whereby some carbon dioxide gas becomes oxygen by splitting. However, even this option has its drawbacks: natural genetic drift should instigate and possibly foster an uncontrolled parapatric speciation in these near-microscopic organisms; free-roaming, technically unmonitored, these creatures— linearly or coincidently—may also evolve by an induced staspatric speciation. All in all, these anticipated biological event-processes are extremely complex because they will be occurring in a very extreme initial planetary environment undergoing induced "catastrophic" geo-evolutionary changes. Interestingly, the Techno-Artist Robert Morris (born 1931) provides an interesting contrast. Between 1948 and1950 Morris studied engineering at the University of Kansas, followed during 1951-1952 by a term of service in Arizona and Korea with the United States Army Corps of Engineers. By 1979 Morrise pondered outloud about such gigantic embellishments "...whether it would be easier in the future [in the Earth-biosphere] to disfigure the landscape to get rid of every last particle of non-renewable energy, knowing that there would be some artist somewhere 'to transform the devastation into a modern and inspired work of art. Or, perhaps, into a recreational place of human leisure? Or else, at a pinch, a nice clean park where you would not run any risks'" (Gilles A.Tiberghien *IN* Reifenscheid, 2011, at page17)! [Circa 2010, about one-third of the the world's population (those with a High Standard of Living) used approximately two-thirds of all energy produced and utilized by humanity (Lawrence et al., 2013); some scientists have, therefore, imagined a depleted-energy Earth (Valero et al., 2011).] In summary, as Harry Donald Goode proposed many decades ago, planetary-scale geologic event-processes may (for better or worse) transform environmental changes enough to develop new geologic event-processes!

Ships, submarines and yachts may one day ply the sparkling Neo-Oceanus Borialis. "At least since Noah's maritime ventures in live-stock transport ships have been symbols of salvation, but admiration for them has always been tempered by awareness of the daily risks posed by seafaring..." (Quartermaine, 1996, page 28). So, Geographos asks silently, what types of watercraft can tolerate such odd planetary environmental conditions as those of an anthropo-centric Mars and yet still work efficiently and profitably? Speculating, it seems logical to assume there would be no need for the construction of super-ports; very possibly there won't be any maritime industry requirement for piers and ports. Could there ever be a fishing industry on Mars? Two types of ships can safely transit the Neo-Oceanus Borialis with relative impunity: (I) horizontally very big ships—that is, slow-cruising ships of great length and shallow displacements or shorter, deeper draft vessels using dynamic-support induced by high-speed propulsion—and (II) submarines that are always sealed for safety reasons whilst underway, possibly even when anchored. (Such mobile air-tight watercraft, with immensely spacious interiors, won't promote a "confinement syndrome" often predictable or associated with human occupancy of small enclosures for long periods of time. That must be the case for yachts and boats too.) Since very little industrial activity may eventuate on the North Pole's highlands marked by a few scenic fiords, it is probable that Mars's settlers will see the new ocean as a picturesque playground or tantalizing tourist-spied region, a psychic relief from the monotony of Mars's ocher sandy and hard-rock geologies and its pinkish sky coloration. And, since air pressure may be rather low compared to Earth's, is it not very likely that all surface vehicles will be air-tight and compartmented for extra-comfortable accommodation? Very big vessels, as for example a "Freedom Ship" (**Figure 8**) that will have the hydrostatic behavior of large keel-less rafts or flat-bottomed buoyant mats—like "Mega-Float" (Hara et al., 2004), (**Figure 9**), and sealed glass or concrete submarines seem most practical to Geographos. These large vessels might also be fabricated with pykrete, shapeable frozen water containing strengthening shredded materials of various kinds (Cross, 2012)

Figure 8. Proposed "Freedom Ship" underway. Illustration is self-explanatory! (Image: Google Images.)

Figure 9. Mega-Float, movable and anchored in this presentation. (Image: Google Images.)

In the ultimate of escorted tourism, imagine a Mars "Love Boat" romantically circuit-cruising (about the planet's natural axis of rotation and one of two geodetic reference points for most measurements of direction and location made anywhere) without visiting any entrepot-ports or exotic ports-of-call (Cockell and Horneck, 2006). Since microbial decomposition of plants grown by settlers will last intact for along time, colorful and interesting shaped and shaded "perpetual" dead plants might be used like artfully applied dramatic Hollywood movie-set decoration outside of the windowed domed cities. The Neo-Oceanus Borialis's northern shoreline may have a few spectacular fiords but its southern shoreline, nearest the Valles Marineris delta, will mostly feature fiards, certainly rather dull viewing for diversion-hunting tourists and distracted honeymooners! Sky-gazing is essentially a visual and mental relation of humans to the Universe which usually stimulates thought and, most importantly, macro-imgaination; still, there is the possibility of illuminated outdoor artistic earthworks—can a Mars artificial outdoor lighting "skyglow" (Duriscoe, 2013) be expected to arise subsequently?—decorating a spooky Alien-themed parkland. Indeed, since people seem to enjoy the sight of the colossal (Mason, 2013), why not carve a suitable rock formation, or cast in concrete, a real "Face on Mars", (**Figure 10**) to be glimpsed by a passing tour ship's excited, thrilled gawker-passengers? Already a human "Ear" on Earth has been done (Switek, 2006; Plait, 2007) Such a duly shaped landscape form would not be any more ridiculous than a realization of the 4th Century BC Dinocratic myth about Greece's Mount Athos sculpted to honor a life-size replica of Alexander the Great (Dora, 2005)! Such fraudulent sight-seeing is already prevalent on Earth where, for example, in Central America a spurious "Tropic Zone Paradise" of phoney Mayan ruins studs a Carribbean Sea island (Bawaya, 2014)!

Figure 10. The controversial natural landscape feature once taken to be artificial. Might humans carve or cast-in-place a massive artificial Martian inselberg equivalent to Australia's famous "Uluru" [Ayers Rock]? (Image: Google Images).

Replicated and Green-approved "Garden of Eden" facilities aboard gigantic vessels—which ought to be veritable floating arboretums since each traveler aboard requires 2.5 kilograms of drinking water and 4.85 kilograms of technical water per day—should be capable of purifying all expelled fluids and other wastes during trips by all the machines and all the living Martian merrymakers carried by these ecologically-friendly ships! On Earth, the collective human respiration alters the climate in a small way by their collective exhalation of carbon dioxide gas. [Seismologists (Ben-Menahem, 1995) first recorded the motions of Earth's crust—earthquakes seemed like violent symptoms of some inner dis-ease—using devices that were, literally, spin-offs from the

Marey pneumographic device that Paul-Marie-Leon Regnard (1850-1927) used to chart the respiratory actions of hysterical patients! Even Mars's pulsating soils have been compared to a planetary gas pump that has a gas flow rate greater than global Nature's gas diffucsion rate (Beule et al., 2013)]

In a new, Technogenic Mars-biosphere, movable floating Astythromes on a Neo-Oceanus Borialis and Ice-domed Astythromes affixed to the Southern Hemisphere's highlands can become nuclear urban-type infrastructures that are so physically extensible and socially intensive that they will foster a second planetary "evolution of global human consciousness time-line (Thompson, 2000). [Once Mars has multiple fixed cities, the old-time mobile cities could become parked theme cities populated by Martians having the same social qualities and inequalities as the Bedouin-themed, tourist-attracting expositions established by Arab governments to collected hard currencies from Europeans and North Americans!] Thus, planetary science and transformative planetarian planner-constructors (Macro-Imagineers/Macro-Engineers: all Geo-engineers plus Terraformers) may successfully merge their meritorious technical skills and their geophilosophical outlooks efficaciously to the ultimate benefit of all humans on Earth and Mars. Both Techno-Artistic groups of planet changers owe their inspiration, in part to, large-scale Artworks of the past, especially since about 1960 (Lippincott, 2010; Viction:ary, 2013).

Martians of (Earth-) human descent might come prefer to adopt the attitude and faith of Australia's first Aborigines who saw themselves as inseperable from global Nature and who located their Creation Myth, "...not remotely in the heavens..." (Hynes, 2013, page 121), but from within the planet itself, the produced superficial landforms permeated by the Aboriginal Dreaming presence of ineffable sacred Ancestral Beings. Geographos wonders, a 40,000 year-old prefiguration, at least philosophically, of Mars Terraforming? Geographos harbors no doubts that, in future, human landscape modification and induced climate change activities in Mars ultimately will produce life-supportive infrastructures worthy of such glorious pictorial representations as *Terra Maxima: The Records of Humankind* (2013), edited by the likes of a Wolfgang Kunth. Our future awaits our arrival and determined efforts! Given adequate planetary conditions, humans can create a metabiosphere in our Solar System—a metabiosphere combines Ecology at an astronomical scale with interstellar biology (Mendonca, 2014). The first birth of a human conceived on Mars will start the first native baby's giving and receiving experience as it draws its first breath and begins, in a natural way, the truly anthropocized circulatory system making Mars a desirable place to dwell.

References Cited

Anker, P. (2005) *The closed world of ecological architecture.* The Journal of Architecture 10: 527-552.

Anker, P. (April 2007) *The ecological COLONIZATION OF SPACE.* Environmental History 10: 1-26.

Arkes, H.R. and Ayton. P. (September 1999) *The sunk cost and Concorde effects: Are humans less rational than lower animals?* Psychological Bulletin 125: 591-600.

Badescu, V. et al. (2009) *Ecopoiesis and Liquid Water Transportation on Mars.* Chapter 26, pages 662-682 *IN* Badescu, V. (Ed.) MARS: Prospective Energy and Material Resources. (The Netherlands: Springer) 695 pages.

Bawaya, M. (4 January 2014) *Land of Make-Believe.* New Scientist 221: 34-37.

Ben-Menahem, A. (1995) *A Concise History of Mainstream Seismology: Origins, Legacy, and Perspective.* Bulletin of the Seismological Society of America 85: 1202-1225.

Beule, C. et al. (2013) *The martian soil as a planetary gag pump.* Nature Physics 9: 822.

Cabrol, N.A. and Grin, E.A. (Eds.) (2010) *Lakes on Mars.* (New York: Elsevier) 410 pages.

Carroll, M. and Lopes, R. (Eds.) (2013) *Alien Seas: Oceans in Space.* (The Netherlands: Springer) 119 pages.

Cathcart, R.B. (1998) *Taming Mars with a Tent and a Tunnel.* Speculations in Science and Technology 21: 117-131.

Cockell, C.S. and Horneck, G. (2006) *Planetary parks—formulating a wilderness policy for planetary bodies.* Space Policy 22: 256-261.

Costanza, R. et al. (2011) *Sustainability of Collapse? An Integrated History and Future of People on Earth.* (Massachusetts: MIT Press) 517 pages.

Cross, L.D. (2012) *Code Name HABBAKUK: A Secret Ship Made of Ice.* (Vancouver, Canada: Heritage138 pages.

Deyong, S. (Summer 2001) *Planetary habitat: the origins of a phantom movement.* The Journal of Architecture 6:113-128.

Dora, V.D. (2005) *Alexander the Great's Mountain.* The Geographical Review 95: 489-516.

Duriscoe, D.M. (2013) *Measuring Anthropogenic Sky Glow Using a Natural Sky Brightness Model.* Publications of the Astronomical Society of the Pacific 125: 1370-1382.

Edwards, P.N. (2012) *Entangle histories: Climate science and nuclear weapons research.* Bulletin of the Atomic Scientists 68: 28-40.

Faunce, T.A. (2012) *Towards a global solar fuels project—Artificial photosynthesis and the transition from Anthropocene to Sustainocene.* Procedia Engineering 49: 348-356.

Gellert, P.K. and Lynch, B.D. (2003) *Mega-projects as displacements.* International Social Science Journal 55: 15-25.

Gourrronc, M. et al. (2014) *One million cubic kilometers of fossil ice in Valles Marineris: Relicts of a 3.5 Gy old glacial land system along the Martian equator.* Geomorphology 204: 235-255.

Gray, W. (23 November 2013) *Abandon Ship!* New Scientist 220: 48-51.

Hara, S. et al. (2004) *At-Sea towing of a Mega-Float Unit.* Journal of Marine Science and Technology 8: 138-146.

Haqq-Misra, J. (2012) *An Ecological Compass for Planetary Engineering.* Astrobiology 12: 985-997.

Haynes, R.D. (2013) *Desert: Nature and Culture.* (London: Reaktion Books) 245 pages.

Hooke, R.L. (September 2000) *On the history of humans as geomorphic agents.* Geology 28: 845.

Jones, S.B. et al. (February 2012) *Beyond Earth: Designing Root Zone Environments for Reduced Gravity Conditions.* Vadose Zone Journal 11: 1-11.

Kemp, C. (16 November 2013) *Rewind, Erase, Rerun.* New Scientist 220: 34-38.

Kleidon, A. et al. (2000) *A Green Planet versus a Desert World: Estimating the Maximum Effect of Vegetation on the Land Surface Climate.* Climatic Change 44: 471-493.

Komatsu, G. (2007) *Rivers in the Solar System: Water Is Not the Only Fluid Flow on Planetary Bodies.* Geography Compass 1: 480-502.

Kronenburg, R. (2013) *Architecture in Motion: The History and Development of Portable Building.* (London: Routledge) 318 pages.

Lawrence, S. et al. (2013) *Global Inequality in Energy Consumption from 1980 to 2010.* Entropy 15: 5565-5579.

Leconte, J. et al. (2013) *Increased insolation threshold for runaway greenhouse processes on Earth-like planets.* Nature 504: 268-271.

Lenton, T.M. and von Bloh, W. (2001) *Biotic feedback extends the life span of the biosphere.* Geophysical Research Letters 43: 991-1002.

Lippincott, J.D. (2010) *Large Scale: Fabricating Sculpture in the 1960s and 1970s.* (New York: Princeton Architectural Press) 256.

Lorenz, R.D. et al. (2013) *Dunes on planet Tatooine: Observations of Barchan Migration at the Star Wars film set in Tunisia.* Geomorphology 201: 264-271.

Marx, B., Stoker, T. and Suri, T. (Fall 2013) *The Economics of Slums in the Developing World.* Journal of Economic Perspectives 27: 187-210.

Mason, P. (2013) *The Colossal: From Ancient Greece to Giacometti.* (Cornwall, UK: Reaktion Books) 208 pages.

Mendonca, M.S. (online 3 February 2014) *Spatial ecology goes to space: metabiospheres.* Icarus. DOI: 10.1016/j.icarus.2014.01.027.

Minter, A. (2013) *Junkyard Planet: Travels in the Billion-Dollar Trash Trade.* (New York: Bloomsbury Press) 285 pages.

Mole, R.A. (1995) *Terraforming Mars with Four War-Surplus Bombs.* Journal of the British Interplanetary Society 48: 321. [See also: Muscatello, A.C. and Houts, M.G., *Surplus Weapons-Grade Plutonium—A Resource for Exploring and Terraforming Mars.* Report Number: LA-UR-96-4463, Los Alamos National Laboratory, New Mexico, USA.]

Morehead, A.V. et al. (2014) *The meteoroid fluence at Mars due to Comet C/2013 A1 (Siding Spring).* Icarus 231: 13-21.

Nordlund, C. (2001) *"On Going Up in the World": Nation, Region and the Land Elevation Debate in Sweden.* Annals of Science 58: 17-50.

OMA, Koolhaas, R. and Mau, B. (1995) *S.M.L.XL.* (New York: Monacelli Press, Inc.) pages 494-516.

Palaia, J.E. et al. *Economics of Energy on Mars.* Pages 369-400 **IN** Badescu, V. (Ed.) *Mars: Prospective Energy and Material Resources.* (The Netherlands: Springer) 695 pages.

Parker, T.J. and Currey, D.R. (2001) *Extraterrestrial coastal geomorphology.* Geomorphology 37: 303-328.

Phillippe, J-M. (1996) *Towards a work of art on planet Mars.* Planetary and Space Science 44: 1463-1470.

Plait, P. (February 2007) *The Face on Mars.* Sky & Telescope 113: 28-29.

Priemus, H. and Wee, B. van (Eds.) (2013) *International Handbook on Mega-Projects.* (London: Edward Elgar Publishing Ltd.) 520 pages.

Quartermaine, (1996) *Building on the Sea: Form and Meaning in Modern Ship Architecture.* (United Kingdom: Academy Editions).

Reifenscheid, B. (2011) *Die Letzte Freiheit: The Last Freedom.* (Milan, Italy: SilvanaEditoriale) 192 pages.

Rey, P.F. (2013) *Opalisation of the Great Artesian Basin (central Australia): An Australian story with a Martian twist.* Australian Journal of Earth Sciences 60: 291-314.

Roach, M. (2010) *Packing for Mars: The Curious Science of Life in the Void*. (New York: W.W. Norton & Company) 334 pages.

Sagan, C. and Fox, P. (1975) *The Canals of Mars: An Assessment after Mariner 9*. Icarus 25: 602-612.

Salisbury, F.B . (6 April 1962) *Martian Biology: Accumulating evidence favors the theory of life on Mars, but we can expect surprises*. Science 136: 17-26.

Sax, D.F. and Brown, J.H. (2000) *The paradox of invasion*. Global Ecology and Biogeography 9: 363-371.

Swanson, K.S. et al. (2013) *Nutritional Sustainability of Pet Foods*. Advances in Nutrition 4: 141-150.

Switek, G.B. (2006) *Project Mars: the nature of macroanthropos*. The Journal of Architecture 11: 485-496.

Valero, A. et al. (2011) *The crepuscular planet: A model for the exhausted atmosphere and hydrosphere*. Energy 36: 3745-3753.

Van Bemmelen, R.W. (1967) *The Importance of the Geonomic Dimensions for Geodynamic Concepts*. Earth-Science Reviews 3: 79-110.

Viction:ary (2013) *OVER S!ZE: The Mega Art & Installations*. (Hong Kong, China: Victionary) 216.

Watson, P. (2012) *The Great Divide: Nature and Human Nature in the Old World and the New*. (New York: Harper) 610 pages.

Wilde, P. and Quinby-Hunt, M.S. (1997) *Collisions with ice/volatile objects: Geological implications—A qualitative treatment.* Palaeogeography, Palaeoclimatology, Palaeoecology 132: 47-63.

Wood, R. et al. (2013) *Climatic change special issue: geoengineering research and its limitations.* Climatic Change 121: 427-430.

Yandle, B. (Fall 1999) *Grasping the Heavens: 3-D Property Rights and the Global Commons.* Duke Environmental Law & Policy Forum 10: 13-44.

www.ingramcontent.com/pod-product-compliance
Lightning Source LLC
Chambersburg PA
CBHW021418170526
45164CB00001B/3